$55.01 UCF

CMOS MEMORY CIRCUITS

CMOS MEMORY CIRCUITS

by

Tegze P. Haraszti
Microcirc Associates

KLUWER ACADEMIC PUBLISHERS
Boston / Dordrecht / London

Distributors for North, Central and South America:
Kluwer Academic Publishers
101 Philip Drive, Assinippi Park
Norwell, Massachusetts 02061 USA
Telephone (781) 871-6600
Fax (781) 871-6528
E-Mail <kluwer@wkap.com>

Distributors for all other countries:
Kluwer Academic Publishers Group
Distribution Centre
Post Office Box 322
3300 AH Dordrecht, THE NETHERLANDS
Telephone 31 78 6392 392
Fax 31 78 6546 474
E-Mail <orderdept@wkap.nl>

Electronic Services <http://www.wkap.nl>

Library of Congress Cataloging-in-Publication Data

Haraszti, Tegze P.
  CMOS memory circuits / by Tegze P. Haraszti.
   p. cm.
  Includes bibliographical references and index.
  ISBN 0-7923-7950-0 (alk. paper)
   1. Semiconductor storage devices--Design and construction. 2. Metal oxide semiconductors, Complementary--Design and construction. 3. Electronic circuit design.
  I. Title.

TK 7895.M4 H37 2000
621.39'732--dc21

                                                                00-058757

Copyright © 2000 by Kluwer Academic Publishers.

All rights reserved. No part of this publication may be reproduced, stored in a retrieval system or transmitted in any form or by any means, mechanical, photo-copying, recording, or otherwise, without the prior written permission of the publisher, Kluwer Academic Publishers, 101 Philip Drive, Assinippi Park, Norwell, Massachusetts 02061

Printed on acid-free paper.

Printed in the United States of America

Contents

Preface xi

Conventions xvi

Chapter 1. Introduction to CMOS Memories 1

    1.1 Classification and Characterization of CMOS Memories 2
    1.2 Random Access Memories 10
        1.2.1 Fundamentals 10
        1.2.2 Dynamic Random Access Memories (DRAMs) 11
        1.2.3 Pipelining in Extended Data Output (EDO) and
              Burst EDO (BEDO) DRAMs 20
        1.2.4 Synchronous DRAMs (SDRAMs) 25
        1.2.5 Wide DRAMs 31
        1.2.6 Video DRAMs 33
        1.2.7 Static Random Access Memories (SRAMs) 36
        1.2.8 Pseudo SRAMs 40
        1.2.9 Read Only Memories (ROMs) 41
    1.3 Sequential Access Memories (SAMs) 42
        1.3.1 Principles 42
        1.3.2 RAM-Based SAMs 44
        1.3.3 Shift-Register Based SAMs 45
        1.3.4 Shuffle Memories 48
        1.3.5 First-In-First-Out Memories (FIFOs) 51
    1.4 Content Addressable Memories (CAMs) 54
        1.4.1 Basics 54
        1.4.2 All-Parallel CAMs 56

		1.4.3	Word-Serial-Bit-Parallel CAMs	58

	1.4.3	Word-Serial-Bit-Parallel CAMs	58
	1.4.4	Word-Parallel-Bit-Serial CAMs	59
1.5	Special Memories and Combinations		61
	1.5.1	Cache-Memory Fundamentals	61
	1.5.2	Basic Cache Organizations	65
	1.5.3	DRAM-Cache Combinations	70
	1.5.4	Enhanced DRAM (EDRAM)	70
	1.5.5	Cached DRAM (CDRAM)	72
	1.5.6	Rambus DRAM (RDRAM)	73
	1.5.7	Virtual Channel Memory (VCM)	76
1.6	Nonranked and Hierarchical Memory Organizations		80

Chapter 2. Memory Cells 85

2.1	Basics, Classifications and Objectives		86
2.2	Dynamic One-Transistor-One-Capacitor Random Access Memory Cell		89
	2.2.1	Dynamic Storage and Refresh	89
	2.1.2	Write and Read Signals	92
	2.1.3	Design Objectives and Trade-offs	96
	2.1.4	Implementation Issues	97
		2.1.4.1 Insulator Thickness	97
		2.1.4.2 Insulator Material	98
		2.1.4.3 Parasitic Capacitances	103
		2.1.4.4 Effective Capacitor Area	105
2.3	Dynamic Three-Transistor Random Access Memory Cell		110
	2.3.1	Description	110
	2.3.2	Brief Analysis	111
2.4	Static 6-Transistor Random Access Memory Cell		113
	2.4.1	Static Full-Complementary Storage	113
	2.1.2	Write and Read Analysis	116
	2.1.3	Design Objectives and Concerns	121
	2.1.4	Implementations	122
2.5	Static Four-Transistor-Two-Resistor Random Access Memory Cells		125
	2.5.1	Static Noncomplementary Storage	125
	2.5.2	Design and Implementation	128
2.6	Read-Only Memory Cells		132
	2.6.1	Read-Only Storage	132
	2.6.2	Programming and Design	134
2.7	Shift-Register Cells		136
	2.7.1	Data Shifting	136
	2.7.2	Dynamic Shift-Register Cells	138
	2.7.3	Static Shift-Register Cells	143
2.8	Content Addressable Memory Cells		146
	2.8.1	Associative Access	146
	2.8.2	Circuit Implementations	148
2.9	Other Memory Cells		151
	2.9.1	Considerations for Uses	151
	2.9.2	Tunnel-Diode Based Memory Cells	152

2.9.3 Charge Coupled Device 154
2.9.4 Multiport Memory Cells 156
2.9.5 Derivative Memory Cells 158

Chapter 3. Sense Amplifiers 163

3.1 Sense Circuits 164
    3.1.1 Data Sensing 164
    3.1.2 Operation Margins 166
    3.1.3 Terms Determining Operation Margins 171
        3.1.3.1 Supply Voltage 171
        3.1.3.2 Threshold Voltage Drop 171
        3.1.3.3 Leakage Currents 173
        3.1.3.4 Charge-Couplings 176
        3.1.3.5 Imbalances 179
        3.1.3.6 Other Specific Effects 181
        3.1.3.7 Precharge Level Variations 182
3.2 Sense Amplifiers in General 184
    3.2.1 Basics 184
    3.2.2 Designing Sense Amplifiers 187
    3.2.3 Classification 190
3.3 Differential Voltage Sense Amplifiers 192
    3.3.1 Basic Differential Voltage Amplifier 192
        3.3.1.1 Description and Operation 192
        3.3.1.2 DC Analysis 193
        3.3.1.3 AC Analysis 196
    3.3.2 Simple Differential Voltage Sense Amplifier 200
        3.3.2.1 All-Transistor Sense Amplifier Circuit 200
        3.3.2.2 AC Analysis 201
        3.3.2.3 Transient Analysis 203
    3.3.3 Full-Complementary Differential Voltage Sense Amplifier 207
        3.3.3.1 Active Load Application 207
        3.3.3.2 Analysis and Design Considerations 209
    3.3.4 Positive Feedback Differential Voltage Sense Amplifier 211
        3.3.4.1 Circuit Operation 211
        3.3.4.2 Feedback Analysis 213
    3.3.5 Full-Complementary Positive-Feedback Differential
            Voltage Sense Amplifier 217
    3.3.6 Enhancements to Differential Voltage Sense Amplifiers 220
        3.3.6.1 Approaches 220
        3.3.6.2 Decoupling Bitline Loads 221
        3.3.6.3 Feedback Separation 224
        3.3.6.4 Current Sources 226
        3.3.6.5 Optimum Voltage-Swing to Sense Amplifiers 229
3.4 Current Sense Amplifiers 232
    3.4.1 Reasons for Current Sensing 232
    3.4.2 Feedback Types and Impedances 236
    3.4.3 Current-Mirror Sense Amplifier 238
    3.4.4 Positive Feedback Current Sense Amplifier 240

      3.4.4 Positive Feedback Current Sense Amplifier 240
      3.4.5 Current-Voltage Sense Amplifier 243
      3.4.6 Crosscoupled Positive Feedback Current Sense Amplifier 245
      3.4.7 Negative Feedback Current Sense Amplifiers..............................249
      3.4.8 Feedback Transfer Functions...250
      3.4.9 Improvements by Feedback...252
      3.4.10 Stability and Transient Damping 256
  3.5 Offset Reduction 257
      3.5.1 Offsets in Sense Amplifiers...257
      3.5.2 Offset Reducing Layout Designs...259
      3.5.3 Negative Feedback for Offset Decrease 260
      3.5.4 Sample-and-Feedback Offset Limitation....................................263
  3.6 Nondifferential Sense Amplifiers..265
      3.6.1 Basics 265
      3.6.2 Common-Source Sense Amplifiers 266
      3.6.3 Common-Gate Sense Amplifiers..269
      3.6.4 Common-Drain Sense Amplifiers 273

Chapter 4. Memory Constituent Subcircuits 277

  4.1 Array Wiring 278
      4.1.1 Bitlines 278
          4.1.1.1 Simple Models..278
          4.1.1.2 Signal Limiters 283
      4.1.2 Wordlines 287
          4.1.2.1 Modelling 287
          4.1.2.2 Signal Control 290
      4.1.3 Transmission Line Models 296
          4.1.3.1 Signal Propagation and Reflections.............................296
          4.1.1.2 Signal Transients 301
      4.1.4 Validity Regions of Transmission Line Models...........................308
  4.2 Reference Circuits 311
      4.2.1 Basic Functions 311
      4.2.2 Voltage References..311
      4.2.3 Current References..318
      4.2.4 Charge References...321
  4.3 Decoders...323
  4.4 Output Buffers...328
  4.5 Input Receivers..336
  4.6 Clock Circuits..341
      4.6.1 Operation Timing 341
      4.6.2 Clock Generators...344
      4.6.3 Clock Recovery 347
      4.6.4 Clock Delay and Transient Control 352
  4.7 Power-Lines 355
      4.7.1 Power Distribution 355
      4.7.2 Power-Line Bounce Reduction...359

Chapter 5. Reliability and Yield Improvement 365

- 5.1 Reliability and Redundancy 366
  - 5.1.1 Memory Reliability 366
  - 5.1.2 Redundancy Effects on Reliability 369
- 5.2 Noises in Memory Circuits 373
  - 5.2.1 Noises and Noise Sources 373
  - 5.2.2 Crosstalk Noises in Arrays 374
  - 5.2.3 Crosstalk Reduction in Bitlines 379
  - 5.2.4 Power-Line Noises in Arrays 382
  - 5.1.5 Thermal Noise 385
- 5.3 Charged Atomic Particle Impacts 388
  - 5.3.1 Effects of Charged Atomic Particle Impacts 388
  - 5.3.2 Error Rate Estimate 390
  - 5.1.3 Error Rate Reduction 398
- 5.4 Yield and Redundancy 402
  - 5.4.1 Memory Yield 402
  - 5.1.2 Yield Improvement by Redundancy Applications 406
- 5.5 Fault-Tolerance in Memory Designs 412
  - 5.5.1 Faults, Failures, Errors and Fault-Tolerance 412
  - 5.5.2 Faults and Errors to Repair and Correct 415
  - 5.5.3 Strategies for Fault-Tolerance 420
- 5.6 Fault Repair 421
  - 5.6.1 Fault Repair Principles in Memories 421
  - 5.6.2 Programming Elements 423
  - 5.1.3 Row and Column Replacement 428
  - 5.1.4 Associative Repair 434
  - 5.1.5 Fault Masking 436
- 5.7 Error Control Code Application in Memories 438
  - 5.7.1 Coding Fundamentals 438
  - 5.7.2 Code Performance 442
  - 5.7.3 Code Efficiency 446
  - 5.7.4 Linear Systematic Codes 453
    - 5.7.4.1 Description 453
    - 5.7.4.2 Single Parity Check Code 453
    - 5.7.4.3 Berger Codes 455
    - 5.7.4.4 BCH Codes 457
    - 5.7.4.5 Binary Hamming Codes 457
    - 5.7.4.6 Reed-Solomon (RS) Codes 461
    - 5.7.4.7 Bidirectional Codes 462
  - 5.8 Combination of Error Control Coding and Fault-Repair 464

Chapter 6. Radiation Effects and Circuit Hardening 469

- 6.1 Radiation Effects 470
  - 6.1.1 Radiation Environments 470
  - 6.1.2 Permanent Ionization Total-Dose Effects 471
  - 6.1.3 Transient Ionization Dose-Rate Effects 475

　　　　　6.1.4　Fabrication-Induced Radiations and Neutron Fluence 477
　　　　　6.1.5　Combined Radiation Effects 478
　　6.2　Radiation Hardening 481
　　　　　6.2.1　Requirements and Hardening Methods 481
　　　　　6.2.2　Self-Compensation and Voltage Limitation in Sense Circuits 486
　　　　　6.2.3　Parameter Tracking in Reference Circuits 491
　　　　　6.2.4　State Retention in Memory Cells 493
　　　　　6.2.5　Self-Adjusting Logic Gates 495
　　　　　6.2.6　Global Fault-Tolerance for Radiation Hardening 499
　　6.3　Designing Memories in CMOS SOI (SOS) 501
　　　　　6.3.1　Basic Considerations 501
　　　　　　　　6.3.1.1　Devices 501
　　　　　　　　6.3.1.2　Features 505
　　　　　6.3.2　Floating Substrate Effects 509
　　　　　　　　6.3.2.1　History Dependency, Kinks and Passgate Leakages 509
　　　　　　　　6.3.2.2　Relieves 516
　　　　　6.3.3　Side- and Back-Channel Effects 520
　　　　　　　　6.3.3.1　Side-Channel Leakages, Kinks and Breakdowns 520
　　　　　　　　6.3.3.2　Back-Channel- and Photocurrents 524
　　　　　　　　6.3.3.3　Allays 526
　　　　　6.3.4　Diode-Like Nonlinear Parasitic Elements and Others 527

References 531

Index 541

Preface

Staggering are both the quantity and the variety of complementary-metal-oxide-semiconductor CMOS memories. CMOS memories are traded as mass-products world wide, and are diversified to satisfy nearly all practical requirements in operational speed, power, size and environmental tolerance. Without the outstanding speed, power and packing-density characteristics of CMOS memories neither personal computing, nor space exploration, nor superior defense-systems, nor many other feats of human ingenuity could be accomplished. Electronic systems need continuous improvements in speed performance, power consumption, packing density, size, weight and costs; and these needs spur the rapid advancement of CMOS memory processing and circuit technologies.

The objective of this book is to provide a systematic and comprehensive insight which aids the understanding, practical use and progress of CMOS memory circuits, architectures and design techniques. In the area of semiconductor memories, since 1977 this is the first and only book that is devoted to memory circuits. Besides filling the general void in memory-related works, so far this book is the only one that covers inclusively such modern and momentous issues in CMOS memory designs as sense amplifiers, redundancy implementations and radiation hardening, and discloses practical approaches to combine high performance, and reliability, with high packing density and yield by circuit-technological and architectural means.

For semiconductor integrated circuits, during the past decades, the CMOS technology emerged as the dominant fabrication method, and CMOS became the almost exclusive choice for semiconductor memory designs also. With the development of the CMOS memory technology numerous publications presented select CMOS memory circuit- and architecture-designs, but these disclosures, sometimes for protection of intellectual property, left significant hollows in the acquainted material and made little attempt to provide an unbiased global picture and analysis in an organized form. Furthermore, the analysis, design and improvement of many memory-specific CMOS circuits, e.g. memory cells, array wiring, sense amplifiers, redundant elements, etc., required expertness not only in circuit technology, but also in semiconductor processing and device technologies, modern physics and information theory. The prerequisite for combining these diverse technological and theoretical sciences from disparate sources made the design and the tuition of CMOS memory circuits exceptionally demanding tasks. Additionally, the literature of CMOS technology made little effort to give overview texts and methodical analyses of some significant memory-specific issues such as sense amplifiers, redundancy implementations and radiation hardening by circuit-technical approaches.

The present work about circuits and architectures aspires to provide knowledge to those who intend to (1) understand, (2) apply, (3) design and (4) develop CMOS memories. Explicit interest in CMOS memory circuits and architectures is anticipated by engineers, students, scientists and managers active in the areas of semiconductor integrated circuit, general microelectronics, computer, data processing and electronic communication technologies. Moreover, electronic professionals involved in developments and designs of various commercial, automotive, space and military systems, should also find the presented material appealing.

The presentation style of the material serves the strong motivation to produce a book that is indeed read and used by a rather broad range of technically interested people. To promote readability, throughout the entire book the individual sentences are chained by key words, e.g. a specific word, that is used in the latter part of the sentence, is reused again in the initial part of the next sentence. Usability is enhanced by developing the material from the simple to the complex subjects within each topic-section and topic-to-topic throughout the book. Most of the sections are devoted to circuits and circuit organizations, and for each of them a section describes what it is, what it does and how it operates, thereafter, if it is appropriate, the section provides physical-mathematical analyses, design and improvement considerations. In the analyses, the equations are brought to easy-to-understand forms, their interpretations and derivations are narrated. The derivations of the equations, however, may require the application of higher mathematics. Most of the physical-mathematical formulas are approximations to make plausible how certain parameters change the properties of the circuit, and what and how variables can be used in the design. Knowledge in the use of the variables, allows for efficient applications of computer models and simulation programs, for shortening design times and for devising improvements for the subject circuits. The simple-to-complex subject composition makes possible to choose an arbitrary depth in studying the material. A considerable amount of the material was presented by the author on graduate and extension courses at the University of California Berkeley, and the response vitally contributed to the organization and expressing style used in the book.

This book presents the operation, analysis and design of those CMOS memory circuits and architectures which have been successfully used and which are anticipated to gain volume applications. To facilitate convenient use and overview the material is apportioned in six chapters: (1) Introduction to CMOS Memories, (2) Memory Cells, (3) Sense Amplifiers, (4) Memory Constituent Subcircuits, (5) Reliability and Yield Improvement, and (6) Radiation Effects and Circuit Hardening. The introductory description of CMOS memory architectures serves as a basis for the discussion of the memory circuits. Because memory cells make a memory device capable to store data, the memory cell circuits are detailed in the next chapter. Sense amplifier circuits are key elements in most of the memory designs, therefore, in the third chapter a comprehensive analysis reveals the intricacies of the sense circuits, voltage-, current- and other sense amplifiers. Subcircuits, beside memory cells and sense amplifiers, which are specific to CMOS memory designs are treated in the fourth chapter. Reliability and yield improvements by redundant designs evolved to be, important issues in CMOS memory designs, because of that, an entire chapter is devoted to these issues. The final chapter, as a recognition of modern requirements, summarizes the effects of radioactive irradiations on CMOS memories, and describes the radiation hardening techniques by circuit and architectural approaches. Since the combination of radiation hardness and high performance was the incipient stimulant to develop CMOS silicon-on-insulator SOI and silicon-on-sapphire SOS memories, this closing chapter devotes a substantial part to the peculiarities of the CMOS SOI (SOS) memory circuit designs.

The circuits and architectures presented in this original monograph are specific to CMOS nonprogrammable write-read and read-only memories. Circuits and architectures of programmable memories, e.g. PROMs, EPROMs, EEPROMs, NVROMs and Flash-Memories are not among the subjects of this volume, because during the technical evolution programmable memories have become a separate and extensive category in semiconductor memories. Yet, a multitude of programmable and other semiconductor memory designs can adopt many of the circuits and architectures which are introduced in this work.

As an addition to this work, a special tuitional aid for CMOS memory designs is under development. The large extent of memory specific tuitional details, which may be read only by a limited number of students, indicates the book-external presentation of this assistance.

The author of this book is grateful to all the people without whom the work could not have been accomplished. The instruction and work of Vern McKenny in memory design, Karoly Simonyi in theoretical electricity and James Bell in technical writing, provided the basis; the inspiration received from Edward Teller, Richard Tsina and Richard Gossen, gave the impetus; the constructive comments of the reviewers, especially by Dave Hodges, Anil Gupta, Ranject Pancholy, and Raymond Kjar, helped to improve the quality; the simulations and modelings by Robert Mento generated many of the graphs and diagrams; the outstanding word processing by Jan Fiesel made possible to compile the original manuscript; the computing skill of Adam Barna gave the edited shape of the equations; the exertion in computerized graphics by Gabor Olah produced the figures; the mastery in programming of Tamas Endrody encoded the graphics; and the editorial and managerial efforts by Carl Harris resulted the publication of the book.

This book has no intention to promote any particular design, brand, business, organization or person. Any nonpromotional suggestion and comment related to the content of this publication are highly appreciated.

Tegze P. Haraszti

Conventions

1. **Basic Units**

    Voltage [Volts]
    Current [Amperes]
    Charge [Coulombs]
    Resistance [Ohms]
    Conductance [Siemens]
    Capacitance [Farads]
    Inductance [Henrys]
    Time [Seconds]

2. **Schematic Symbols**

    Circuit Block

    Data Path

Address Information

Other Connections

Inverter

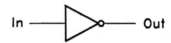

AND Gate

NAND Gate

OR Gate

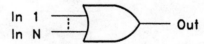

NOR Gate

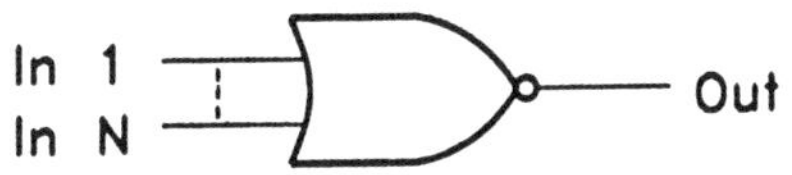

In 1 — ⟩— Out
In N

XAND Gate

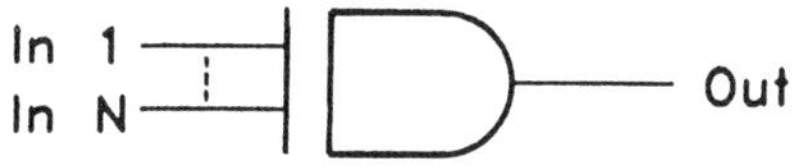

In 1 —|⟩— Out
In N

XOR Gate

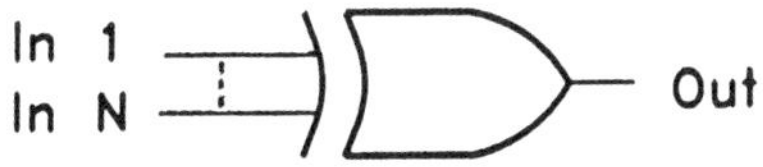

In 1 —⟩⟩— Out
In N

Linear Amplifier

In —[<]— Out

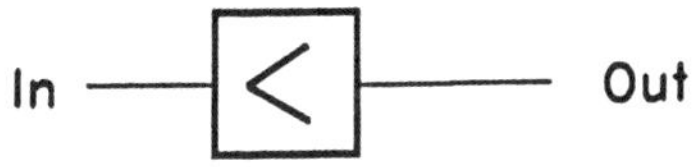

NMOS Transistor Device

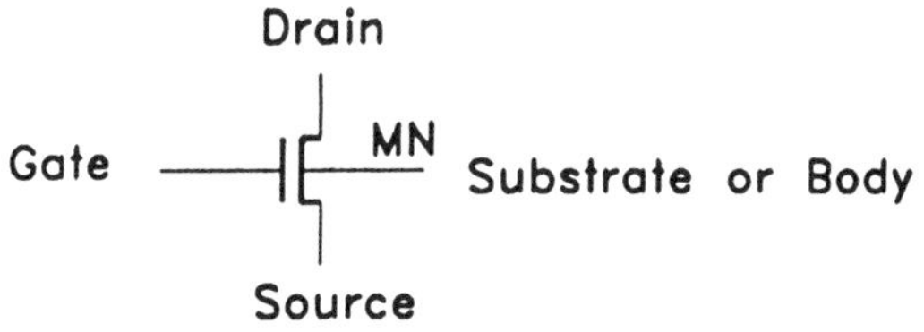

Drain
Gate —|[MN Substrate or Body
Source

PMOS Transistor Device

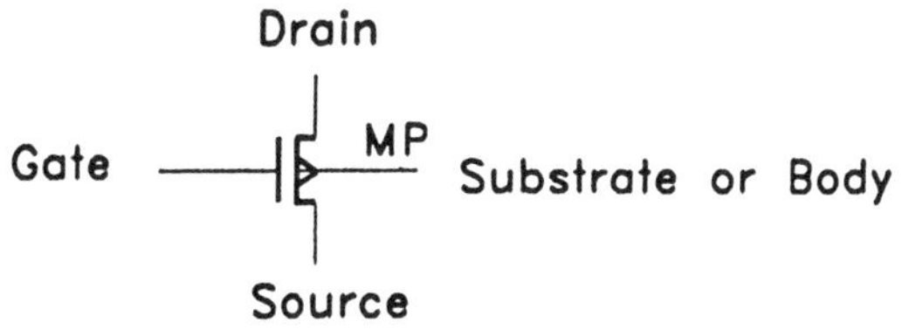

Drain
Gate —|◁ MP Substrate or Body
Source

Bipolar Transistor Device

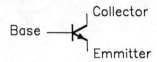

Diode

Anode ——▷|—— Cathode

Tunnel Diode

Anode ——▷⊤—— Cathode

Complex Impedance

Resistor

⌇ R

Capacitor

╪ C

Conventions   xxi

Inductor

$\text{\Large \Epsilon}$ L

Voltage Signal Source

 $v, v(t), V$

Current Signal Source

 $i, i(t), I$

Fuse

$\otimes$

Antifuse

$\ominus$

Positive Power-Supply Pole

V_{DD}

Negative Power-Supply Pole or Ground

"Only God has all the knowledge…"

1

Introduction to CMOS Memories

CMOS memories are used in a much greater quantity than all the other types of semiconductor integrated circuits, and appear in an astounding variety of circuit organizations. This introductory chapter describes concisely the architectures of the circuit organizations which are basic, have been widely implemented and have foreseeable future potentials to be applied in memory designs. The architectures of the different CMOS memories reveal what their major constituent circuits are, and how these circuits associate and interact to perform specific memory functions. Furthermore, the examinations of memory architectures aid to understand and to devise performance improvements through organizational approaches, and lay the foundation to the detailed discussion of the CMOS memory circuits delivered in the next chapters.

1.1 Classification and Characterization of CMOS Memories

1.2 Random Access Memories

1.3 Sequential Access Memories

1.4 Content Addressable Memories

1.5 Special Memories and Combinations

1.6 Nonranked and Hierarchical Memory Organizations

1.1 CLASSIFICATION AND CHARACTERIZATION OF CMOS MEMORIES

CMOS memories, in a strict sense, are all of those data storage devices which are fabricated with a complementary metal-oxide-semiconductor (CMOS) technology. In technical practice, however, the term "CMOS memory" designates a class of data storage devices which (1) are fabricated with CMOS technology, (2) store and process data in digital form and (3) use no moving mechanical parts to facilitate memory operations. This specific meaning of the term "CMOS memory" results from the historical development, application and design of semiconductor data storage devices [11].

In general, data storage devices may be classified by a wide variety of aspects but most frequently they are categorized by (1) fabrication technology of the storage medium, (2) data form, and (3) mechanism of the access to stored data (Table 1.1). From the variety of technologies which may be applied to create data storage devices, the semiconductor integrated circuit technology, and within that, the CMOS technology (Table 1.2) has emerged as the dominant technology in fabrication of system-internal memories (mainframe, cache, buffer, scratch-pad, etc.), while magnetic and optical technologies gained supremacy in production of auxiliary memories for mass data storage. The dominance of CMOS memories in computing, data processing and telecommunication systems has arisen from the capability of CMOS technologies to combine high packing density, fast operation, low power consumption, environmental tolerance and easy down-scaling of feature sizes. This combination of features provided by CMOS memories has been unmatched by memories fabricated with other semiconductor fabrication technologies. Applications of semiconductor memories, so far, have been cost prohibitive in the majority of commercial mass data storage devices. Nevertheless, the design of mass storage devices, which operate in space, military and industrial environments can require the use of CMOS memories, because of their good environmental tolerance.

Introduction to CMOS Memories 3

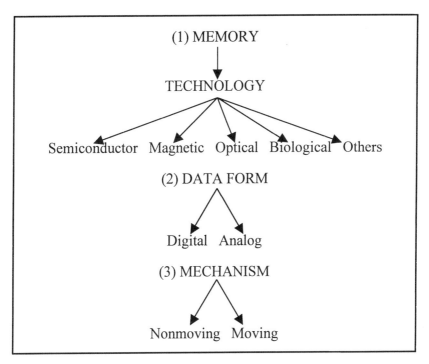

Table 1.1. Memory Classification by Technology, Data Form and Mechanics.

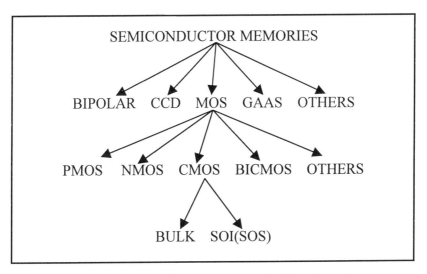

Table 1.2. Semiconductor memory technology branches.

Historically, system requirements in data form, performance, environmental tolerance and packing density, have dictated the use of digital signals in CMOS memories. With the evolution of the CMOS memory technology, data storage in digital form has become dominant and self-evident without any extra statement, and the alternative analog data storage is distinguished by using the expression "CMOS analog memory." Similarly, because all CMOS memories operate without mechanically moving parts, an added word for mechanical classification would be redundant. A plethora of subclasses indicates the great diversity of CMOS memories, and includes classification by (1) basic operation mode, (2) storage mode, (3) data access mode, (4) storage cell operation, (5) storage capacity, (6) organization, (7) performance, (8) environmental tolerance, (9) radiation hardness, (10) read effect, (11) architecture, (12) logic system, (13) power supply, (14) storage media, (15) application, (16) system operation (Table 1.3) and by numerous other facets of the memory technology.

The vast majority of CMOS memories are designed to allow write and read and, in a much less quantity, read-only basic operation modes. In mask-programmed read-only memories the data contents cannot be reprogrammed by the user. User-programmable (reprogrammable, read-mostly) memories can also be, and are, made by combining programmable nonvolatile memory cells (which retain data when the power supply is turned off) with CMOS fabrication technologies. During the advancement of the memory technology, nevertheless, user-programmable nonvolatile memories emerged as a separate main class of memory technology that has its own specific subclasses, circuits and architectures. Therefore, the circuits and architectures of the user-programmable nonvolatile memories are discussed independently from the write-read and the mask-programmed read-only CMOS memories elsewhere in other publications, e.g., [12].

In CMOS memory technology the classification by access mode and by storage cell operation is of importance, because these two categories can incorporate the circuits and architectures of all other subclasses of CMOS memories. Consistently with the categories, this work first provides a general introduction to the CMOS random-, serial- and mixed-access and content-addressable memory architectures and, then, it presents the dynamic, static and fixed type of memory cells and the other compo-

1.	**Basic Operation Modes:**	Write-Read, Read-Only, User-Programmable.
2.	**Storage Mode:**	Volatile, Nonvolatile.
3.	**Access Mode:**	Random, Serial, Content-Addressable, Mixed.
4.	**Storage Cell Operation:**	Dynamic, Static, Fixed, Programmable.
5.	**Storage Capacity:**	Number of Bits or Storage Cells in a Memory Chip.
6.	**Organization:**	(Number of Words) X (Number of Bits in a Word).
7.	**Performance:**	High Speed, Low-Power, High-Reliability,
8.	**Environmental Tolerance:**	Commercial, Space, Radiation, Military, High-Temperature.
9.	**Radiation Hardness:**	Nonhardened, Tolerant, Hardened.
10.	**Read Effect:**	Destructive, Nondestructive.
11.	**Architecture:**	Linear, Hierarchical.
12.	**Logic System:**	Binary, Ternary, Quaternary, Other.
13.	**Power Supply:**	Stabilized, Battery, Photocell, Other.
14.	**Storage Media:**	Semiconductor, Dielectric, Ferroelectric, Magnetic.
15.	**Application:**	Mainframe, Cash, Buffer, Scratch-Pad, Auxiliary.
16.	**System Operation:**	Synchronous, Asynchronous.

Table 1.3. CMOS Memory Subclasses.

nent circuits which are specific to CMOS memories. The discussion of component circuits constitutes the largest part of this book, and includes the analysis of operation, design and performance-improvement of each circuit type. Improvements in reliability, yield and radiation hardness of CMOS memories by circuit technological means are provided separately, after the discussion of the component circuits.

CMOS memory integrated circuits are characterized, most commonly and most superficially, by memory-capacity per chip in bits and by access-time in seconds or by data-repetition rate in Hertzes. Generally, the access time indicates the time for a write or read operation from the appearance of the leading edge of the first address signal or from that of a chip-enable signal to the occurrence of the leading edge of the first data signal on the data outputs. The data repetition rate represent the frequency of the data change on the inputs and outputs at repeated writing and reading of the memory. Operational speed versus memory-capacity at a certain state of the industrial development (Figure 1.1) is of primary importance in choosing memories to a specific system application. access and content-addressable memory architectures and, then, it presents the dynamic, static and fixed type of memory cells and the other component circuits which are specific to CMOS memories. The discussion of component circuits constitutes the largest part of this book, and includes the analysis of operation, design and performance-improvement of each circuit type. Improvements in reliability, yield and radiation hardness of CMOS memories by circuit technological means are provided separately, after the discussion of the component circuits.

For system applications, a CMOS memory is succinctly described by a quadruple of terms, e.g., CMOS 32Mb x 8 25nsec DRAM, or CMOS-SOI 0.5Gbx1 200MHz Serial-Memory, etc. The first term states the technology; the second term indicates the memory storage capacity and organization; the third term marks the minimum access time or the maximum data rate; and the fourth term includes the access mode and often, also the storage cells' operation mode. The capability to operate in extreme environments, or, a specific performance or feature, are frequently identified by added terms, e.g., radiation-hardened, low-power or battery-operated, etc. Attached terms, of course, may emphasize any class, subclass or important property of the memory, e.g., synchronized, second-level cache, hierarchical, etc., which may be important to satisfy various system requirements.

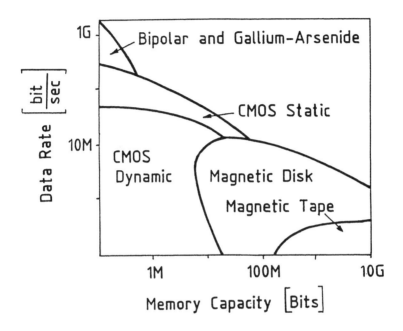

Figure 1.1. A data-rate versus memory-capacity diagram indicating application areas.

Performance requirements may also be expressed by cycle times in addition to access times. Commonly, cycle times are measured from the appearance of the leading edge of a first address signal, or that of a chip enable signal, to the occurrence of the leading edge of a next address or next chip-enable signal, when a memory performs a single write, or a single read or one read-modify-write operation.

Many of the memory applications in computing systems require data transfer in parallel-serial sets. For data-set transfers the data-transfer rate f_D [bit/sec, byte/sec], data-bandwidth BW_D [bit/sec, byte/sec] and the so-called fill-frequency FF [byte/sec/byte, Hertz], rather than access and cycle times are important. The fill-frequency is the ratio of the data-bandwidth and the memory-granularity MG [byte, bit], and it indicates the maximum frequency of data signals that fills or empties a memory device completely [13]. Memory-granularity, here, designates the minimum increment of memory-capacity [byte, bit] that operates with a single data-in and data-out terminal, and the data terminals of the individual memory-granules can simultaneously be used in a memory and in a computing system.

For a 16Kbyte memory that consists of two 16Kbit RAMs performing a maximum data-rate of 2.5 MHz the granularity is 16,384 bit and the fill-frequency is 152.59 Hz. The fill-frequency of the memory should exceed that of the computing circuits to obtain economic systems which are competitively marketable.

During the evolution of computing systems, the gap between the operational frequency of the central computing unit (f_{CPU}) and the data-transfer rate of CMOS memories f_D has continually increased (Figure 1.2). Since, at a given state of CMOS technology, $f_{CPU} \gg f_D$, high performance systems attempt to narrow the speed-gap by augmenting the bandwidth of the data communication between central computing units and CMOS memory devices, and by exploiting spatial and temporal relationships among data fractions which are stored in the memory and to be processed by the computing unit. A burgeoning variety of CMOS memory architectures have been developed for applications in high-performance systems. These performance-enhancing architectures are comprehensively described and analyzed in the literature of computing systems, e.g., [14], and memory applications, e.g., [15]. Furthermore, CMOS memories, which are designed specifically for low power consumption, have been developed to allow for packing-density increase and for applications in battery- and photocell-powered portable systems. Low-power systems and circuits use some special techniques which are widely published, e.g., [16], and the publications include the applications of the special techniques to low-power CMOS memory designs also, e.g., [17]. In this book, design approaches to both high-performance and low-power memories are integrated to the discussions of specific memory circuits and architectures rather than treated as separate design issues.

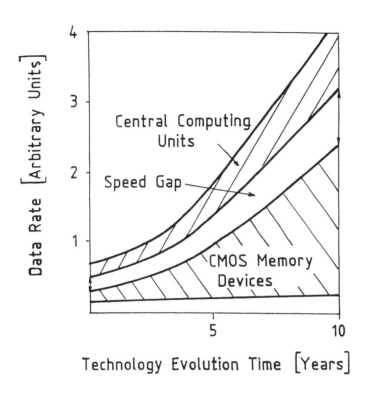

Figure 1.2. Widening performance gap between central processing units and CMOS memory devices.

In subject matter, this work focuses on write-read and mask-programmed read-only CMOS memories which either have established significant application areas or have foreseeable good potentials for future applications. The material presented here, nevertheless, can well be applied to the understanding, analysis, development and design of any CMOS memory.

1.2 RANDOM ACCESS MEMORIES

1.2.1 Fundamentals

In random access memories the memory cells are identified by addresses, and the access to any memory cell under any address requires approximately the same period of time. A basic CMOS random access memory (RAM) consists of a (1) memory cell array, or matrix, or core, (2) sensing and writing circuit, (3) row, or word address decoder, (4) column, or bit address decoder, and an (5) operation control circuit (Figure 1.3).

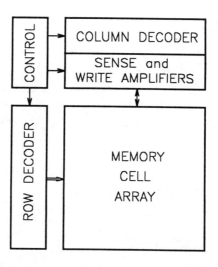

Figure 1.3. Basic RAM architecture.

Generally, the operation of a write-read RAM may be divided into three major time segments: (1) access, (2) read/write, and (3) input/output. The access segment starts with the appearance of an address code on the inputs of the decoders. The N-in/2^N-out row decoder selects a single wordline out of the 2^N wordlines of the memory-cell array. In an array of $2^N \times 2^N$ memory cells, this wordline renders the data input/output terminals of 2^N cells to 2^N bitlines, and an N-in/2^N-out column decoder selects S number of bitline. S is also the number of the sense and write amplifiers, and S may be between one and 2^N. In the second read/write operation segment, the sense and write amplifiers read, rewrite or alter the data content of the selected memory cells. During the input/output time segment, the sensed or altered data content of the memory cells are transferred through datalines to logic circuits, and to one or more output buffer circuits. The output buffer is either combined or separated from the input buffer. A data input is timed so that the write data reaches the sense and write amplifiers before the sense operation commences. Of course, no write can be performed in read-only memories. Every memory operation, e.g., write, read, standby, enable, data-in, data-out, etc., is governed by RAM internal control circuits.

Most frequently, CMOS RAMs are categorized by the operation of the storage cells into four categories: (1) dynamic RAMs (DRAMs), (2) static RAMs (SRAMs), (3) fixed program or mask-programmed read-only memories (ROMs), and (4) user-programmable read-only memories (PROMs). The following sections introduce the architectures of CMOS DRAMs, SRAMs and ROMs. CMOS PROM architectures are not discussed here, as stated beforehand, because PROMs have emerged as a distinct and extensive class of memories, and their architectures are developed to accommodate and exploit the unique properties of the nonvolatile memory cells (Preface).

1.2.2 Dynamic Random Access Memories (DRAMs)

Write-read random access memories, which have to refresh data in their memory cells in certain time periods, are called dynamic random access memories or DRAMs. CMOS DRAMs along with microprocessors evolved to be the most significant products in the history of solid-state circuit technology. Among all the various solid-state circuits, CMOS DRAMs are manufactured and traded in the largest volumes.

The attractiveness of DRAM devices is attributed to their low costs per bit, which stems from the simplicity and minimum area requirement of their fundamental elements, the dynamic memory cells. In a DRAM cells a binary datum is represented by a certain amount of electric charge stored in a capacitor. Because the charges inevitably leak away through parasitic conductances, the data must periodically be rewritten, or with other words "refreshed" or "restored", in each and all memory cells. Refresh is provided by sense and write amplifiers associated with each individual bitline. Commonly, the number of individual sense amplifiers is the same as the number of the bitlines in the array, or, with other words, same as the number of bits in a DRAM internal word. A DRAM refreshes the data in all bits which connect to a selected wordline at all three operations, at write, read and refresh, with the exception of the data of those bits which are modified at a write operation.

A typical architecture of a DRAM (Figure 1.4) includes a refresh controller, a refresh counter, buffers for row and column addresses and for data input and output, and clock generators, in addition to the constituent circuits of the basic RAM. The refresh controller and counter circuits assure undisturbed refresh operation in all operation modes, i.e., it addresses sequentially and provides timing for the refresh of each row of memory cells. The DRAM-internal refresh control simplifies the DRAM applications in systems. Applicability is facilitated also by the address and data buffers, while clocks are fundamental to the internal operation of the DRAM.

The general operation of the DRAM has little deviation from the operation of the basic RAM (Section 1.2.1). Initially, the DRAM is activated by a chip-enable signal CE. CE and the row and column address strobe signals RAS and CAS generate control signals. Some of these control signals allow the flow of the address bits to the decoders either simultaneously or in a multiplexed mode. Multiplexing can reduce pin-numbers and, thereby, costs without compromises in memory access and cycle times. In multiplexed memory addressing, first the row address and, thereafter, the column address is transferred to the row and column address buffers. Next, the row decoder selects a single wordline from 2^N wordlines. The selected wordline activates all 2^N memory cells in the accessed row, and 2^N memory cells put a 2^N-bit data set to 2^N bitlines. On the bitline terminals 2^N sense amplifiers read and rewrite, or just write, the data in accordance with the state of the write/read control signal W. From the 2^N-bit data set, the column decoder selects a single bit or a multiplicity of bits,

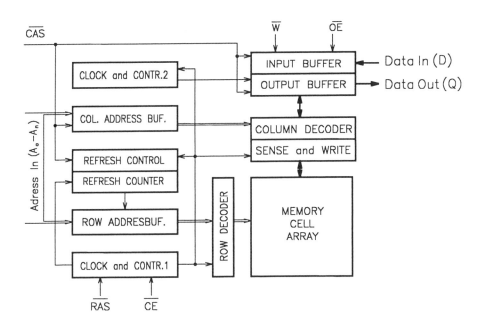

Figure 1.4. Typical write-read DRAM architecture.

and these data bits are passed to the output buffer and to the data output Q. The time when the data are valid, i.e., they can be used to further processing, is controlled by the output enable signal OE. The input data are stored in the input buffer, and timed to reach the sense amplifiers before the memory cells are activated. Timing is crucial in DRAM operations, and a DRAM may apply over a hundred chip-internal clock impulses to keep the wiring and circuit caused delays under control, and to synchronize the operation of the different part-circuits which are distributed in various locations of the memory chip. In this sense, almost all DRAMs are internally synchronous designs. Historically, DRAM designs with self-timed (asynchronous) internal operation have provided significantly slower and less reliable operations than DRAMs of internally synchronous designs have done. Asynchronous interfaces may be applied between synchronously operating blocks in a large-size DRAM chip to recover from the effects of clock-skews. In CMOS technology, nevertheless, the term "synchronous DRAM" reflects that the DRAM is designed for application in synchronous system, and the DRAM operation requires a system master clock and eventually other control signals which are synchronized to the master clock.

Similarly to other memory devices, a DRAM is characterized by features, absolute maximum ratings, direct current (DC) and alternating current (AC) electrical characteristics and operation conditions. AC electrical characterization and operating conditions of DRAMs, however, involves rather specific timing of clocks. The use of some external clock signals such as row-address-strobe RAS, column-address-strobe CAS, write control W, output enable OE and chip enable CE or address change detection is required to most of the applications, and their timing determines several performance parameters, e.g., the access times from the leading edge of the RAS signal t_{RAC}, from the leading edge of the CAS signal t_{CAC}, and from the appearance of the column address t_{AA}, and the read-modify-write cycle time t_{RWC} (Figure 1.5) and others.

The access and cycle times of a memory is determined by the longest delay of the address and data signals along the critical path. In DRAMs, a greatly simplified critical path includes the (1) row address buffer, (2) row address decoder, (3) wordline, (4) bitline, (5) sense amplifier and (6a) output buffer or (6b) precharge circuit (Figure 1.6). The output buffer delay is a segment of the access times, while the precharge time is a portion of the cycle times. Neither access nor cycle times are influenced by the data input buffer delay, because the data buffer operation may be timed simultaneously with the column address decoder or even sooner, e.g., as in an "early write" operation mode. Furthermore, cycle times may be shortened by performing precharge during column addressing and data-output time.

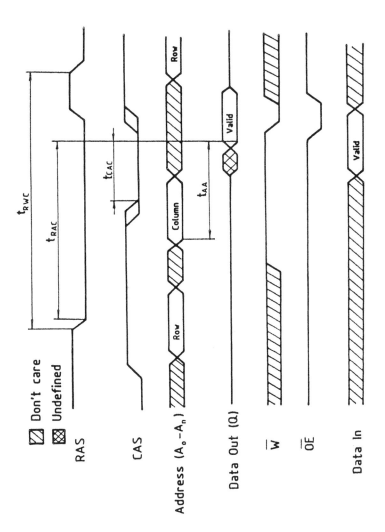

Figure 1.5. Some commonly used DRAM timing signals and parameters.

16   CMOS Memory Circuits

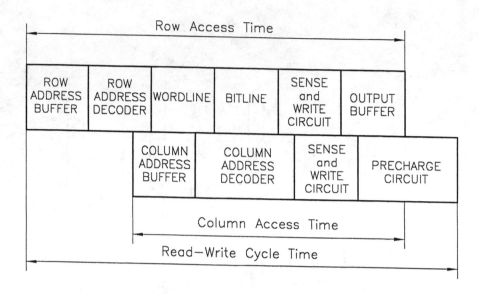

Figure 1.6. Simplified critical paths in a DRAM

Access and cycle times may be reduced by giving up some randomness in the data access, e.g., by exploiting that the bits within a selected row or column are sooner available than bits randomly from the whole array. That type of randomness limitation is utilized in page mode (PM) and static column mode (SCM) operations. For the accommodation of the page and static column operation modes all DRAM array designs (Figure 1.7) are inherently amenable. In page mode, after the precharge and row activation, 2^N data bits are available in the 2^N sense amplifiers. Any number of these 2^N data bits can be transferred to the output, or rewritten, in the pace of the rapidly clocking CAS signal when an extended RAS signal keeps the wordline active for t_{RASP} time (Figure 1.8). The data rate of the CAS signal clocks $f_{CD} = 1/t_{PC}$ is fast because $t_{PC} = t_{CAS} + t_{CP} + 2t_T$. Here, t_{CAS} is the width of the CAS pulse, t_{CP} is the exclusive time to precharge the bitline capacitance, and t_T is an arbitrary signal transition time. Throughout the duration of t_{RASP} column addresses may change randomly, but the row address remains the same. A row address introduces a latency time $t_{LP} = t_{RAC} + t_{RP} + 2t_T$, where t_{RP} is the RAS

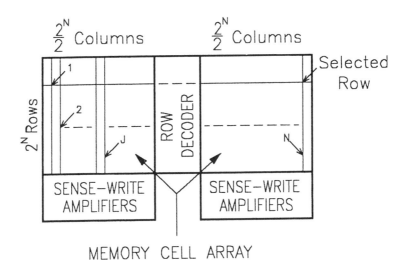

Figure 1.7. Accommodation of page and static-column operation modes in a DRAM array.

precharge time that appears after a sense operation is completed and before the row address buffer is activated.

Furthermore, before the column address reaches the memory cell array the precharge of the sense amplifier has to be concluded, and every column address needs a setup time t_{ASC} and hold time t_{CAH}. Precharge, setup and hold time periods are inherent to traditional DRAM operations, and constrain the possibility of obtaining gapless changes in address and data signals in page mode or in its improved variations, e.g., in the enhanced or fast page mode (FPM). The static column operation mode attempts to make the fast page mode faster by keeping the CAS signal "statically" low rather than pulsated, and the bits in the sense amplifiers are transferred to the output or rewritten simultaneously with the appearance of the column address signals. Thus, after a single preparatory period, that consist of a row address, a setup, a hold and a transition time, gapless column address changes (Figure 1.9) can be obtained during the time when RAS signal keeps a single wordline active. Because the CAS

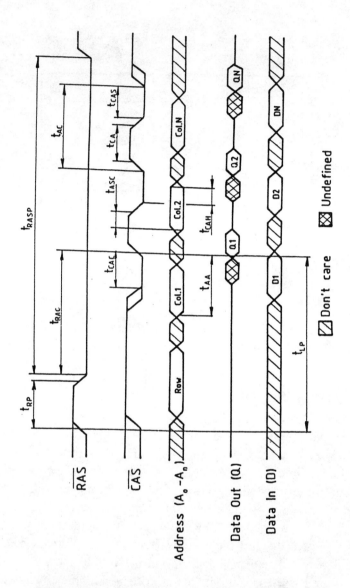

Figure 1.8. An example of page-mode timing.

signal is unpulsated, one t_T transient time can be eliminated at each column access, but the lack of defined CAS clocks for data transfers and

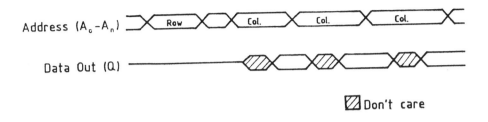

Figure 1.9. Address and read-data signals in a static column operation mode.

the rather high implementation costs make the application of static column mode less attractive than that of the page mode.

Apart from fast page and static column mode implementations, a large variety of architectural approaches can increase DRAM data rates by the accommodating operation modes which hide or eliminate some of the internal DRAM operations for the time of a set of accesses. By little extension in the DRAM architecture, e.g., by adding a nibble selector and M-bit (traditionally M=4) input and output registers, a nibble mode operation may be implemented. In a nibble mode read, the nibble selector chooses M bits from the contents of the sense amplifiers, these bits are transferred parallel to the output register which clocks its content rapidly into the output buffer. In a write operation, the input buffer sequentially loads the input register which transfers the M-bit data to the write and sense amplifiers. Because most of DRAM-sense-amplifiers can latch data, the input and output registers may be eliminated from the design. In designs where the memory cell array is divided into M blocks, M bits can be addressed simultaneously and used for fast data input and output in a fast nibble mode. No extra input and output data register is needed in the multiple I/O configuration, but M simultaneously operating input and output buffers transfer the data parallel into and out of the memory chip. The parallel buffer operation dissipate high power, and may result in signal ringing or bounces in the ground and supply lines, which limit the implementability and performance of wide I/O architectures.

Generally, DRAM performance in computing systems can be improved through an immense diverseness of both architectural and circuit technological approaches. Minor architectural modifications to accom-modate page, fast page, static column and nibble modes provide only temporary solutions for the demands of increasing DRAM data rates and bandwidths. To keep up with the aggrandizing demand for higher performance (Section 1.1), DRAM architectures need to implement extensive pipelining and parallelism in their operations, and to minimize the data, address and other signal delays. The mostly applied architectural approaches to improve DRAM data rates and bandwidths are described next (Sections 1.2.3-1.2.6) and under the special memories and combinations (Section 1.5).

1.2.3 Pipelining in Extended Data Output (EDO) and Burst EDO (BEDO) DRAMs

Pipelined architecture and operation in DRAMs are usually implemented to increase the data transfer rate for column accesses, although pipelining could increase the data rate at for row accesses as well. Column access and cycle times are usually shorter than row access and cycle times. An access to a particular row and consequent rapid column-address changes within the single accessed row, can greatly enhance the data rate of traditional DRAMs.

The effect of pipelining on the data transfer rate can be made plausible by a greatly simplified chart of successive critical paths (Figure 1.10). Here, in a critical path, the time period from the appearance of address change to the access of a memory cell is $A(N)$; the time from the end of $A(N)$ to the accomplished data sensing is $S(N)$; the delay from the end of $S(N)$ to the valid data output is $O(N)$; and the total precharge time is $P(N)$. N designates the operations and signal delays associated with column address N, while N+1 indicates the address that follows N in time. The time period between $A(N)$ and $A(N+1)$ is t_{CD}, and t_{CD} indicates the efficiency of the pipelining schema. Some pipelining may inherently appear in fast page mode, when the addressing phase $A(N+1)$ follows the output delay O(N) rather than the precharge or initiation phase P(N). Namely, in many designs P(N) = P(N+i) is longer than O(N) = O(N+i), where i is an integer, and the time difference between P(N) and O(N) allows for shortening the data repetition time t_{CD}.

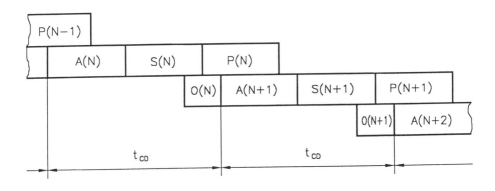

Figure 1.10. Covert pipelining in a fast page mode implementation.

Reduction in t_{CD} can easily be achieved by using extended-data-out (EDO) architecture. EDO DRAM architecture connects a static flip-flop (FF) directly to the common output of a row of sense amplifiers (Figure 1.11). Since FF provides the data of the address N for the time the data travels to the output, the next column addressing phase A(N+1) may appear as soon as the data transfer from the sense amplifier into FF is accomplished (Figure 1.12). FF allows CAS signal to go higher while waiting for the data to become and stay valid on the output node, and perform the precharge simultaneously with O(N) and A(N+1). This simultaneousness shortens $t_{CD} = t_{CP}$ in fast-page mode. The EDO fast-page mode, also called hyper-page mode, may be controlled by OE or W signals to turn the output buffers into high impedance states after the appearance of valid output data.

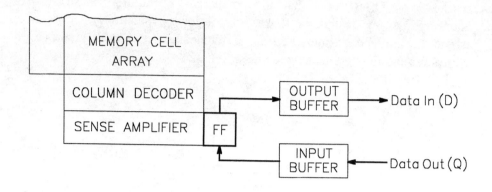

Figure 1.11. Flip-flop placed into the data path to facilitate EDO operation mode.

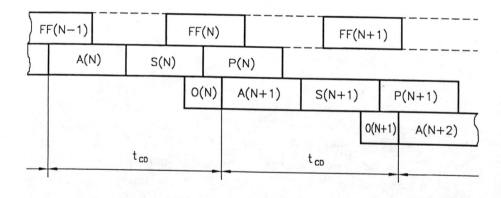

Figure 1.12. Pipelined timing in an EDO DRAM.

As a further improvement, the data transfer from the sense amplifier to a digital storage stage FF may start as early as the output signal reaches the level which can change the state of the storage stage. Moreover, the correct level of precharge or initiation can much sooner be provided on the sense amplifier nodes than on the bitlines, because the bitline-capacitances are much smaller than the sense amplifier node-capacitances. Separating bitline precharge from the sense-input precharge, the bitline precharge can be taken out from the critical path and hidden. Thus, in the critical path a reduced precharge or initiation time PR(N) and a shorter sense time SS(N) occur.

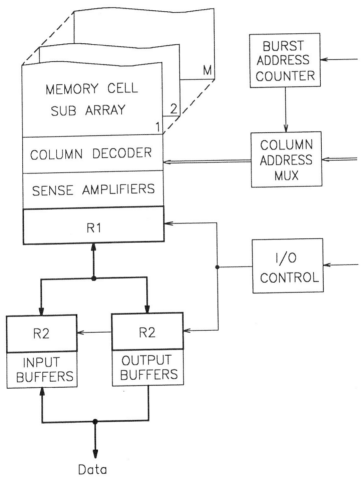

Figure 1.13. A pipelined BEDO implementation in a DRAM.

Pipelined burst EDO (BEDO) DRAMs take advantage of quick sense SS(N) and precharge PR(N) operation by replacing FF by a two-stage register R1 and R2 so that the second R2 is placed close to the data output buffer (Figure 1.13). In pipelined BEDO timing (Figure 1.14) R1 receives the data from the sense amplifier, stores data during the precharge or initiation time of the sense amplifier PR(N), and transfers data to R2. While the data of address N is in R1 and R2, most of A(N+1) and SS(N+1) can be accomplished. The data from address N reaches the outputs only after a follower CAS signal initiates an access to the data stored on address N+1. A data burst is created by parallel access of one memory cell in each of M subarrays, when only a single row and a single column address is used with the RAS and CAS signals and the rest of the column addresses are generated DRAM-internally. Column addressing, in most BEDO implementations, is also pipelined through the column address buffer, a multiplexer and the decoder. This column address pipeline allows a new random column address to start consequent bursts without a gap. A change in row address, however, requires longer latency time than that in EDO or fast-page mode DRAM.

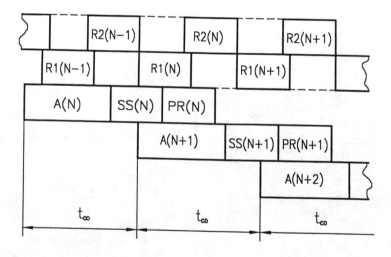

Figure 1.14. Simplified timing structure of a pipelined BEDO operation.

Neither the EDO nor the BEDO with pipelining can operate with such data output and input signals which have no gaps between the periods of their validity. Gaps among the valid output signals appear, because only the data of one row can be sensed at a time by one set of sense amplifiers and during the precharge time of the sense amplifiers data can not be sensed, and because within a particular memory the delays for addressing, sensing, and precharging are usually unequal.

1.2.4 Synchronous DRAMs (SDRAMs)

In RAMs pipelining can reduce the data repetition time t_D to the time period of the required minimum for output signal validity, and can allow for a gapless input and output signal sequence (Figure 1.15). Gapless input and output signal sequences can be provided in DRAMs which are designed with synchronous data and address interfaces to the system and which are controlled by one or more DRAM-external clock signals.

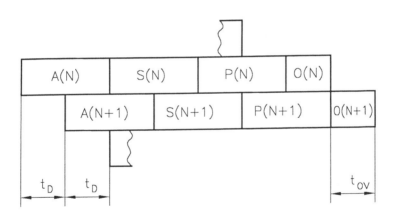

Figure 1.15. Possible data-repetition time reduction in pipelined synchronized DRAMs.

In an SDRAM, an external clock synchronizes the DRAM operation with the system operation, and, therefore, such a clock controlled DRAM is called synchronous-interface DRAM, or synchronous DRAM or

SDRAM. Because of the synchronized operation a full period of the master clock can be used as unit of time, e.g., a 4-1-1-1-1 SDRAM uses four clock periods of RAS latency time and after that it generates a series of four valid data output signals in each clock period (Figure 1.16). Here, the access and RAS latency times take four clock periods, the CAS latency time lasts two clock periods and, thereafter, the data change back-to-back in every single clock period. The number of back-to-back appearing data bits can be as high as the number of the sense amplifiers, when the CAS data burst mode is combined with a so-called wrap feature.

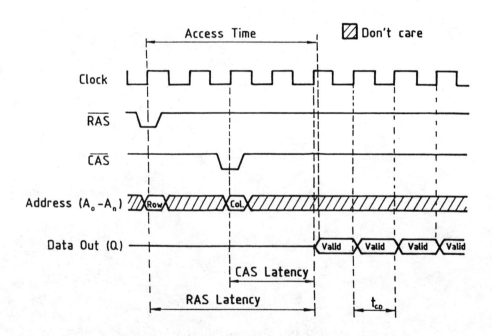

Figure 1.16. Timing in a hypothetical SDRAM design.

A wrap instruction allow to access a string of bits located in a single row regardless of the column address of the initially accessed memory cell.

Synchronous memories may have either single data rate (SDR) or double data rate (DDR) type of interface signals. In a single clock period, an SDR interface uses a single valid input or output datum, while a DDR interface accommodates two valid data which are synchronized with the rising and falling edges of the clock signal.

Synchronization by external clock signals can be designed into any memory, also in DRAMs which have single or multibank architecture. A bank, here, includes the memory-cell array and its row- and column-decoder, sense-, read- and write-amplifiers, read- and write-register, data input and output buffer circuits. Single-bank SDRAM circuits have been noncompetitive with dual-bank SDRAMs, because the dual-bank architecture allows for significantly faster operation than the single-bank approach does, and the chip size of both single- and dual-bank SDRAMs can approximately be the same at same bit-capacities for both the single- and dual-bank designs.

Although all dual-bank SDRAMs can exploit pipelining in some forms, the nomenclature distinguishes so-called prefetched (Figure 1.17a) and pipelined (Figure 1.17b) dual-bank SDRAM data-interface structures. The prefetch technique brings a data word alternatively from each bank to a multiple-word input-output register during each clock cycle. A word, here, is the number of columns in an array or of the sense amplifiers which serve one bank. A prefetched dual-bank structure allows to run the data inputs and outputs of the memory faster than the operational speed of the individual banks, but disallows back-to-back column CAS addressing during a word-wide data burst. In the pipelined structure, two separate address registers provide addressing alternatively to the two banks, no added register for data input and output is needed, and back-to-back column CAS addressing is permitted. Nevertheless, the clock-frequency of the data inputs and outputs is the same as that in the banks. Whether a dual bank memory can achieve minimum output valid time in gapless operation, it depends on the combination of its internal delays. Thus, the addressing delays may limit the performance of those dual-bank DRAMs which are designed with one set of address input. Internal time multiplexing of the input address to each row and column access and a combination of prefetch and pipelined architectures may be used to further improve cycle times in numerous designs.

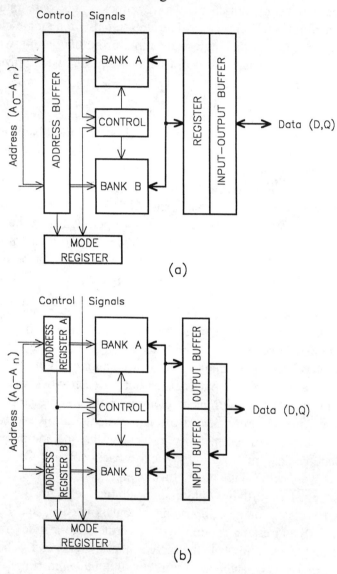

Figure 1.17. Prefetched (a) and pipelined (b) dual-bank DRAM architectures.

Introduction to CMOS Memories 29

Cycle times can be minimized to minimum output valid times in architectures which have more than two banks (Figure 1.18). Multibank DRAM (MDRAM or MSDRAM) architectures may include data first-in-first-out (FIFO) registers (Section 1.3.5), data formatter, address FIFO register, timing register and a phase locked loop, in addition to the DRAM banks. The data and the address FIFO registers serve as interface buffers between the memory and the memory-external circuits which may operate

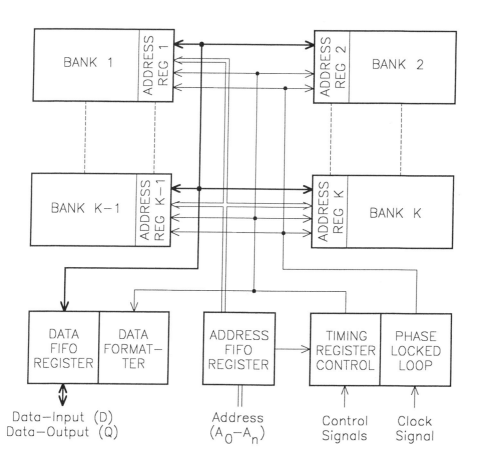

Figure 1.18. A multi-bank SDRAM architecture.
(Derived from a product of MoSys Incorporated.)

with different clock-frequencies. With the frequency of the system clock a data formatter regulates burst lengths, burst latencies, data-masking and their timings. The timing register may accommodate clock enable CKE, master clock MC, mask data output DQM, function F as well as the usual RAS, CAS, W and CE signals or other control signals to enhance system applicability. Usually, a skew of the system clock is corrected and the memory operation is synchronized by a phase-locked-loop circuit (Section 4.6.3).

In synchronous memories pipelining may be designed by applying the staged circuit or the signal wave technique [17]. The staged circuit technique requires the insertion of signal storage elements between the end and the beginning of two pipelined subcircuits, e.g., a flip-flop set between the sense amplifiers and the output buffers, or latches between the column-address predecoder and decoder circuit, etc. Stage-separator storage circuits buffer the variations in subcircuit delays and make possible to synchronize memory-internal operations with the system's master clock. Subcircuit delays may significantly differ, and the longest subcircuit delay or part-delay and the largest part-delay variation restrict the maximum data-rate achievable by staged pipelining. Wave pipelining divides the total delay, from the inception of the address-change to the appearance of the valid output data, into even time intervals, and in each interval an address-change can occur. The duration of the intervals and the maximum obtainable data-rate are limited, here, by the spread or dispersion of the total delay time rather than by the sum of the largest part-delay and part-delay variation. Since the maximum dispersion of the total delay is smaller than the amount of the longest part-delay and part-delay spread, higher data-rates are achievable by wave pipelining than by staged pipelining.

Generally, implementations of pipelined multibank SDRAM architectures provide very attractive means to greatly increase the input/output data rates of traditional DRAMs, without improvements in DRAMs' circuit and processing technologies. Multibank SDRAM architectures, nevertheless, can not multiply the data rate and the bandwidth of traditional DRAM proportionally with the number of pipelined memory banks. The timing of DRAM and SDRAM operations (Figure 1.16), namely include a row access strobe, or RAS latency time, which may last some, e.g., four, clock-period long. At low frequency operations, the RAS latency is only a small percentage of the write and read times, but at high frequencies the RAS latency may several times be longer than the write and read times. To decrease the influence of the RAS latency in the write/read data rates, an increasing number of banks are required. At a certain number of banks, however, the increasing number and lengths of long chip internal interconnects, and the increasing complexity of the circuits, degrade the speed, power and packing density parameters to unacceptable values.

1.2.5 Wide DRAMs

The bandwidth of communication between a memory and other circuits can effectively be enhanced by the increase of the number of simultaneously operating write and read data inputs and data outputs in the memory chip. A random access memory that has a multiplicity of data inputs and outputs is called wide RAM, and if dynamic and static operation is particularly indicated, they are called wide DRAMs and wide SRAMs, respectively. Wide DRAMs feature significantly higher packing densities than wide SRAMs do. Most of the wide DRAM designs are required to provide a number of inputs and outputs which are integer-factors of a byte, e.g., 8, 16, 32, or 64 bits, but some designs need to add a few inputs and outputs, e.g., 1, 2, or 4 bits, for error detection and correction. The multiple data inputs and outputs can most speed-efficiently be supported by an architecture that divides the memory cell array into blocks, which are simultaneously accessed.

The architecture of a wide DRAM (Figure 1.19) comprises Z number of X x Y memory cell subarrays, an X-output row decoder, a Y-output column decoder, N-bit input and output data buffers, and a mask register in addition to the traditional clock generators, refresh counter and controller and the address buffers. Usually, in the data buffers $N = i \times 8$, where $i = 1,2,4...$, and in the memory cell array $Z=N$, e.g., for a subarray of $X \times Y = 1024$ an $i=2$ and a $Z=N=16$ are designed. During a read operation, Z blocks of X sense amplifiers, e.g., $X \times Z = 1024 \times 16$ sense amplifiers, are active. Each sense amplifier bank holds temporarily the data from the addressed X-bit row of its XxY bit memory cell array after the same rows in all Z arrays are accessed. When the same columns in Z arrays are selected, the data of each individual one of Z columns are moved simultaneously from Z sense amplifiers to Z=N output terminals. In a write operation, the input data flow simultaneously from Z=N input terminals to Z sense amplifiers. The data content of the sense amplifiers may be masked. If the mask changes the data patterns during the appearance of the consequent write enable W signals or of row address strobe RAS signals, the masked write is called nonpersistent, otherwise the masked write is persistent. Write enable may separately be provided for lower and higher byte sets, e.g., through WEL and WEH control signals.

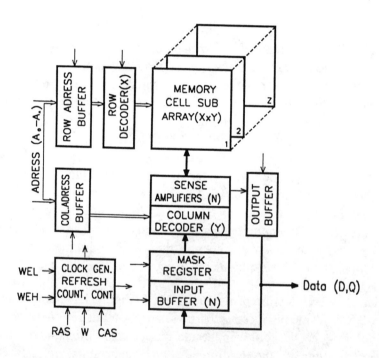

Figure 1.19. Data and address paths in a wide DRAM.

In systems, very high bandwidth and data rate can be obtained by applications of wide DRAMs. Nonetheless, wide DRAMs dissipate great amounts of power due to the high number of simultaneously activated data output buffers and sense amplifiers. Furthermore, the simultaneously operating output buffers and sense amplifiers cause large current surges, which may degrade the reliability of wide DRAMs mainly by hot-carrier emissions in the constituent transistors. Both the power dissipation and reliability of wide DRAMs can substantially be improved by using special circuit techniques and coding in the design of the output buffers (Section 4.9).

1.2.6 Video DRAMs

A video random access memory (VRAM) is a high-speed wide-bandwidth DRAM, that is designed specifically for applications in graphics systems. In graphics systems a display memory stores the data representing pixels to be displayed, a buffer memory provides timing interface and parallel-serial data conversion between the display memory and a monitor, and a video monitor shows the pattern or picture assembled from the pixel data (Figure 1.20).

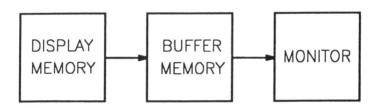

Figure 1.20. Simplified graphics system.

High data rates and wide bandwidths of VRAMs are achieved by combining the display and buffer memories in a single chip, by constructing the display memory of a multiplicity of simultaneously accessible arrays made of two-port or triple-port DRAM cells, and by implementing the buffer memory in SRAMs. A VRAM architecture is either dual-port with one random-access and one serial-access port, or triple-port with one random-access and two serial-access ports. A dual-port VRAM comprises only one buffer memory that restricts the sequential data flow to one direction at a time. Data can move in both directions in and out of a triple-port VRAM simultaneously, because it has two buffer memories and one display memory. In essence, the buffer memory is a serial access memory SAM, that can move the columns or rows of bits step-by-step sequentially, and in that SAM either a column, or a row, or both can be written and read parallel.

To data writing and reading the design has to provide three typical VRAM operations: (1) asynchronous parallel access of a DRAM data port, (2) high-speed sequential access of one or two SAM ports, and (3) data transfer between an arbitrary DRAM row and one or two SAMs. DRAM and SAM ports must be accessible independently at any time except during a data transfer between the DRAM and a SAM. Some design requirements comprise data transfer only from DRAM to SAM, others encompass bidirectional data moves between the DRAM and a SAM. Not all VRAMs perform sequential data write, but all VRAMs have serial read capability.

In a triple-port VRAM (TPDRAM) architecture (Figure 1.21) the data transfer between the DRAM and both SRAMs is very quick, because these memories are placed in close proximities to each other in the chip. Each (X x Y)-bit memory cell array has Y number of sense amplifiers, all the Z number of arrays are addressed simultaneously. The operation of Y x Z number of sense amplifiers allow for minimizing both RAS and CAS latency times and for speedy parallel data transfers from and to all of the DRAM arrays. Data transfer rates in the SAMs are much faster than those in the DRAMs, because SRAMs have inherently short access times and because the size of each SAM (Y x Z)/2 bits is much smaller than that of the DRAM (X x Y x Z) bits. Both SAMs may operate synchronously or asynchronously and independently from each other. For independent operation each of both SAMs is controlled its individual address counter. While an address counter can point any word address in a SAM, a bit-mask register controls the location of the bits in the word, which is to be written and, in some designs, read also. A column mask register is also used to modulate the DRAM's column address by the content of the color register. Otherwise, the DRAM constituents and operation are the same as

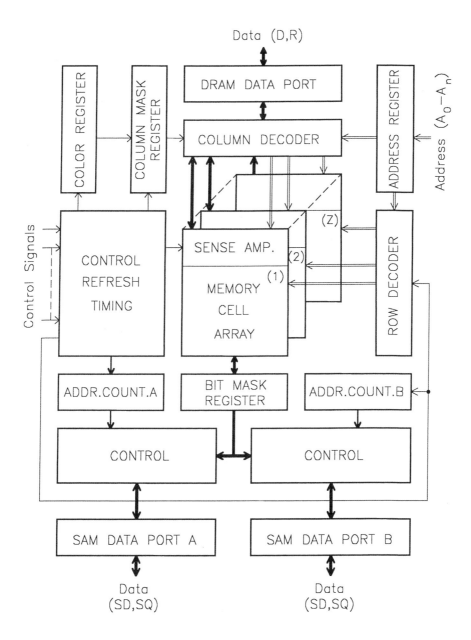

Figure 1.21. A triple-port VRAM architecture.
(Derived from a product of Micron Semiconductor Incorporated.)

those for other DRAMs with the exception of the memory cells, which are, in the depicted architecture, triple-port dynamic memory cells.

Dual-port VRAMs, of course, apply dual-port memory cells, and their architecture is similar to that of triple-port VRAMs, but dual-port VRAMs have only one SAM array, one address counter and one serial data port. Both the dual- and triple-port VRAMs allow for substantial increase in system performance without placing stringent requirements on DRAM access and cycle times.

1.2.7 Static Random Access Memories (SRAMs)

Random access memories which retain their data content as long as electric power is supplied to the memory device, and do not need any rewrite or refresh operation, are called static random access memories or SRAMs. CMOS SRAMs feature very fast write and read operations, can be designed to have extremely low standby power consumption and to operate in radiation hardened and other severe environments.

The excellent speed and power performances, and the great environmental tolerances of CMOS SRAMs are obtained by compromises in costs per bit. High costs per bits are consequences of the large silicon areas required to implement static memory cells. A CMOS SRAM cell includes four transistors, or two transistors with two resistors, to accommodate a positive feedback circuit for data hold, and one or two transistors for data access. The positive feedback between two complementary inverters provides a stable data storage, and facilitates high speed write and read operations. The data readout is nondestructive, and a single sense amplifier per memory cell array or block is sufficient to carry out read operations.

SRAM architectures and operations are very similar to that of the generic RAM (Section 1.2.1), but an SRAM architecture comprises also row and column address registers, data input-output control and buffer circuits, and a power down control circuit, in addition to the constituent parts of the generic RAM (Figure 1.22). The operation of the SRAM starts with the detection of an address change in the address register. An address change activates the SRAM circuits, the internal timing circuit generates the control clocks, and the decoders select a single memory cell. At write, the memory cell receives a new datum from the data input buffers; at read, the sense amplifier detects and amplifies the cell signal and transfers the datum to the output buffer. Data input/output and write/read are controlled

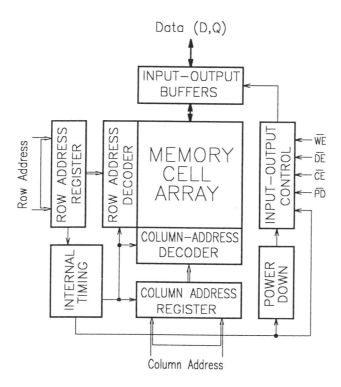

Figure 1.22. An SRAM architecture.

by output enable OE and write enable WE signals. A chip enable signal CE allows for convenient applications in clocked systems, and system power consumption may be saved by the use of the power down signal PD. The power down circuit controls the transition between the active and standby modes. In active mode, the entire SRAM is powered by the full supply voltage; in standby mode, only the memory cells get a reduced

38   CMOS Memory Circuits

supply voltage. In some designs, the memory-internal timing circuit remains powered and operational also during power down.

In SRAM operations, the read and write cycle times t_{RC} and t_{WC} are specified either as the time between two address changes or as the duration of the valid chip enable signal CE, the address access time t_{AA} is the period between the leading edge of the address change signal and the appearance of a valid output signal Q, the chip enable access time t_{ACE} is the time between the leading edge of the CE signal and the appearance of Q, t_{AW} is the time between address change and the end of CE, and t_{CW} is the period from CE to the write signal or the duration of CE (Figure 1.23). In certain

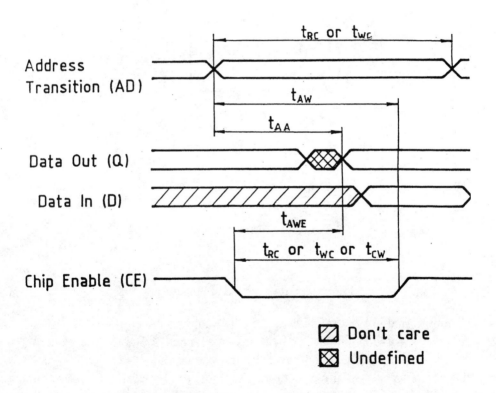

Figure 1.23. Cycle and access times in an SRAM.

SRAM designs, the output enable signal (OE) may also be used as reference for determining various cycle and access times. The critical path determining cycle times comprises the delays through the (1) row address buffer, (2) row address decoder, (3) wordline, (4) bitline, (5) sense amplifier and (6) output buffer circuits (Figure 1.24). Precharge and initiation times for sensing as well as column address buffer and decoder delays can well be hidden in the critical timing of an SRAM.

For internal timing, SRAMs apply a high number of clock impulses generated from an address change or chip enable signal in asynchronous systems, or and from the systems' master clock in synchronous systems. Synchronous and asynchronous operation, here also, relates to memory-system interface modes rather than to chip-internal operation modes.

The cycle times of SRAM operations can significantly be decreased by pipelined designs, similarly to the pipelining in DRAMs. (Sections 1.2.3 and 1.2.4). Synchronized SRAMs or SSRAMs, designed with multiple bank architectures, can serve very fast computing and data processing circuits. Some SRAM circuit designs allow to implement pipelining without using registers to separate the intervals of the individual delays in the so-called wave pipelining schema (Section 1.2.4).

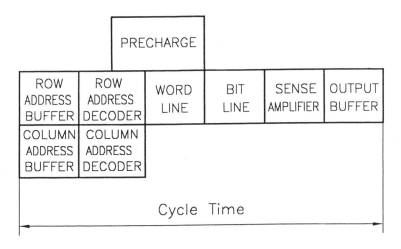

Figure 1.24. Critical timing path in an SRAM.

1.2.8 Pseudo SRAMs

Pseudo-SRAMs are actually DRAMs, which provide chip-internal refresh for their stored data and in effect appear as SRAMs for the users. Like many DRAMs, pseudo-SRAMs also, allow three methods for refresh: (1) chip-enable, (2) automatic, and (3) self-refresh. A chip-capable refresh cycle operates during the appearance of the CE signal, the rows have to be externally addressed, and the row accesses have to be timed so that none of the data stored in the DRAM can degrade to an undetectable level. Automatic data refresh cycles through a different row on every output enable OE or refresh RFSH clock impulse without the need to supply row addresses externally. Self-refresh does not need any address input nor any external clock impulse. Without any external assistance, a timer circuit keeps track with the time elapsed from the last memory access, and

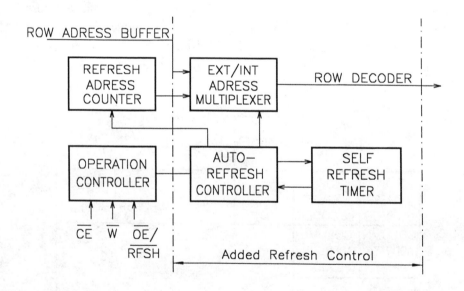

Figure 1.25. Refresh controller and timer in a pseudo-SRAM architecture.

generates signals to refresh each individual row of the pseudo-SRAM sequentially.

The architecture of a pseudo-SRAM is the same as that of a DRAM but for internal refresh some logic to the operation control an automatic refresh controller and a self-refresh timer are added to the basic DRAM configuration (Figure 1.25). Pseudo-SRAMs' characterizations and applications are also similar to those for SRAMs. Pseudo-SRAMs operate some slower than their SRAM counterparts due to the hidden refresh cycles, but their smaller memory cell size makes possible to produce them at low costs.

1.2.9 Read Only Memories (ROMs)

Random access memories, in which the data is written only one time by patterning a mask during the semiconductor processing and cannot be rewritten nor programmed after fabrication, are the read only memories or ROMs. CMOS ROMs can combine very large memory bit capacity, fast read operation, low power dissipation and outstanding capability to operate in radiation hardened and other extreme environments at low manufacturing costs.

The combination of high performance and packing density with low costs is attributed to the use of the one-transistor NMOS or PMOS ROM cells. The sole transistor in the ROM cell can be programmed to provide either high or low drain-source conductance at a certain gate-source voltage. Furthermore, the programming can be performed by one of the last masks in the processing sequence, which results in short turn-around times in ROM manufacturing. ROM cells may be arranged in NOR or NAND configurations depending on the particular design objectives.

Both the architectural design (Figure 1.26) and the operation of ROMs are basically the same as those of the SRAMs (Section 1.2.7) with the exception that ROMs have no write-related circuits, signals and controllers. The critical signal path in a ROM includes the delays of the (1) row address buffer, (2) row address decoder, (3) wordline, (4) bitline, (5) sense amplifier, and (6) output buffer. Sense amplifiers may be very simple and fast operating because the hard-wired data storage provides large differences in signal levels. Pipelining of addressing and data signals are commonly used in ROM designs, which along with synchronous operation can greatly improve output data rates.

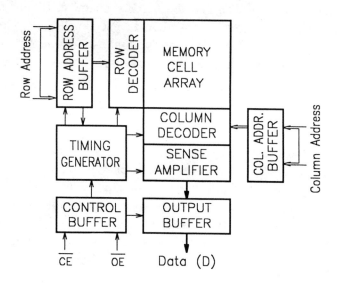

Figure 1.26. ROM architecture.

1.3 SEQUENTIAL ACCESS MEMORIES (SAMS)

1.3.1 Principles

In a sequential or serial access memory (SAM) the memory cells are accessed each after the other in a certain chronological order. The locations of the memory cells are identified by addresses, and their access time depends strongly on the address of the accessed memory cell. In a generic SAM (Figure 1.27), an address pointer circuit selects the requested address and controls the timing of the data in- and outputs, a sequencer circuit controls the order of succession of addresses in time, and a memory cell array contains the spatially dispersed data bits.

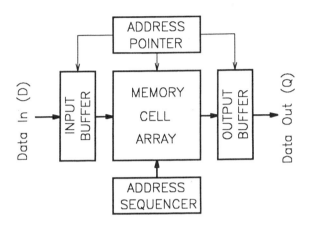

Figure 1.27. Generic SAM structure.

A SAM array is combined either of static or dynamic random access memory cells, or static or dynamic shift register cells. Theoretically, nonvolatile data storage cells may also be used in sequential memory designs. In characterization of sequential memories the input/output data rate or frequency f_D and the bandwidth BW indicate the speed performance more unambiguously than the write and read access and cycle times do. Namely, SAM access and cycle times are functions of the memory bit-capacity, the location of the addressed memory cell, the used processing technology and design approach. By design approach CMOS SAMs may be categorized in three major groups: (1) RAM-based, (2) shift-register (SR) based and (3) shuffle memories.

SAM circuits based on CMOS RAM designs provide lower frequency data rates than their SR based counterparts do, because the length of each input/output period is determined by the access times of the RAMs. To circumvent the access time limitations RAMs in SAMs may feature fast page, fast nibble, fast static column, extended data out or burst extended data out modes, or they are synchronous, extended, cached or

Rambus DRAM or SRAM designs. RAM based SAMs, by all means, have the fundamental advantages against SR-based approaches, that RAM-cell sizes are much smaller than the corresponding SR-cell sizes, and that the power dissipation of a RAM is a small fraction of the power consumption of a corresponding SR memory. The small cell size and power consumption allow for designs of high-packing density large-bit-capacity SAMs which can operate with rather high speed. Very high data rates, exceptionally low power dissipation and high packing density can be combined in shuffle memories.

1.3.2 RAM-Based SAMs

A RAM-based SAM consists of a complete DRAM or an SRAM (Sections 1.2.2 and 1.2.7), an address counter, an address register and an address comparator (Figure 1.28). The address counter generates the add-

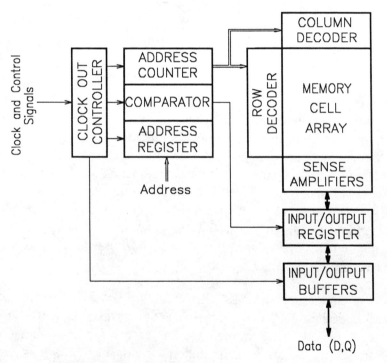

Figure 1.28. A RAM based SAM configuration.

resses from 0 to N in an N-capacity SAM, and enables write or read, when the content of the address register and the counter are the same. The address comparator and the clock and controller unit provide the timing signals, and the combination of the counter, comparator and register circuits has the role of an address pointer circuit.

1.3.3 Shift-Register Based SAMs

High-speed sequential data access can be provided by shift register (SR) based SAMs. A single SR stage, however, contains at least two data storage elements and two data transmission devices, and needs at least two clock phases for operation. For a write or read operation, and in dynamic SRs during storage also, the entire data content of the SAM has to be moved simultaneously. The simultaneous transfer of all data results in large power dissipation. Both the rather large SR-cell size and the excessive power consumption limit the bit-capacity of SR-based SAMs. Nevertheless, the application of SRs makes possible to eliminate the decoders, sense amplifiers, and the bitlines from the storage array. In a CMOS SR-based array, the cells drive very little capacitive loads and, therefore, very high input/output data rates can be achieved. Furthermore, with SR-based memory designs the achievement of very high radiation hardness is also possible, because SRs do not need sense amplifiers.

A typical SR-based SAM (Figure 1.29) includes a word-or-block-wide barrel shift-register array, a clock generator, an address counter and input/output control and buffer circuits. Input/output data may be transferred serially bit-by-bit or parallel word-by-word depending on the state of the SPD signal. A W/R signal controls write or read operations, an M signal selects masked or unmasked write or read, and an R signal allows data recycle in the array. In an N x M bit array, with clock phase ϕ_1 the data from the input/output register, or from the last N-th M-bit column, are transferred into a first M-bit column, and simultaneously the data content from every second column storing complement data are transferred into their neighbored odd numbered column 1,3,...N-1. During clock phase ϕ_2 the data content of all columns storing true data are transferred simultaneously into their neighbored even numbered column, and so forth with consecutive ϕ_1 and ϕ_2 clock phases. Always two neighbored columns contain a single one of each data word after the data

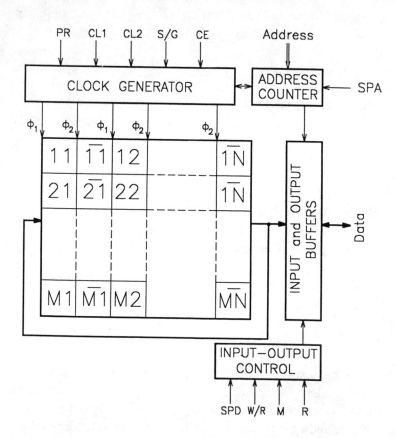

Figure 1.29. A typical SR-based SAM architecture.

transitions are accomplished. Thus, only N/2 number M-bit words can be stored and shifted in an N-column by M-bit SR-based SAM. The data shift in an SR-based SAM may be made plausible in a 4 x 4 bit array, i.e., in an 8 x 4 cell array (Table 1.4), in which the initial data content of the storage cells is represented by numbers, and the storage cell locations correspond to positions in the matrix. The location of a word is pointed by an address counter, which can be controlled for serial or parallel address reception by

Introduction to CMOS Memories 47

an SPA signal. PR signals program the speed of two clock signals CL1 and CL2. CL1 and CL2 provide the two basic operation phases ϕ_1 and ϕ_2, and serve as references to various clock signals required for the operation of the memory frequencies. The chip-enable CE signal keeps the memory active, clocking may be stopped and started any time by an S/G signal, or the stop and start can be determined by the content of the address counter in most SR-based SAM designs.

11	$\overline{11}$	12	$\overline{12}$	13	$\overline{13}$	14	$\overline{14}$
21	$\overline{21}$	22	$\overline{22}$	23	$\overline{23}$	24	$\overline{24}$
31	$\overline{31}$	32	$\overline{32}$	33	$\overline{33}$	34	$\overline{34}$
41	$\overline{41}$	42	$\overline{42}$	43	$\overline{43}$	44	$\overline{44}$

Phase 1

14	$\overline{11}$	11	$\overline{12}$	12	$\overline{13}$	13	$\overline{14}$
24	$\overline{21}$	21	$\overline{22}$	22	$\overline{23}$	23	$\overline{24}$
34	$\overline{31}$	31	$\overline{32}$	32	$\overline{33}$	33	$\overline{34}$
44	$\overline{41}$	41	$\overline{42}$	42	$\overline{43}$	43	$\overline{44}$

Phase 2

14	$\overline{14}$	11	$\overline{11}$	12	$\overline{12}$	13	$\overline{13}$
24	$\overline{24}$	21	$\overline{21}$	22	$\overline{22}$	23	$\overline{23}$
24	$\overline{34}$	31	$\overline{31}$	32	$\overline{32}$	33	$\overline{33}$
44	$\overline{44}$	41	$\overline{41}$	42	$\overline{42}$	43	$\overline{43}$

Table 1.4. Barrel-shifting in a 4 x 4 bit (8 x 4 memory cell) array.

The SR operation allows for gapless streams of input and output data which appears synchronized with the applied master clock without a latency time. Between the appearance of the clock and a read signals the delay is very small, while the write data may be required to somewhat precede the clock at high frequency operation.

The maximum clock frequency, at which an SR-based SAM can operate, depends on the design of the SR-cell, the SR array and the data input/output circuits. Since each word of data is transferred between two neighborhood columns, no small sensing signal is required, and the whole memory can be designed of digital logic gates. CMOS circuits designed of digital logic gates provide considerable larger noise and operation margins than CMOS RAMs do. Therefore CMOS SR-based SAMs can be designed for better environmental tolerance and for higher radiation hardness than CMOS-DRAM and SRAM-based circuits.

In comparison to RAM-based SAMs, SR-based SAMs can provide 10-30 times higher data rates, but they need 2-6 times more silicon area and operate at 10-100 times higher power dissipations than RAM-based SAMs do, if the same technology and same bit capacity are assumed, for both RAM- and SR-based designs. The bit capacity per chip in an SR-based SAM is limited primarily by the excessive power dissipation of the SR-array, which increases in accordance with the increase in operational speed and in bit-capacity per chip. Approximately a ten-fold increase in radiation hardness may also be expected in SAM designs which apply SR- rather than RAM-cell arrays.

1.3.4 Shuffle Memories

Shuffle memories [18] combine the advantages of both RAM- and SR-based designs including very high operational frequency, packing density and radiation hardness, and extremely low operating power consumption. The operation principle of shuffle memories rests on a data transfer algorithm in an array, which allows the application of RAM-cells without the need for bitlines and sense amplifiers in contrast to RAM-based SAM designs, and without the need of moving the entire data content of the array of write and read accesses in opposite to SR-based SAM designs. In shuffle SAMs during one clock cycle, the data content of only one column is transferred into the next column, and both the input and output registers step only one bit to allow the input and output of one bit datum.

A basic shuffle memory array (Figure 1.30) comprises input and output registers, shuffle signal generator and a memory-cell array. A memory cell consists of one storage- and one transfer element. The in- and output nodes of N transfer elements couple N storage elements in a chain, and the control nodes of N transfer elements are coupled to a single wordline. The N x N memory cell array is terminated by the registers for data input and output. Data shuffling in the array and from and to the registers are controlled by the shuffle signal generator.

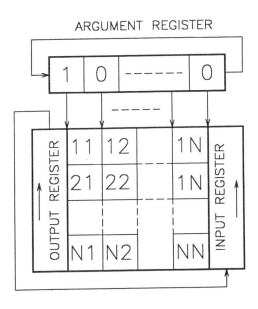

Figure 1.30. Shuffle-memory architecture.
(Derived from a product of Microcirc Associates.)

The shuffle signal generator may be implemented in a shift-register which circulate a single log. 1 and N-1 log.0, or in a $n = \log^2 N$-bit counter and an N-output decoder, or in other digital circuits.

In SAM operation the shuffle signal generator circuit activates one, and only one, word of N bits. In an initial clock period, the content of a first N-bit word or a set of external data is dumped into the output register, and the datum of from the first stage of the output register is shifted into the input register. At this time the output register holds a copy of the data content of the first N-bit word, and the data in none of the N-1 other words are moved or copied. In a next clock period, the content of a second N-bit word is transferred into the storage cells of the first N-bit word, and the second bit of the first N-bit word is shifted from the output register into the input register buffer. Now, the output register holds a copy of N-1 bits of the first N-bit word, the input register has the first bit of the first N-word, and the storage cells of the first N-word contain a copy of the data content of the second N-bit word, and none of the other N-1 words is moved or copied. In similar steps, N^2 clock periods recycles the N^2 bits of the memory cell array. The shuffle array allows for both bit-sequential or word-parallel write or read when the addressed bit or word appears in the input or in the output register. The internal operation of a shuffle memory may be illustrated in a 4 x 4 matrix (Table 1.5), where the initial content of each storage cell is represented by numbers, and the allocation of numbers in the matrix corresponds to the locations of the storage cells in the memory cell array and in the registers.

Shuffle memories, like SR-based SAM, provide continuous strings of data which appear in synchrony with the master clock without latency, and may feature dual and stop-start clocking, serial and parallel data and address transfer, data recycling, masked write and read as well as other memory operations. The operational speed is as fast as the speed of SR-based SAMs due to the small capacitive loads on the memory cells. Nevertheless, the bit-capacitance per chip can be as high as that of a RAM, because the shuffle memory may use RAM cells rather than SR cells in the design. Much less than the power dissipation of RAMs can also be achieved in shuffle memories, since only one word is activated at a time, and the activated memory cells are not coupled to highly capacitive bitlines and sense amplifiers. Moreover, the lack of sense amplifiers eliminates the most difficult obstacles for radiation hardened designs.

Compared to RAM-based SAM designs, shuffle memories can provide 10-30 times faster data rates, about 10-1000 times less power consumption and approximately 10 times higher radiation hardness than RAM-based SAMs do, and the bit-capacity per chip of shuffle memories can near that of RAM-based SAMs.

11	11	12	13	14		
21	21	22	23	24		Step 1
31	31	32	33	34		
41	41	42	43	44	11	

21	12	12	13	14		
31	22	22	23	24		Step 2
41	32	32	33	34	11	
	42	42	43	44	21	

	12	12	13	14		
11	22	22	23	24		Step N^2-1
21	32	32	33	34		
31	42	42	43	44	41	

Table 1.5. Principle of data-shuffling in a 4 x 4 bit (4 x 4 memory cell) array.

1.3.5 First-In-First-Out Memories (FIFOs)

A first-in-first-out (FIFO) memory is a sequentially accessible write-read data storage device, in that data rate for writing can substantially differ from the data rate of reading while keeping the same sequence of data transfer in both write and read. If the data sequence of read is the opposite of the write data sequence, the device is called last-in-first-out (LIFO) memory. In essence, FIFO and LIFO memories are data-rate transformer devices, but LIFOs reverse also the data sequence. Transformation in both data rate and data sequence may beneficially be applied in designs of digital systems. Therefore, numerous FIFO designs allow for a variety of output data sequences which can be generated from one input data sequence in addition to performing data rate transformations and other features. FIFOs may feature data write and read on both the inputs and outputs, as well as data reset, and flow-through, and expandable word-width and memory-length.

A FIFO memory is applied most often, as an interface between two digital devices D_A and D_B to temporarily store the data received from D_A until D_B is ready to accept. The FIFO's storage capacity must be large enough to accommodate the amount of data that D_A generates between services, and the FIFO has to perform the write operations with the clock frequency of D_A and the read operations with the clock rate of D_B. Either one, or both write and read, may operate synchronously or asynchronously.

FIFO operation can be obtained, most trivially, in shift registers, if there is no requirement for independent writing and reading. Writing is clocked by device D_A, the received information is shifted until the first data bit or word is available at the output, and the data is shifted out with the clock rate of device D_B.

Most FIFO devices, nonetheless, are modified RAMs (Figure 1.31) in which the addressing is word-sequential. Rather than using addresses received from another circuit, a FIFO chip internally generates address sequences in the write and read address pointer circuits. Address pointers are usually digital counters or shift registers used independently for write and read. Thus, writing and reading can be randomly intermixed, but only one operation, write or read, can be performed in one clock cycle, because only one address can be selected at a time. An address generated by the pointer circuit is decoded, and enables a single or word in the memory cell array. The enabled row can receive the content of the input register, or send its stored information to the sense amplifiers and the output register. The input and output register operations are governed by the write and read

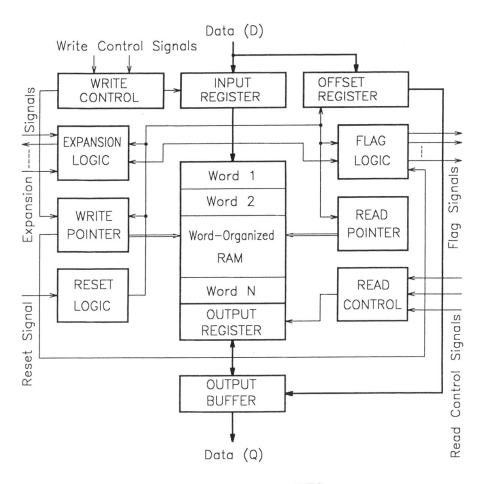

Figure 1.31. A modified RAM as FIFO memory.
(Derived from a product of Quality Semiconductor Incorporated.)

control circuits, respectively. These control circuits accommodate the external clocks, synchronous and asynchronous enable signals for each write and read operation. Flag signals indicate how much of the RAM storage capacity is occupied, e.g., empty, full, almost empty, almost full, etc., and the amount of flagged storage capacity can be programmed in most FIFOs. Both the depth and width of the FIFO storage array may be extended by coupling a multiplicity of FIFOs to each other through the expansion logic. The content of an FIFO may be bypassed in designs where offset registers are implemented.

FIFOs implemented in two-port RAMs, rather than in traditional RAMs, are capable to handle write and read operations simultaneously. In such a FIFO the write and the read address code has independent sets of address inputs. The dual addressability requires the use of complex and large memory cells, which increases chip size and costs of FIFOs. A FIFO architecture that is based on dual-port RAMs is similar to the FIFO architecture using single-port RAMs, except the difference in write and read addressing.

1.4 CONTENT ADDRESSABLE MEMORIES (CAMS)

1.4.1 Basics

In associative or content addressable memories (CAMs) [19] data are identified and accessed by the data content of a multiplicity of memory cells rather than by addresses. Associative memories address data elements, or words, in their memory cell array by associating input data words, or keys, with some or all the stored words (Figure 1.32). The input word is stored in an argument register. A mask register can exclude arbitrary "don't care" bits from the associative search. The result of the associative search is a flag signal that indicates match or mismatch between the masked search argument and an interrogated word located in the CAM array, and this flag signal is transferred into a response store. When more than one flag signals indicate matches, a multiple response resolver establishes a sequence for the further processing of the matching words. Further processing may also require the use of both the matched words and the addresses of the matched words, therefore output buffer and an address encoder are also parts of an associative memory.

Associative search can determine not only an exact match, i.e., all bits are identical in the compared words, but also the degree of similarity in a partial match. For similarity measure CMOS CAMs use mostly the Hamming distance, which simply gives the number of bits the compared words differ from each other, but CMOS CAMs may also apply Euclidean

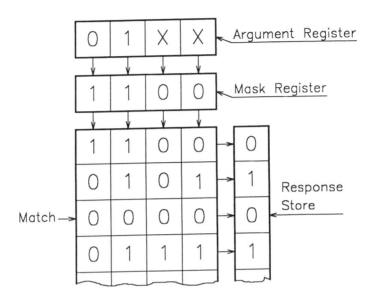

Figure 1.32. Associative-search in a CAM.

distance, Minkowsky metric, Tanimoto-measure and various algebraic and logical correlation measures. The measure of similarity may provide the basis for the operation-strategy of the multiple response resolver. In practice, however, multiple responses are resolved mostly by using spatial or timing orders rather than by similarity measures. To establish similarity magnitudes repeated search for the same search argument at systematic masking is required in all CAMs.

From the wide variety of potential CAM applications data content associations, magnitude search, catalog memory and information retrieval are only the most apparent examples of use. CAM circuits are often used also in cache memory implementations. Moreover, entire computer implementations are based on associative circuits [110]. Partly-associative computations are routinely applied in object and voice recognition and in numerous other military, government and forensic systems. Generally, for systems which would be unacceptably slow or complex with traditional Von Neumann type of computations, the application of CAM-based imputing systems may be attractive design alternatives.

CAMs, most commonly, are categorized as (1) all-parallel, (2) word-serial-bit-parallel, and (3) word-parallel-bit-serial classes. The implementation of all types of CAMs, except the all-parallel CAM, can be based on any type of random or sequentially accessible CMOS memory designs, including all types of RAM, shift-register and shuffle memory arrays. The dominance of RAM technology and the excellent performance-cost parameters of RAMs, nevertheless, gives little chance to justify the use of any other approach than the RAM-based design to CAM implementation. What is more, in many designs, CAM functions are provided by software, e.g., hash coding, TRIE, etc., so that traditional RAM or SAM hardware can be used.

1.4.2 All-Parallel CAMs

The all-parallel CAM compares the search argument with all words residing in the CAM simultaneously. The simultaneous comparison is made possible by combining a RAM cell with an exclusive-OR (EXOR) gate in a single CAM cell, and the outputs of the EXOR gates are tied to an interrogation line in each word. In practical designs, the elements of the RAM cell and the EXOR gate are blended with each other to decrease complexity and size of CAMs.

An all-parallel associative memory consist of a (1) content addressable memory cell array, (2) search argument register, (3) mask register, (4) response store, (5) multiple response resolver, (6) write and read control, (7) address decoder, (8) address encoder, and (9) output buffer (Figure 1.33). Addressing allows to access CAM cells by word addresses, and addresses are needed to facilitate and control a variety of CAM operations. Initially, a CAM cell is addressed word-by-word, and each data word held in the search argument register is written into the CAM cell array. At an associative search, the data content of the search argument register is compared with all or with a required fraction of the words stored in the CAM cell array. Where the search argument word matches the masked content of a word stored in the array, match signals occur. The match signals are put in the response store, prioritized by the multiple response resolver, and the addresses of the matching words are encoded for further nonassociative operations.

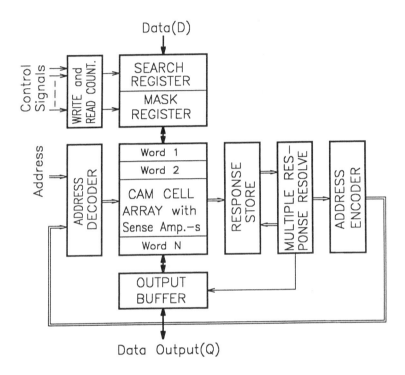

Figure 1.33. All-parallel CAM architecture.

The all-parallel associative memory executes the data search in one short cycle. The length of the search cycle t_S is determined by the consecutive delays occurring in the search argument and mask registers t_i, in the CAM array t_{CAM}, in the response store and multiple response resolver (t_{res}), and in the data output buffer t_o as:

$$t_S \approx t_i + t_{CAM} + t_{res} + t_o.$$

The longest data propagation delay appears in the CAM array, in which the delay time depends mainly on the bit-capacity of the array and on the CAM-cell design. Since full featured CAM cells are much more complex and larger than RAM-cells are; for CAMs the practical limit of bit capacity per chip is much smaller and the power dissipations per chip is much larger than those for RAMs.

1.4.3 Word-Serial-Bit-Parallel CAMs

Any RAM array can be used to implement word-serial-bit-parallel CAMs. In such a CAM, a parallel-operating digital comparator is placed between the mask-register and the RAM-cell array, preferably next to the sense amplifiers, and an address counter generates a sequence of word-

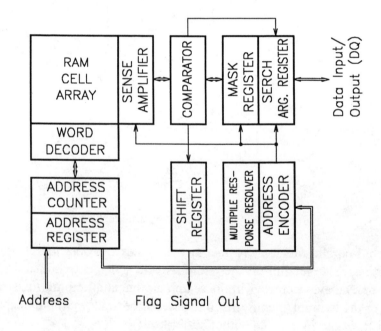

Figure 1.34. A word-serial-bit-parallel CAM configuration.

addresses to the RAM-cell array (Figure 1.34). One-by-one each word is compared to the search argument and the responses are transferred in a shift register. The shift register content preserves the time sequence of matches, and, therefore, it may also be used as a multiple response resolver. Since the responses appear serially, the maximum search cycle time t_S to associate N-words with a search argument, comprises N times the wordline, bitline and sense amplifier delays in the RAM array t_{RAM}, and N times the comparison delays t_{com} in addition to t_i, t_{res} and t_o:

$$t_S \approx t_i + Nx\,(t_{RAM} + t_{com}) + t_{res} + t_o.$$

Very short search times can be combined with exceptionally small power dissipations and high packing densities in designs which apply shuffle memory arrays in place of the RAM-cell array. The RAM-cell array may also be substituted by a shift-register array, but shift register arrays are plagued with low packing density and hefty power dissipations at high frequency operations.

1.4.4 Word-Parallel-Bit-Serial CAMs

Word-parallel-bit-serial CAMs may also be implemented by application of RAM-cell arrays similarly to the previously introduced word-serial-bit parallel CAMs. Word-parallel-bit-serial CAM designs apply either modified RAM cells in the array or the arrays have to be subdivided into subarrays, and each individual subarray has to include a specific decoder circuit. The high circuit complexity, and the resulting low packing density and moderate performance, make RAM based world-parallel-bit-serial CAMs less attractive than other CAM implementations. Excellent speed performance can also be obtained by application of a barrel shift-register array, but because of packing density and power considerations shift-register arrays should be kept in small bit-capacity. Large bit-capacity can be combined with very low power consumption and very high data rate by implementations in shuffle memories.

Since both shift-register and shuffle memory clock data bits in and out of the circuit one-by-one in the same order, the operations of both shift-register- and shuffle-memory-based CAMs may be illustrated in the same block diagram (Figure 1.35). In this word-parallel-bit-serial CAM array, the word, search argument and mask data may be written either in series or in parallel mode into the registers. All registers are the same, and in all

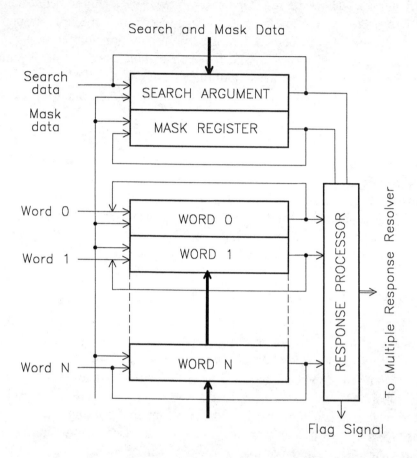

Figure 1.35. A word-parallel-bit-serial array.

registers the data are moved and circulated by the same clock simultaneously. In the response processor, one bit of each word is compared simultaneously to the masked search argument, and the result is stored and evaluated as the data circulates in the registers. Except the data storage and associative search circuits, a word-parallel-bit-serial CAM can have the same constituents as the all-parallel and word-serial-bit-parallel CAMs do.

Introduction to CMOS Memories 61

In this word-parallel-bit-serial CAM, the maximum search cycle t_S for N words of M bits, includes M-times the data advance time for one bit t_D and the signal delay in the response processor t_{RP} in addition to the input, output and resolver time t_i, t_o and t_{res}:

$$t_S \approx t_i + M \times (t_D + t_{RS}) + t_{res} + t_o.$$

The magnitude of t_S greatly depends on the speed of the response processor and the shift-register or of the shuffle memory.

1.5 SPECIAL MEMORIES AND COMBINATIONS

1.5.1 Cache-Memory Fundamentals

A cache-memory, or cache, is a short-access-time small-bit-capacity memory that temporarily stores a fraction of the data and instruction content from the overall memory content of a computing system. Cache memories are applied in traditional Von Neumann type of computers to improve their performance, i.e., to narrow the gap between the high operational speed of the computing unit and the low input/output data rates of the main memory (Section 1.1). Generally, the greater the storage capacity of a memory is, the longer are the access, cycle and data transfer times and the slower are the memory input/output data rates. Slower memory operations cause more idle runs in the central computing unit (CPU). Placed between the CPU register and the main memory, a cache or a complex of cache memories (Figure 1.36) can decrease the time the CPU receives instruction and data and, thereby, it can improve the speed and efficiency of computing systems. System performance improvement by cache applications, nevertheless, requires increase in system complexity, i.e., the addition of cache memory and controller.

One level of cache may not be sufficient to provide a required system performance. In systems using multi-level cache hierarchy the cache closest to the CPU is denoted as primary or level-one (L1) cache, while the cache coupled to L1 is the secondary or level-two (L2) cache and so forth. Caches on an arbitrary i level (Li) can contain either the instructions (LiI), or the data (LiD) or both (LiID).

62   CMOS Memory Circuits

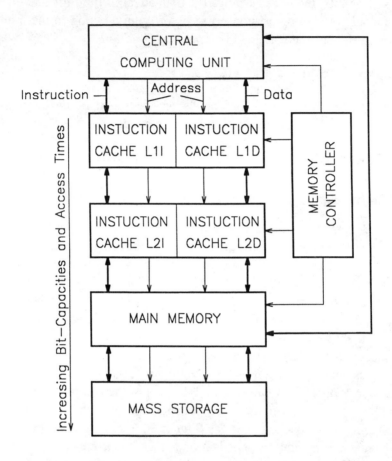

Figure 1.36. Cache memory application in computing systems.

The applicability of any cache LiI, LiD and LiID are based on the (1) temporal, and (2) spatial locality of the instruction and data items. Temporal locality means that items used in the recent past are likely to be used in the near future, while spatial locality implies that items placed physically near to each other are likely to be used in consecutive operations. The efficiency of cache operations depends on the probability that the item when requested by the CPU is in the cache or not, and the cache performance is characterized by the hit rate and miss rate and by the memory cycle times. A hit occurs when the item is found in the cache when the CPU requests it, and miss appears when the cache does not contain the referenced items during the time of their request. The number of CPU references which result hits divided by the total number of CPU references gives the hit rate H_R, and the miss rate M_R may be expressed as $M_R=1-H_R$. An increasing hit rate decreases the total number of CPU cycles needed to execute an instruction, and improves system performance. Nevertheless, the performance of a system, that includes cache, depends also on the number of CPU cycles required to fetch an item from the cache at a hit and from the main memory in case of a miss. Therefore, both the cache and main memories should be designed so that they operate with the shortest possible memory cycle times. What is more, the signal propagation times between the main and cache memories, between the CPU register and the cache and also between the CPU and the main memory should also be short.

A memory address code contains bits to determine the location of words within blocks, blocks within sections, and sections within the cache memory. A block or a line includes a certain number of words, as well as a number of tag bits which uniquely determine the location of the block in the main memory. An address tag is applied to indicate which portion of the main-memory content is present in the cache, and it may contain either a part of the address bits, usually the most significant address bits, or all of the address bits. The number of tag bits depend on the size of the blocks and the bit capacity of the main memory. To reduce the number of tag bits the block sizes may be increased, or the blocks may be grouped in sections and only the sections are tagged. A collection of blocks or sections, for which the tags are checked simultaneously or parallel in a cache memory, is a set. The number of blocks in a set is the degree of associativity or the set size of the cache.

Performance improvements by cache memory applications have become an important part of the system design and are discussed in publications extensively, e.g., [111]. The following discussion focuses on the architectures of CMOS cache memories.

64   CMOS Memory Circuits

Cache hit rates can be improved by optimizing both organizational and strategical variables. Organizational variables include the data storage capacity C, number of blocks in a set or degree of associativity A, number of sets S, and block size in bytes or bits B of the cache. Increasing cache capacity $C = A \times S \times B$, increases the probability of hits. A simple increase in cache size C would result in longer cache operation times. In a large cache, yet, the cycle time can be kept short by increasing the number of blocks per set A and by decreasing the amount of data stored per tag B. Clearly, variables C, A, S, and B can be optimized in a system. The performance of the system may also be ameliorated by careful choice of strategies in replacement of blocks in the cache (last recently used LRU, first-in-first-out FIFO, random, etc.), in data writing (copy back to main-memory, write through cache to main memory, buffer several writes, etc.), in data fetching (number of blocks, speculative prefetching, order of data return, etc.), and in workload management (adjusting block sizes, request buffering, system timing, etc.). System strategies for optimum performance may substantially vary system to system, and the chosen strategies determine a great part of the overhead circuit design in the cache memory. Since the optimum cache memory size, associativity, block-size, set-size as well as strategies of replacement, writing, fetching and workload management are actually system dependent, the optimization of these parameters are not discussed here, but they are available in the literature of computing systems, e.g., [112].

CMOS caches are not only fast-operating small-size memories with the ability to determine quickly whether the requested instruction or data are stored in the cache (hit or miss), but they are also capable to replace the stored items by items fetched from a main memory if the requested item is not found in the cache memory. Since the cache is a temporary buffer for a portion of the main-memory content, the data in the cache have to be coherent with the data in the main-memory. Common coherency strategies are the copy-back and write-through policies.

Under the copy-back policy, the cache records writes and reads, and the cache can operate without using the circuits of the main-memory. An update of the main-memory occurs when the data block that contains the write address in the cache is replaced. No replacement can be performed in cache locations which are occupied and flagged by a "dirty" signal indicating that the information must be written into the main-memory, otherwise they would get lost. With the copy-back strategy, the main-memory is updated far less frequently than with other coherency strategies, but the replacement of information in the cache requires the transfer of large amount of data and, thereby, rather long time for each transfer event.

In the write-through policy the reads are cached, but the writes are stored in the main-memory, and in write cycle the main-memory is accessed. This ensures coherency between the cache and the main-memory operations, but the large number of slow accesses to the main-memory decreases the system performance. In the majority of computing systems the use of the write-through policy is less efficient than the application of the copy-back strategy.

1.5.2 Basic Cache Organizations

CMOS cache memory designs, like other cache memory implementations, may be fully-associative, direct-mapped and set-associative.

In a fully-associative cached computing system, the main memory and the cache are divided into storage blocks, and the cache stores one set of blocks. Any block in the main memory can be mapped into any block in a fully-associative cache memory (Figure 1.37). The fully-associative cache compares simultaneously each bit in every tag stored in the cache to each bit of the effective or physical address generated by the computing unit, to determine a cache hit or miss. The cache performs a fully associative tag comparison, and the tag memory is designed as an all-parallel content addressable memory. Fully-associative cache memories are capable to keep the most frequently accessed data and instructions, no matter in what location of the main memory these data and instructions are stored. However, to find the desired data and instructions the entire cache must be searched, and this search-time compromises the performance of the fully-associative cache.

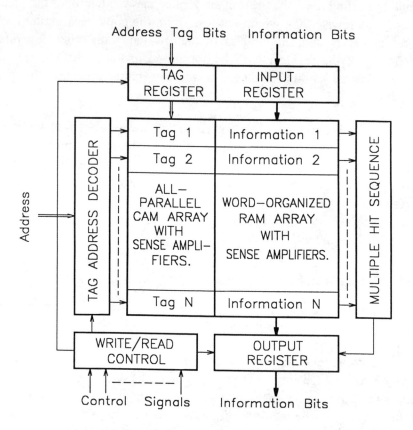

Figure 1.37. A fully associative cache organization.
(Derived from a product of Microcirc Associates.)

In a direct-mapped cached system each storage block of the cache is linked to several predetermined blocks of the main memory. The direct-mapped cache memory has a tag and information store, and a comparator for matching the tag bits of the addresses (Figure 1.38). The block address is broken into tag and index bits. The index bits are the addresses in the cache memory and determine the depth of the cache. Cache locations correspond to predetermined main-memory locations which are identified

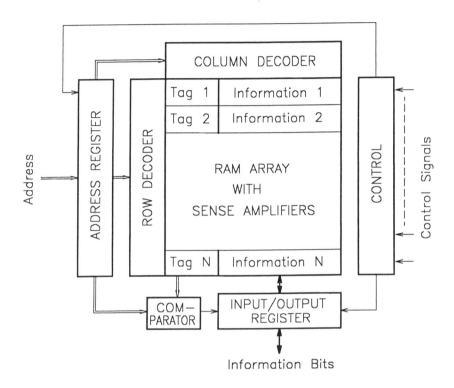

Figure 1.38. A direct mapped cache configuration.

by the same index bits. Because the use of index bit reduces the number of bits to be compared, the detection of a cache hit or miss is fast. A direct-mapped cache, however, can not maintain a nearly optimal collection of blocks, because the new block, that replaces an old one, determines which one of the old blocks has to depart. Furthermore, if the block required to be in the cache is the same as the one used in the preceding program step except this block is requested from a different set, the block has to be fetched from the main-memory and written into the cache. In such cases, the direct-mapped cache has frequent misses and operates inefficiently.

A set-associative cache operation alleviates the contentions of the fully-associative and the direct mapped systems, keeps the tag-memory small and the tag comparison simple and speedy, by dividing the cache's memory capacity into N direct-mapped sets, and by using N number of simultaneously operating comparators. In a two-way (N=2) set-associative cache (Figure 1.39) both of the RAM arrays, decoders, comparators and input/output registers are identical. Furthermore, identical are the predetermined locations in both RAM arrays in which the main-memory blocks can directly be mapped, and blocks from a certain main memory location can be copied into two cache locations. If the tag-bits of the address match either one or both of the tags residing in the cache, then a hit occurs. At a miss, the set-associative cache can maintain a favorable set of blocks, because the cache user is free to decide which one of the blocks within a set should be replaced. The application of four-way (N=4) set-associative caches represents a good balance between performance and cost in memory systems.

Whether a set-associative, or a direct-mapped or a fully-associative cache memory provides a required hit rate and the most economical solution, it depends on specific system parameters. Therefore, many application-specific cache designs are available, and many designs allow to change between set associative and direct mapped organizations, or to use fully-associative caches also in direct-mapped in set-associative operation modes by one or two external control signals.

Cache operation may also include error detection and correction by the application of simple error correcting codes and a few redundant bits for each block. Additional status bits may also be required to record the time-order of the tag-usage, or to store other information which assist the replacement of the cache's information content. Cache designs often comprise multiplexers to avoid repetitions of significant circuit parts in the layout and, thereby, to keep the chip-size of the cache small. Many caches use multiplexing also for accommodating certain input and output data formats. Data formats, block- and tag-sizes may be programmable to allow for adjustments in system optimizations.

Requirements for fast system operation and ease in applications point to the use of SRAMs in cache designs. Nevertheless, cache designs can apply any type of fast operating memory circuits, e.g., small-size DRAMs which have the potential to approach the access times of SRAMs.

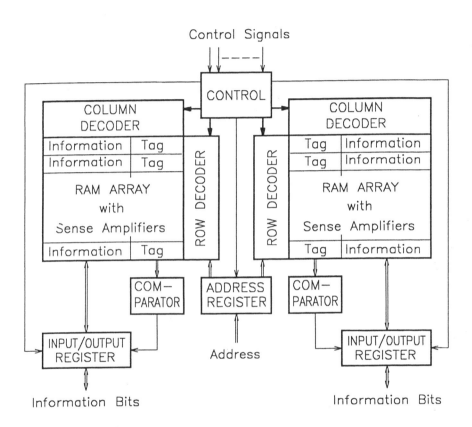

Figure 1.39. A two-way set associative cache architecture. (Derived from a product of Micron Incorporated.)

1.5.3 DRAM-Cache Combinations

The integration of a DRAM with a cache memory in a single CMOS chip is aimed to combine the low cost per bit of DRAMs with the high bandwidth of cache memories. A DRAM-cache combination improves the operational speed of computing systems by both means: (1) by the wide bandwidth of the integrated cache and (2) by the greatly reduced data transfer time between the DRAM and cache memory. Small SRAM caches have wide band widths due to their inherently high speed write and read operations, and due to their large word size designs. Furthermore, the continuity of the write and read operations are infrequently halted, as a result of the high probability that the requested information is stored in the cache at the time of the request. If the requested information is not stored in the cache, a rapid instruction and data transfer is required to update the cache. In DRAM-cache combinations, this transfer of instructions and data can be very fast due to the lack of chip-to-chip interfaces, and due to the very short chip-internal wire-lengths between the DRAM and the cache.

The most publicized DRAM-cache combinations are the (1) enhanced DRAM (EDRAM), (2) cached DRAM (CDRAM) and (3) Rambus DRAM (RDRAM) and (4) virtual channel memory (VCM).

1.5.4 Enhanced DRAM (EDRAM)

The EDRAM (Figure 1.40) blends a primitive one-row (X bit) wide cache directly into the column decoder of the DRAM. It caches one row at a time, but in a three dimensional X x Y x Z memory array organization it may cache X x Z bits. On every cache miss, the EDRAM loads a new row into the cache, most frequently, on a last row read (LRR) basis. Thus, an LRR register and a comparator are also added to the generic DRAM design.

An EDRAM writes directly to the DRAM, but reads from the cache memory. During a burst read the EDRAM provides data burst $B \ll X$, and hides the precharge cycle of the DRAM. The precharge is completed before a consecutive burst-read can start. To read the cache no row enable signal is needed, when the cache has hits. At cache misses the EDRAM accesses the DRAM. The DRAM cell can be refreshed without any waiting time during write, read and burst-read operations.

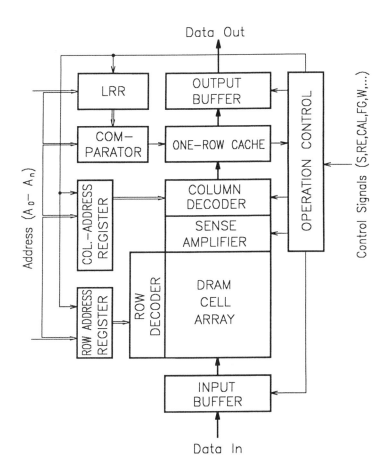

Figure 1.40. EDRAM organization.
(Derived from a product of Ramtron Incorporated.)

The differences between DRAM and EDRAM operations are manifested in differing operation controls. An EDRAM may include chip select S, row enable RE, column address latch CAL, refresh F and separate output enable G control terminals which enhance the application flexibility of the built-in cache. Since the cache is a part of a DRAM architecture, most EDRAMs operate asynchronously, although synchronous EDRAM may also be designed with little added effort.

1.5.5 Cached DRAM (CDRAM)

A CDRAM integrates a complete cache memory and a generic DRAM, so that the cache and DRAM can operate independently. In a typical design (Figure 1.41), between a (2m x D)-bit DRAM and a (2n x D)-bit cache a row-wide buffer of D2 bit capacity is applied, and each of the DRAM and the cache has a separate control circuit. The cache is directly addressed by n addressing bits, while the DRAM addressing is multiplexed in a ratio of m/n. Between the DRAM and the cache a buffer facilitates communication.

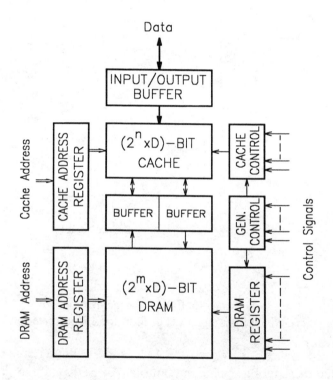

Figure 1.41. A CDRAM architecture.
(Derived from a product of Mitsubishi Electronics America Incorporated.)

A half of the buffer-bits serves read-data transfers, and the other half of the bits serves write-data operation transfers from and to the DRAM. All buffer bits are used for data transfers between the buffer and the cache memory.

The cache is segmented into $D^2/2$ cache lines with D words per line, and can be either direct-mapped or set-associative depending on the implementation of the external cache controller. At a cache hit, D bit of data is transferred through the buffer to the output. At a cache miss the buffer receives $D^2/2$ bits simultaneously from the DRAM either in write-through or write-back replacement mode. The write-back cycle may be hidden by buffering. At a write miss, the buffer data can be posted during a complete DRAM write cycle.

A CDRAM may use a synchronous clock to control all operations. All control and address signals should be set up before the appearance of the clock signal, and access and cycle times are from and to the appropriate edges of the synchronizing clock signals. The synchronous output register may operate in transparent, cached, registered and masked modes.

1.5.6 Rambus DRAM (RDRAM)

The RDRAM unifies a synchronous DRAM, a row-cache and a complete data-bus interface circuit (Figure 1.42) [113]. The on-chip interface circuits greatly reduces the communication time between the DRAM and the other parts of the computing system, in addition to the performance benefits of integrating the cache and the DRAM in a single chip. RDRAM application and operation in systems are determined by specific protocols, which allow to combine the addressing, data commun-ication and control signals through a single bus complex.

In RDRAMs write and read operations include bus request, data and acknowledgement of data arrival. At read, positive acknowledgement means that the requested data are in the cache, negative acknowledgement appears if the data are not in the cache or a refresh cycle is in progress. At a second cache miss, upon a consecutive read-request, the RDRAM issues a write request to the cache from the DRAM. A write into the RDRAM commences also with a request for data package. When the write data can be accepted by the RDRAM, positive acknowledgement occurs. Negative acknowledgement may appear when the write address is not found in the cached page, or when a refresh cycle is in progress during the write request. After the first write request the controller waits a DRAM cycle and repeats the write request for the same data and address.

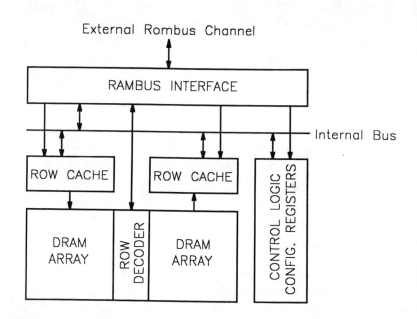

Figure 1.42. RDRAM scheme. (Inferred from [113].)

In the proprietary Rambus systems (Figure 1.43), the data, address, control and clock signals can be transferred at very high frequencies, e.g., at 800 MHz. At high frequencies the bus, clock and other interconnect lines behave as transmission lines (Sections 4.1.3 and 4.1.4), in which the propagating signals have significant delays or flight times, and can greatly be distorted by reflections on the interfaces and, in small degrees, by attenuations within the lines. Different flight times for the data, address, control and clock signals, and distorted signal forms, would make a system inoperable and, in unfavorable combinations, even damaged. Rambus systems counteract the effects of transmission lines by the use of a folded clock signal path and of resistive line terminals. For flight time compensation the folded clock signal path provides forwards and backwards signals along the bus lines. Forward clock signals are applied in read operations, because the read data propagates from the RDRAMs to the control unit. In write operation backward clock signals are applied, because the write data runs toward the RDRAMs. For minimizing the effects of signal reflections all signal and clock bus lines are terminated by resistors R_T-s which are connected to a terminal voltage. The termination resistors keep reflections under control and determine the signal amplitudes on the bus and clock lines. All bus and clock lines are expandable up to a functional limit, e.g., 10 cm,

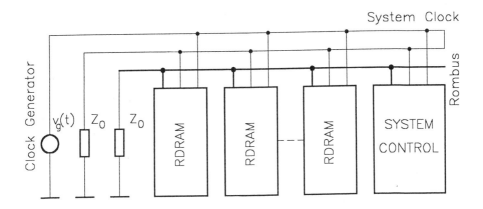

Figure 1.43. RDRAM system with Rambus protocol [114].
(Derived from a design of Rambus Incorporated.)

to accommodate various numbers of RDRAMs. The operation of the RDRAMs are controlled by different protocols in the base, concurrent and direct Rambus systems. Since the Rambus systems use pocket codes, they are referenced as pocket protocol systems.

Other protocol based memory and system architectures, e.g., Synclink DRAM, an implementation of the Ramlink protocol [114], differ little from the RDRAM and the Rambus protocol in concepts. While the Rambus concept applies single ended terminals and linear extension of the bus wires and the number of RDRAMs, the Ramlink concept applies a closed ring bus structure in which the extension of the bus and the number of Synclink DRAMs, or SLDRAMs, are confined in the ring. Synclink, as Rambus, is also a pocket protocol system. Extended performance can be obtained by protocol implementations which use small-signals and differential signal-pairs in data transfers.

1.5.7 Virtual Channel Memory (VCM).

A VCM combines a DRAM, preferably a dual- or a multibank synchronous DRAM, with a multiplicity of cache SRAMs (Figure 1.44) in a single memory chip. The K number of cache SRAMs create K number of virtual channels, and to each channel the computing system assigns a distinguished task. e.g., in a graphics system one channel is for reading display list, another one is for loading texture maps and a third channel is for loading vertice data. Data transfers between the SRAMs and the DRAM have to explicitly be ordered by a VCM-external memory controller in contrast to EDRAMs, SDRAMs and RDRAMs, which manage SRAM-DRAM data transfers chip-internally. The cache SRAMs are placed next the DRAM so that the X number of SRAM columns joins through minimum wire-lengths to the X number of the DRAM's sense amplifiers. In this SRAM-DRAM combination, only the cache SRAMs need an X-output column decoder for access, while a single row- and a segment-decoder support the wordline selection in the DRAM. For the selection of an SRAM a channel selector circuit is used. Each virtual channel has its own interface circuits and dedicated resources to accommodate the operation modes. The operation control of the VCM requires a control logic circuit which is more complex than that of other DRAM RAM designs.

In a VCM design, the circuits which surround the DRAM array are organized to hide the precharge time, or precharge latency, by pipelining the operations of the individual cache SRAMs. The precharge time in a DRAM array is several times longer than the access time for the bits in a certain selected row or in a page (Section 1.2.2). Upon a change in page address, the first write or read of a memory cell can be performed only after the precharge is completed, because all the three write, read and precharge operations use the same bitlines and sense amplifiers. Bits in a page can be addressed with high frequency, e.g., one or two bits with

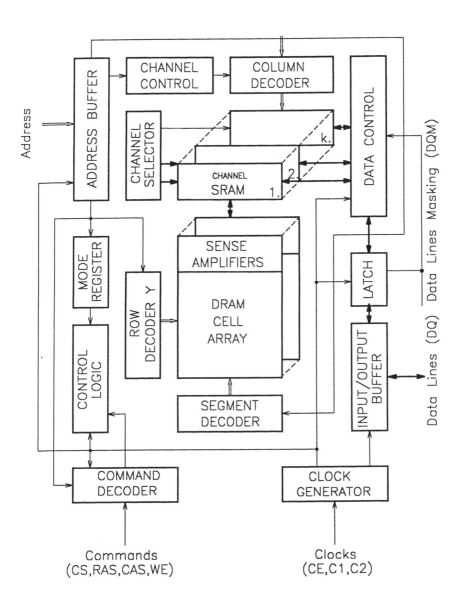

Figure 1.44. A VCM organization.
(Derived from a product of NEC Electronics Incorporated.)

every system clock, without precharge latencies (Section 1.2.4). Although the DRAM precharge times are unchanged by the VCM organization, during the precharge the virtual channels perform operations which do not interfere with the precharge and can be controlled so that their operation times conceal the precharge latency. An intelligent memory controller can assign a well-localized code thread to each of the cache SRAMs and protect them from page-address misses triggered by other memory accesses. Page-address misses may often occur when multitask operations force the memory to write and read information into and from randomly addressed memory locations which are far from each other in the memory space and in time of use. Spatial and temporal coherency are improved, here, by storing the components of a specific task in the same cache SRAM and processing the data of this specific task through the same virtual channel.

Apart from SRAMs the virtual channels may also be implemented in small DRAMs or in parallel-serial registers, and a VCM may be designed for synchronous or asynchronous operations. The precharge operations may also be pipelined for the segments of the DRAM, which may reduce the bit-capacity requirements in the channels of the VCM.

VCM designs may adopt a variety of input and output interface schemes including the traditional unterminated CMOS or the terminated Rambus, Synclink and other interfaces. As an alternative to Rambus, VCMs and numerous other high-speed memory designs may support the bidirectional stub series terminated logic SSTL interface complex [115] (Figure 1.45). The bidirectional SSTL interface applies series resistors R_S-s and two termination resistors R_T-s to approximate a match with the characteristic impedance Z_o of the transmission line. Signal lines are terminated to the terminal voltage V_{TT}, and reference voltage V_{ref} is used to distinguish logic levels. Usually, $V_{TT}=V_{ref}=V_{DD}/2$, where V_{DD} is the supply voltage to the system. In this system, logic levels fluctuate with the variations of V_T and each memory device is activated in accordance with the flight time of the clock impulse from or to the driver/receiver. At write, the driver provides a data-strobe signal toward the memories while at read, the data-strobe signal is driven by the memories toward the driver/receiver to compensate the flight times. This type of data-strobe technique allows to use both the rising and falling edges of the clock signals as references to start write or read operations. Since traditional systems timed only by the leading edge of the clock signal, the use of both clock-impulse edges can double the data rates in compatible memories. Memories which can operate with such double data rates are called double data rate DDR devices. In practice, DDR techniques do not double the achievable data rate in DRAM operations, because each write and read access is preceded by a latency time. Critical is the row latency, that is influenced by the signal delays in the input buffer, decoders, word and bitlines, and sense amplifiers, and the latency time can greatly be reduced by highly segmented and by hierarchical organization of the memory cell arrays.

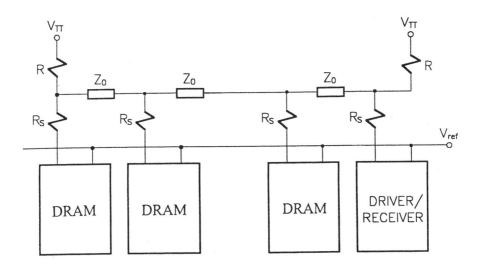

Figure 1.45. Bidirectional stub series terminated logic interface.

1.6 NONRANKED AND HIERARCHICAL MEMORY ORGANIZATIONS

All of the CMOS memory types can be organized in nonranked and hierarchical memory architectures. In nonranked or nonhierarchical architectures each subarray has equal organization ranks, and a one-level decoding for each row and column selection is sufficient to access any of the words and any of the bits in the array. A nonranked array may be simple (Figure 1.46a) or segmented (Figure 1.46b), but in both nonranked array

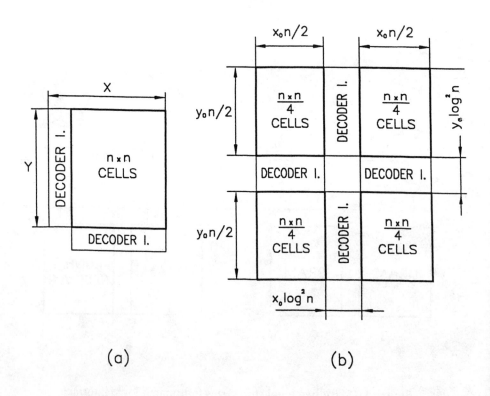

(a)   (b)

Figure 1.46. Simple (a) and segmented (b) nonranked arrays.

types each word and each bit of a word are directly accessible through the use of a single-level array decoder. The hierarchical organization, in a memory chip, partitions the memory cell array into subarrays (Figure 1.47), the subarrays into sub-subarrays, etc., and divides the decoding into a certain number of levels [116]. The division ratio, i.e., the number of subordinate

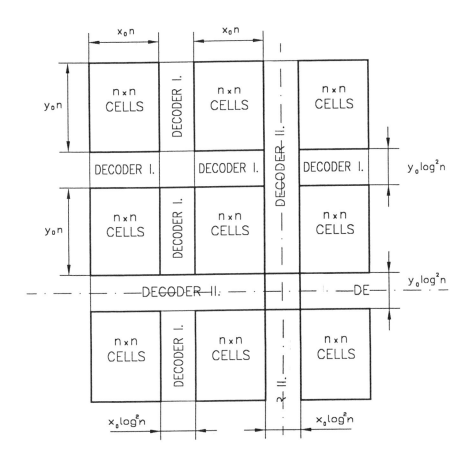

Figure 1.47. A hierarchical memory organization with shared decoders.

arrays contained by one module of the one-level-higher ranked array, may be the same or different for the various organizational levels. In designs the most apparent rationale for using hierarchical organization is to reduce access times by architectural means. A worst case access time appears in a single-array organized memory where the array delay T_a includes both the entire wordline delay t_W and entire bitline delay t_B (Figure 1.48a)

$$T_a \approx t_W + t_B .$$

Both word- and bitline delays can be reduced to approximately t_W/m and t_B/m by organizing the array of memory cells into m x m modules. Although the modularity increases the word decoding time t_X by Δt_X and the bit decoding time t_Y by Δt_Y, nonetheless, the delay in the array organized in m x m modules T_a^1 (Figure 1.48b)

$$T_a^1 \approx \frac{t_W}{m} + \Delta t_x + \frac{t_B}{m} + \Delta t_y$$

is smaller than T_a, because $t_X \ll t_W$ and $t_Y \ll t_B$. Namely, Δt_X and Δt_Y occurs in low-capacitance interconnect lines which run over field-oxide, while t_W

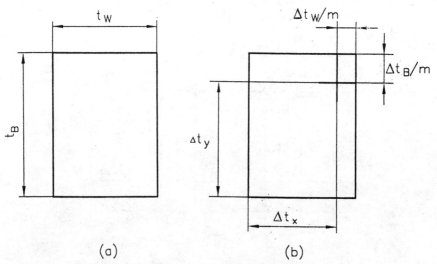

Figure 1.48. Approximate access delay lengths in a nonranked (a) and in a two-level (b) hierarchical architecture.

and t_B are caused mainly by the capacitance of the high number of memory cells coupled to the word- and bitlines.

As a tradeoff for the shorter memory access time, increased layout area is required to accommodate the hierarchical decoding. In most of the designs, each of the row and column decoders is placed between two symmetrical parts of the memory-cell array, because a single decoder can drive n number of memory cells by two buffers much faster than the same decoder could drive 2n memory cells by a single buffer. Thus, a decrease in memory access time virtually without an increase in layout area may be obtained. If every quadrant of this generic bisymmetrical layout includes n x n memory cells, and each memory cell has X_o length and Y_o width, then this module's total width X and length Y may be approached as

$$X = X_o (2n + \log^2 n) \quad \text{and} \quad Y = Y_o (2n + \log^2 n).$$

Mirroring this bisymmetrical module in both direction N times and adding decoder rows and columns for the selection of the increased number of segments, in an N level of hierarchical memory the total enlarged length X^1 and width Y^1 of the enlarged memory may be approximated as

$$X^1 = X_o 2^N (2n + \log^2 n + \log^2 N) \quad \text{and} \quad Y^1 = Y_o 2^N (2n + \log^2 n + \log^2 N),$$

where N=2 for two levels and N=4 for three levels of hierarchy. More than three levels are unlikely to be used in a memory chip.

At the implementation of any level of hierarchy $NXY < X^1 Y^1$ applies, but the area difference $\Delta A^1_{XY} = X^1 Y^1 - NXY$ is very small at high number of memory cells per segment n^2 and at small number of bisymmetrical levels N. Because the number of memory cells per bitline is limited by required extents of operation and noise margins and by speed requirements, many of the large memory designs apply hierarchical architectures. Generally, hierarchical organizations allow for improvements in both operational speed and power consumption at little trade-off in chip-size.

2

Memory Cells

Memory cells are the fundamental components to all semiconductor memories, and their features predominantly effect the chip-size, operational speed and power dissipation of memory devices. This chapter examines the CMOS-compatible memory cells which are extensively applied or have good potentials to be used in CMOS memories. The examination of the memory cells comprises structural, storage-mechanism, write, read, design and improvement issues. The structural and operational characteristics of a memory cell set the primary parameters for the design of sense amplifier, memory-cell, array, reference and decoder circuits.

2.1 Basics, Classifications and Objectives

2.2 Dynamic One-Transistor-One-Capacitor Random Access Memory Cell

2.3 Dynamic Three-Transistor Random Access Memory Cell

2.4 Static Six-Transistor Random Access Memory Cell

2.5 Static Four-Transistor-Two-Resistor Random Access Memory Cell

2.6 Read-Only Memory Cells

2.7 Shift-Register Cells

2.8 Content Addressable Memory Cells

2.9 Other Memory Cells

2.1 BASICS, CLASSIFICATIONS AND OBJECTIVES

Memory cells are the irreducible elementary circuits which are able to store data and allow for addressable data access in a memory, and they are the key elements in determining the characteristics of a memory device. Memory cells are applied in arrays (matrices, cores), and the functions of memory-cell arrays are served by all other (peripheral, overhead) circuits of the memory.

Generally, a memory cell that is applicable to CMOS memory designs, comprises (1) a data storage circuit or circuit element, (2) one or more data access devices and, in some designs, (3) additional circuit elements (Figure 2.1). Nearly in all CMOS memories, one storage circuit or element is capable to hold one bit of binary information, but some storage elements are able to store a multiplicity of binary or nonbinary data. A data access device allows or disallows data read or write from and to the storage circuit part depending on the state of a control signal on the control node of the access device. Additional circuit elements may be used to improve environmental tolerance and to accommodate a variety of functions in a single memory cell.

To CMOS memory applications, memory cells are classified mostly by (1) featured operation modes, (2) data form, (3) logic system, (4) storage mode, (5) storage operation, (6) number of constituent elementary devices, (7) access mode, (8) storage media, (9) radiation hardness and others (Table 2.1). From the immense variety of memory-cell types most of the CMOS write-read memories use dynamic one-transistor-one-capacitor (1T1C), static 6-transistor (6T), and static four-transistor-two-resistor (4T2R) memory cells. These three types of memory cells are applied primarily to implement write-read random access memories, but also to numerous serial and special access memory designs. Small capacity serially-accessible CMOS memories employ also dynamic 6- and 8-transistor shift-register (6T SR and 8T SR) cells as well as static 7-transistor shift register (7T SR) and other serially accessible cells. Static 6T and dynamic 1T1C memory cells combined with a one-bit digital comparator are the favored approaches to construct 10- and 4-transistor content addressable memory cells (10T CAM and 4T CAM).

Read-only memory designs are based on the mask-programmable one-transistor (1T ROM) cells.

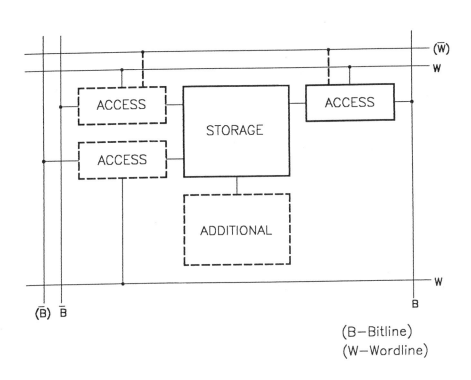

Figure 2.1. General memory cell structure applicable to CMOS designs.

In CMOS memories, the mostly used dynamic 1T1C and static 4T2R cells are generally not full CMOS circuits, but n-channel-only memory cells which operate with the support of CMOS peripheral circuits. Memory cells formed exclusively of n-channel or exclusively of p-channel devices, can be designed in much smaller area, and can be fabricated with less complex processes, than their complementary counterparts do. In designs with single-channel-type of memory cells, no added area is required for isolating the n-wells from the p-wells, and capacitor or resistor devices can readily be placed above and under the transistors and interconnects.

1.	Operation Modes:	Write-Read, Read-Only, User-Programmable
2.	Data Form:	Digital, Analog
3.	Logic Systems:	Binary, Nonbinary
4.	Storage Mode:	Volatile, Nonvolatile
5.	Storage Operation:	Dynamic, Static, Fixed, Programmable
6.	Device Number:	1T1C, 4T2R, 6T, 10T, Others
7.	Access Mode:	Random, Serial, Content-Addressable, Multiple, Mixed
8.	Storage Media:	Dielectric, Semiconductor, Ferroelectric, Magnetic
9.	Radiation Hardness:	Nonhardened, Tolerant, Hardened

Table 2.1. Mostly used classifications of CMOS-compatible memory cells.

Memories, which apply both n- and p-channel devices in full-complementary circuit configurations, are designed to satisfy stringent requirements for high operational speed, low power consumption, and radiation hardness, or to combine extra operational features with the basic write, store and read functions in a single memory cell. Implementations of full-complementary memory cells in traditional CMOS processing technologies take rather large silicon areas. Nevertheless, the emergence of thin-film transistor technologies allow for vertical stacking of n- and p-channel transistors and, thereby, for designs of full-complementary memory cells in very small sizes.

In memory cell designs, the most important objective is to minimize the size, i.e., the semiconductor silicon surface area of the memory cell. Smaller cell area decreases (1) costs per bit, (2) access and cycle times and (3) power dissipation in CMOS and other semiconductor memories.

Cost-per-bit benefits are results of the increased number of bits which can be stored in a memory chip of given size. Improvements in memory operational speed and power are consequences, chiefly, of the reduced capacitances which result from the use of smaller size memory cells. A cell-area enlargement in the design of memory cells may be justified by requirements for such (1) high performance, (2) operation in severe environments, or (3) functional complexity which can not be provided by the use of smaller size memory cells.

This chapter focuses on those write-read and read-only digital binary memory cells which are manufactured with CMOS processing technologies in high volumes, have established a significant application area in CMOS memory technology, and have, most likely, potential to be used in future CMOS memories. Of course, memory cell types other than the ones discussed here, have been and are being developed and applied to CMOS memory designs. Memory-cell research has become a world-wide effort to satisfy the increasing demand for low-cost, small-size, high-performance and low-power CMOS memories for applications in standard, military, space and in other environments. Memory cells applicable to the designs of user-programmable nonvolatile memories are not subjects of this book, because the technology of user-programmable nonvolatile memories has grown to be an independent technology for itself.

2.2 DYNAMIC ONE-TRANSISTOR-ONE-CAPACITOR RANDOM ACCESS MEMORY CELL

2.2.1 Dynamic Storage and Refresh

Most of the CMOS memories apply the dynamic one-transistor-one-capacitor (1T1C) memory cell in their design, because it can be implemented in smaller silicon surface area than other memory-cell types do, its implementation is compatible with CMOS processing technologies, and it is able to provide good performance in memory-cell arrays. Furthermore, the 1T1C memory cell is inherently amenable for coherent down-scaling along with the evolution of the CMOS technology, and for accommodating data not only in binary but also in future nonbinary, multi-level, and analog memories. Principally, CMOS compatible 1T1C memory cells are developed for the designs of cost-effective, high-packing-density, write-read dynamic random access memories (DRAMs). Since the CMOS DRAM technology dominates the semiconductor industry, designs of pseudo-static random-access, sequential-access and specialty memories use also dynamic 1T1C memory cells.

The dynamic 1T1C memory cell (Figure 2.2) employs a single capacitor C_S to store a certain amount of electric charge that represents a datum, uses a single MOS access transistor MA1 to couple and decouple the storage capacitor to and from other circuits, and requires a periodical refresh of the stored datum and a rewrite of the read datum. The charge pocket that represents a datum corresponds to a voltage difference between the storage node voltage V_S and the cell-plate voltage V_{CP}, and $V_{CP}=V_{SS}$ or $V_{CP}=(V_{DD}-V_{SS})/2$ in most of the designs. Here, V_{DD} and V_{SS} are the positive and negative supply voltages and, for convenience, $V_{CP}=V_{SS}$ is assumed in the following discussion of 1T1C DRAM cells.

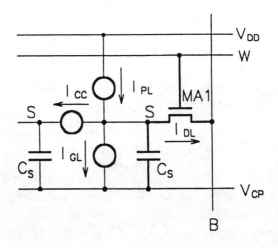

Figure 2.2. Dynamic one-transistor-one-capacitor memory cell circuit with the main leakage-current paths.

In a DRAM cell array, leakage currents through the access device I_{DL}, between the storage node S and the ground I_{GL}, between node S and power-supply pole I_{PL} and between memory cells I_{CC}, alter the charge and the voltage $v_S(t)$ across C_S (Figure 2.3). To avoid great changes in $v_S(t)$ which could destroy the stored data, the capacitor C_S must be rewritten or refreshed to V_S in a certain time period, in the so-called refresh time t_{ref}.

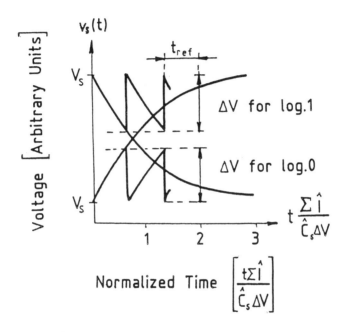

Figure 2.3. Leakage current effects on stored voltage levels.

The maximum allowable $\hat{t}_{ref}$ may be calculated from the time functions of the storage node voltage $v_s(t)$:

$$v_s(t) = V_s e^{-t_{ref} \frac{\hat{C}_s \Delta V}{\Sigma \hat{i}}} \quad \text{if } V_s \to \log 1,$$

$$v_s(t) = (V_{DD} - V_S)(1 - e^{-t_{ref} \frac{\hat{C}_s \Delta V}{\Sigma \hat{i}}}) \quad \text{if } V_s \to \log 0,$$

where V_S is the initial write voltage in worst case, $\check{C}_s$ is the minimum storage capacitance, $\sum \hat{I} = I_{DL} + I_{GL} + I_{PL} + I_{CC}$ is the maximum total leakage current to alter V_s, and ΔV is the voltage change allowed by the operating and noise margins of the sense circuit. From the equations of $v_s(t)$ with a predetermined ΔV, the refresh time may roughly be approximated as

$$t_{ref} \approx 0.07 C_s \Delta V / \sum \hat{I}$$

in most of the sense circuit designs.

2.2.2 Write and Read Signals

Waveforms in the accessed cell and on the bitline may be approximated by the analysis of a simple model circuit (Figure 2.4) that consist of a generator $v_g(t)$, resistor $r(t)$ and capacitor $c(t)$ in the investigations of both write and read operations.

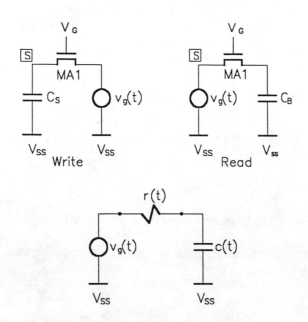

Figure 2.4. Simple model to approximate write and read signal forms.

In write operation mode a write or a sense-write amplifier switches the bitline voltage to a memory-interim standard log.0 or log.1 level. The same voltage level appears on the storage node $\boxed{S}$ after a transient time; because the gate voltage signal, that turns on the n-channel access device MA1, is boosted to $V_G \geq V_{DD} + V_T(V_{BG})$. High $V_S \approx V_{DD}$ is needed to maximize the amount of charge stored on capacitor C_S. A higher amount of stored charge can generate larger and faster signals on the bitline during read operation, and increases the immunity of data-storage against the effects of incident atomic particles. The assumption that a data signal has already reached V_{DD} on the bitline and that $V_S = 0V$ before device MA1 is turned on, permits an approximation of the waveform of the write-data $v_{SW}(t)$ on node S by applying a step-function as generator signal $v_g(t)$ and time-invariant parameters r_{dt} and C_S in the simple write-read model circuit

$$v_g(t) = V_{DD} 1(t)$$

$$r(t) = r_{dt}, \; c(t) = C_S$$

$$\tau_c = r_{dt} C_S$$

With these parameters the analysis of the circuit gives

$$v_{SW}(t) = V_{DD} \left(1 - e^{-\frac{t}{\tau_c}}\right)$$

Here, r_{dt} is the drain-source resistance of the access device in the triode region, and τ_c is the time constant of the cell. From $v_{SW}(t)$ the rise time t_r can be approximated by the well known $t_r = 2.2\tau_c$ formula. For the fall time $t_f = t_r$ and for the propagation delay $t_p = 0.5 t_r$ may be used, because during writes device MA1 operates mainly in the triode region.

In read operation mode the memory cell generates the signal $v_g(t)$ on the bitline capacitance C_B. Before activating device MA1, bitline capacitance C_B is brought to a precharge voltage V_R, and the bitline is disconnected from all the other circuits. When C_S is coupled to the bitline resistance R_B and capacitance C_B through a turned-on MA1, then both the stored voltage $V_S = V_{DD} - \Delta V$ and the bitline voltage $V_B = V_{PR}$ change. For calculation of the voltage-level change on the bitline as a function of time $v_{rB}(t)$ the rudimentary model may also be applied at read with the assumptions

$$v_g(t) = (V_D - \Delta V - V_{PR})\, e^{\frac{-t}{\tau_c}},$$

$$\tau_C = r_{dt} C_s, \qquad \tau_B = R_B C_B,$$

$$r(t) = R_B, \qquad c(t) = C_B,$$

where τ_B is the time constant of the bitline. The Laplace-transformed of $v_g(t)$ and the bitline voltage $v_{rB}(t)$, $V_g(p)$ and $V_{rB}(p)$ respectively, may be expressed as

$$V_g(p) = (V_{DD} - \Delta V - V_{PR}) \frac{1}{p + \dfrac{1}{\tau_c}}$$

and

$$V_{rB}(p) = \frac{1}{\tau_c}(V_{DD} - \Delta V - V_{PR}) \frac{1}{(p + \dfrac{1}{\tau_B})(p + \dfrac{1}{\tau_C})};$$

while the reverse transformation of $V_{rB}(p)$ results

$$V_{rB}(t) = V_0 \frac{\tau_C}{\tau_B - \tau_C}(e^{-\frac{t}{\tau_B}} - e^{-\frac{t}{\tau_c}}),\ V_0 = V_{DD} - \Delta V - V_{PR}.$$

Clearly, both the amplitude and the time-behavior of $v_{rB}(t)$ are functions of the time constants τ_C and τ_B. Depending on the ratio τ_B/τ_C the normalized $v_{rB}(t)$ curves have different voltage-maxima $\hat{V}$ appearing at different time points, and have different switching times for each signal-swing (Figure 2.5). In designs, the read-signal amplitude can most effectively be influenced by adjusting the storage capacitance C_S and bitline capacitance

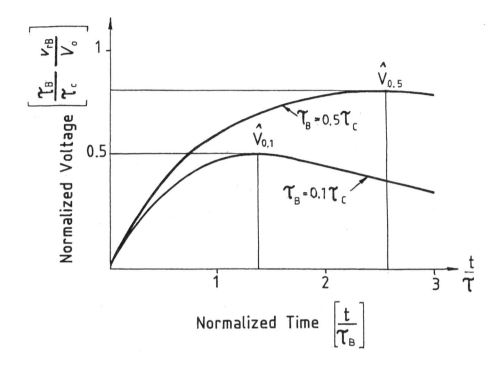

Figure 2.5. Read-signals as functions of time constants.

C_B. A $C_B/C_S < 10$ is required in most of the sense circuits, to provide reasonable read-signal amplitudes. In fast and large size memories, the read-signal amplitude is a function of both time t and bitline location x. Apart from C_B and C_S, the read-signal shape on the bitline strongly depends also on the drain-source resistance of the access transistor r_{dt}, bitline resistance R_B, terminating impedances of the bitline Z_{L1} and Z_{L2}, and the shape and timing of the control and data signals (Sections 4.1.1 and 4.1.3). The switching times of a write signal can also substantially be affected by r_{dt} and C_S. While a small r_{dt} improves both write and read speeds, a small C_s, that can facilitate for fast read, may render the DRAM cell array dysfunctional due to a large C_B/C_S. Yet, for 1T1C cells the read

96 CMOS Memory Circuits

switching times may be reduced by optimizing the amplitude of the read signals (Section 3.3.6.5). Fast write signals can be obtained by boosting the wordline voltage to exceed the supply voltage V_{DD} by the threshold voltage $V_T(V_{BG})$, where V_{BG} is the substrate bias.

2.2.3 Design Objectives and Trade-offs

The design objectives of the dynamic 1T1C memory cell may include (1) small area on silicon surface for implementation, (2) high number of memory cells tying to a single bitline, (3) large operating and noise margins, (4) speedy read operation, (5) fast write operation, (6) long time between refresh operations, (7) low power dissipation, (8) insensitivity to the impact of atomic particles and, in some cases, (9) operation in extreme environments. Normally, the circuit design can engineer only the area of the storage capacitor A_C, and the width W and the length L of the access device, and no change in CMOS processing is allowed. For the extent of A_C and W the design objectives usually dictate opposing requirements in many aspects (Table 2.2), while L may be kept as short as the processing and leakage-current considerations allow. Though the requirement for a small memory cell size is basic, if the small-size cell can not allow sufficient operating and noise margins (Section 3.1), or can not generate signals on the bitline which are large enough for error free detection and fast sense amplifier operation (Section 3.2), or needs too frequent refresh, or operates unreliable because incident alpha particles are capable to upset the stored data (Section 5.3), or can not operate in eventually required radiation environments (Section 6.1), then the cell size has to be compromised. To design a small memory cell that satisfies the variety of requirements, a combined effort of circuit design, layout design, process development, capacitor and transistor device design is needed. Such combined design and development efforts are time consuming and expensive, but the extremely high volume production of CMOS DRAMs and related memory products greatly rewards the efforts.

Paradoxically, the storage capacitance C_S in the 1T1C memory cell has to be increased as the CMOS feature sizes decrease, whereas ideal scaling would lead to reduce all parameters. Usually, the down-scaling results in connecting increasing number of memory cells to a bitline of constant length, which enlarges the bitline capacitance. Moreover, bitline lengths

Objectives	Area A_C	Width W
(1) Small Silicon Surface Area	Small	Small
(2) Many Cells on Bitline	Small	Small
(3) Large Operating and Noise Margins	Large	Small
(4) Speedy Read	Large	Large
(5) Fast Write	Small	Large
(6) Long Time Between Refreshes	Large	Small
(7) Low Power Consumption	Small	Small
(8) Particle Impact Insensitivity	Large	Small
(9) Operation in Extreme Environments	Large	Small

Table 2.2. Objectives and trade-offs in one-transistor-one-capacitor memory cell designs.

extend with the evolution of CMOS technology in fabricating larger and larger chip sizes. Thus, for greater bit-capacities, CMOS memories call for memory cells which combine smaller silicon surface area with larger storage capacitances. Capacitance enlargement may be obtained by (1) reduction of the dielectric thickness t_d, (2) use of insulator with high relative dielectric constant ε_d, (3) abatement of the parasitic capacitances C_{ld} coupled serially to C_S, and (4) expanding the effective area of the capacitor plates A_C.

2.2.4 Implementation Issues

2.2.4.1 Insulator Thickness

In the storage capacitor, the insulator thickness t_d scales down proportionally with the other facets of miniaturizations. In most of the designs, a proportional down-scaling of t_d is insufficient to provide the needed storage capacitance C_S, and further extra thinning of t_d may be required. The thinning of t_d, however, is limited [21] by the effects of the increasing electric field strength $E \approx V_C/t_d$, where V_C is the voltage across the capacitor. With increasing E

(1) the conduction of the dielectric insulator I_d grows dramatically;

$$I_d = I_0 e^{BE^{\frac{1}{2}}},$$

(2) through the insulator the quantum mechanical tunneling current I_{tu} may become significant;

$$I_{tu} = qv_t \frac{\varepsilon_d \varepsilon_0}{\varepsilon_{Si}} \cdot \frac{V_c}{kT} E \{ e^{-\frac{C_k}{E} \phi^{\frac{3}{2}}[1-(1-\frac{V_c}{\phi})]} \},$$

3) the defect density D_D gets greater, i.e., for S_iO_2;

$$D_D \approx D_0 e^{-0.2t_d}.$$

Here, current I_0 and parameter B are constants for the specific dielectric, q is the electron charge, v_t is the electron thermal velocity, ε_S is the relative dielectric constant of silicon, ε_0 is the permittivity of the empty space, k is the Planck's constant, T is the temperature, C_K is a physical constant containing electron effective mass and k, ϕ is the barrier energy between the conduction bands of silicon and dielectric insulator, and D_0 is the defect density before the thinning of t_d. In accordance with the equations, a thinning of the insulator layer may lead to such excessive dielectric and tunneling currents or reliability degradations which can render the 1T1C cell unusable in practical memory designs.

2.2.4.2 Insulator Material

The most widely used insulation material is SiO_2 in 1T1C cell implementations. SiO_2 is the fundamental insulator in all semiconductor integrated circuits, and its material characteristics have been thoroughly examined. SiO_2 is a paraelectric material, i.e., the displacement charge density D is linearly dependent on the external electric field E and in which no spontaneous polarization P occurs (Figure 2.6). If E_0 and E_1 are the electric field strengths generated by the voltages which in the binary

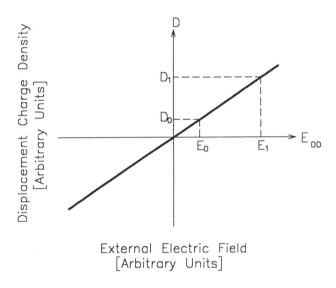

Figure 2.6. Displacement charge density versus external electric field in SiO_2.

logic-system represent log.0 and log.1 respectively, then the charge amount that differentiates the logic levels, i.e., the charge storage capacity of the 1T1C memory cell Q_C is

$$Q_C = (D_1 - D_0)A_C = \Delta D A_C,$$

and the charge density on the capacitor Q'_C is

$$Q'_c = \varepsilon_0 \varepsilon_d E,$$

where A_C is the effective area of the storage capacitor plate and ΔD is the difference in charge density. Increased Q'_C may be obtained by the application of materials with higher ε_d than SiO_2 has at a given E. Higher

ε_d and Q'_c makes possible to reduce A_c and, thereby, the size of the memory cell. Size reductions by increases in ε_d, however, are limited by the raises in currents I_O and by the variations in material constant B (Table 2.3). A raise in I_O, or B, or in both I_O and B, results in higher insulation conductivity I_d and may appear as excessive leakage current through the dielectric material. Yet Ta_2O_5 dielectric insulation in 1T1C cells [22] seems to be attractive despite its very high I_O and B. Si_3N_4 insulator has a high I_O, a low B and a small improvement factor of 1.8 in ε_d in comparison to SiO_2. Popularly applied SiO_2-Si_3N_4 combinative insulators exploit the higher material integrity of Si_3N_4 to improve memory yield rather than taking advantage of the elevated ε_d to increase capacitance per area. Other paraelectric materials with high ε_d, e.g., Y_2O_3, ZrO_2, etc., may also be applied in 1T1C cells, but the emergence of ferroelectric insulators offers also attractive alternatives.

Material	ε_d	I_O	B
SiO_2	3.9	5.1×10^{-28}	17.7
Si_3N_4	7.0	9.2×10^{-18}	11.7
Ta_2O_5	23.0	4.1×10^{-15}	23.4

Table 2.3. Material parameters ε_d, I_O and B of SiO_2, Si_3N_4 and Ta_2O_5 insulators. (Source [21])

Most of the CMOS-applicable ferroelectric materials [23] have a perovskite crystal structure described by the general chemical formula ABO_3. Elements A and B are large and small cations respectively. This cation-oxide crystals possess pyroelectric, piezoelectric and ferroelectric properties. Ferro-electricity means that the material possesses spontaneous electric polarization P that can be reversed by an applied external electric field E. With P and E the displacement charge density D can be expressed as

$$D = \varepsilon_0 \varepsilon_d E + P$$

In the ferroelectrics, which are considered to use in 1T1C cells, $P \gg \varepsilon_0 E$ and, therefore, $D \approx P$. P as a function of E curves a hysteresis loop, and the P(E) curve is nearly identical with the D(E) curve (Figure 2.7). In this curve, E_c is the coercive electric field where the net polarization reverses,

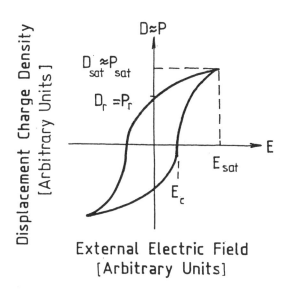

Figure 2.7. Polarization and displacement charge density versus external electric field in a ferroelectric material.

E_{SAT} and P_{SAT} define the saturation point of the polarization, and P_r is the remanent spontaneous polarization that remains aligned with previously applied E. With the slopes of the D(E) curve the ε_d is nearly proportional [24]; $\varepsilon_d \rightarrow \infty$ at E_C, and $\varepsilon_d \rightarrow 0$ at E_{SAT}. Furthermore ε_d rises to anomalously high values near the phase-transition temperature T_o (Figure 2.8). At the characteristic temperature T_o the material changes from ferroelectric phase to paraelectric phase and from paraelectric phase to ferroelectric phase. Thus, the use of ferroelectric insulators to shrink 1T1C cell sizes can be beneficial in both paraelectric and ferroelectric phases near T_o, if T_o is adequately chosen. T_o should be outside the operating temperature range of the memory so that the cell operates only in either para- or ferroelectric phase. In ferroelectric state the charge storage density Q'_c may be obtained from the hysteresis loop

$$Q'_c \approx D_{SAT} - D_r \approx P_{SAT} - P_r \, .$$

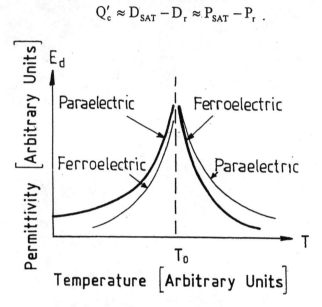

Figure 2.8. Permittivity versus temperature. (Source [24].)

The hystereses in D(E) and P(E) curves are unintended to be used in dynamically operating memory cells; merely the benefits of the high ε_d are exploited. In addition to high ε_d the insulator in a 1T1C memory cell has to satisfy requirements in dielectric leakage-current, breakdown-voltage, defect-density and reliability including the effects of aging and fatigue; and the implementation of the high ε_d must be compatible with the CMOS process compatible. A combination of CMOS-process compatibility and high ε_d may be achieved, e.g., by the use barium-strontium-titanate compounds as capacitor dielectric and platinum for capacitor plates [25]. Experimental processing of memory cells using Pb,Zr,TiO_3 (PZT); Pb,La,TiO_3 (PLT); Pb,La,Zr,TiO_3 (PLZT); $BaTiO_3$; Pb,Mg,NbO_3 (PMN); Pb,Mg,Nb,O_3-$PbTiO_3$ (PMNPT); $SrTiO_3$ and other materials have also shown some encouraging results.

2.2.4.3 Parasitic Capacitances

In the implementations of memory cells, the use of CMOS processing causes to exist parasitic capacitances, which may significantly reduce the effectiveness of the storage capacitors. Dynamic memories of moderate bit capacities may use planar capacitors (Figure 2.9) in which the depletion

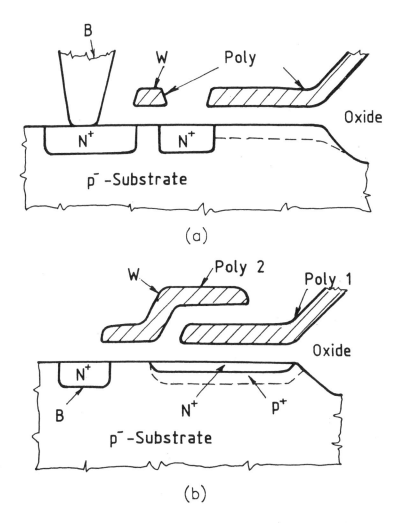

Figure 2.9. Storage and depletion layer capacitances in 1T1C memory cells using lightly doped (a) and heavily doped (b) silicon surface.

layer in the silicon forms a serial capacitance C_{ld} with the storage capacitance C_S. To provide a reasonable $C_{ld}/C_s \leq 0.05$ the doping concentration N_d under the capacitor plate has to be adjusted. A doping adjustment may interfere with other facets of CMOS processing, such as the control of junction breakdown and threshold voltages, and may be constrained by the critical electric field E_{CR} that causes impact ionization. Applying the dependency of E_{CR} from N, a lower bound for doping concentration adjustment can be found [26] as

$$N \geq 1.5 \times 10^{-12} \frac{C_s V_s}{\varepsilon_{Si} \varepsilon_0 A_c}$$

Here V_s is the voltage on C_s, ε_{si} and ε_o is the permitivity of the silicon and vacuum respectively, and A_c is the surface area of the storage capacitor. Storage capacitors which are formed between a pair of polysilicon layers (Figure 2.10) make the design free from the constraint imposed on N.

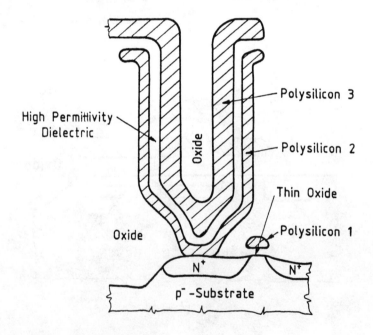

Figure 2.10. Storage capacitance between polysilicon layers.

2.2.4.4 Effective Capacitor Area

To increase the storage capacitance C_S, the extension of the effective area of the capacitor plates A_C has widely been used in CMOS memories. CMOS fabrication technologies allow to include processing steps to create trenches in the silicon bulk, to stack polysilicon devices above transistors and wirings, and to make coarse polysilicon surfaces. Making trench-capacitors (Figure 2.11) [27] seems to be a cost-effective approach; but an

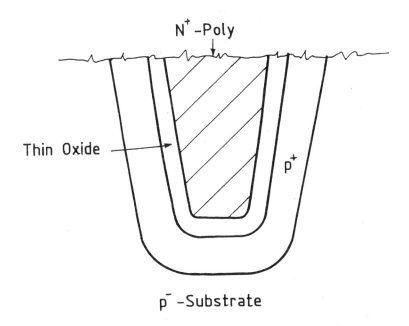

Figure 2.11. Trench-capacitor structure. (Extracted from [27].)

increase in processing complexity is necessary to control the uneven oxide-growths and leakage currents at the surface of the trench. These phenomena in the vicinity of the trench are results of the various crystal-orientations (Figure 2.12) which occur due to the oval or circular shape of the trench on the silicon surface.

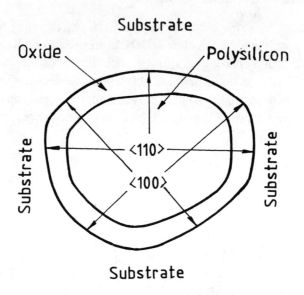

Figure 2.12. Crystal orientations along the contour of an actual trench-capacitor.

Stack capacitors can most effectively be formed between polysilicon layers (Figure 2.13) [28], but the implementation of large capacitor plates above the access transistor and wiring may require high-temperature processing steps which effect the characteristics of the circuits placed under the capacitor.

The effective surface of the capacitor plates may also be enlarged by shaping grains, textures or other forms of granuality [29] into the surface of the polysilicon storage nodes. Granulation, of course, is the most effective in extending capacitance when the layers can follow each other's surface shape (Figure 2.14).

The most efficient silicon surface utilization may be achieved by designing memory cells, which can be placed in the area determined by the crossover of a minimum-width bitline and a minimum-width wordline,

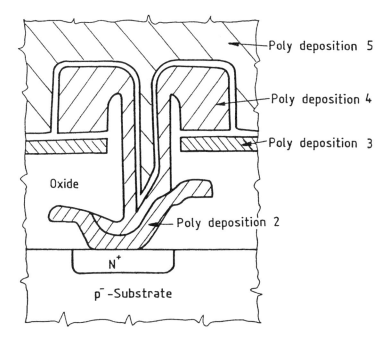

Figure 2.13. A stack capacitor formation. (Derived from [28].)

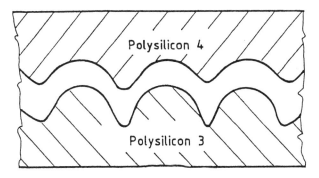

Figure 2.14. Granulation of polysilicon surface.

108   CMOS Memory Circuits

and in a distance from each other established by the metal-to-metal minimum spacing (Figure 2.15). Designs to approach this minimum cell-area require to place perpendicularly to the silicon surface both the capacitor and the access transistor (Figure 2.16) [210,211]. Nevertheless, access devices formed on the surfaces of polysilicon slabs or of trenches in silicon-crystals may have characteristics which differ substantially from those of conventional CMOS transistors. Degradations in transistor parameters e.g., in drain-source and drain capacitances, subthreshold drain-source leakage currents, gain factors, and others, and interactions among neighbored transistors and capacitors may seriously constrain the

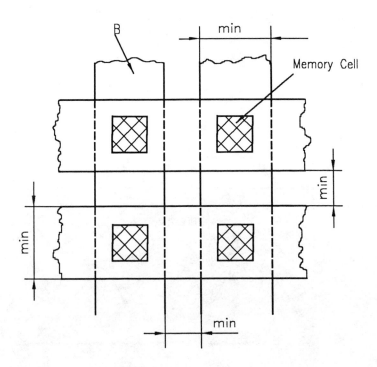

Figure 2.15. Efficient surface utilization.

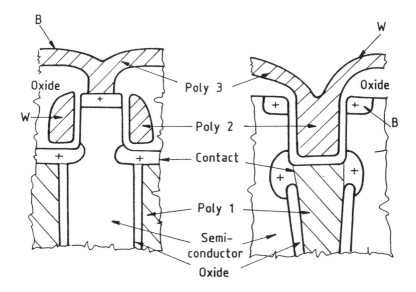

Figure 2.16. One-transistor-one-capacitor memory cell designs perpendicular to the silicon surface. (Derived from [210] and [211].)

applicability of 1T1C memory cells with structures perpendicular to the silicon surface. Yet, the mitigation of these transistor-parameter degradations is within the grasp of the CMOS technology, and the resulting improvements further strengthens the dominance of 1T1C cells in CMOS memory designs.

2.3 DYNAMIC THREE-TRANSISTOR RANDOM ACCESS MEMORY CELL

2.3.1 Description

Three-transistor (3T) dynamic cells have applications in integrated circuits which combine CMOS digital logic and memory functions, but they are seldom used in designs of memory-only chips. 3T memory cells can be implemented with inexpensive processing technologies which are developed for fabrication of digital circuits, because their implementation does not require the use of sophisticated processing steps. In stacked configuration the circuit of a 3T memory cell may be designed in a silicon surface area that is smaller than the area requirement of any static memory cell design, but larger than that of the 1T1C memory cell. Yet, memories using 3T memory cells provide faster write and read operations than those applying 1T1C memory cells. As a drawback, the data stored in 3T memory cells can easily be upset by the impacts of low-energy atomic particles, because the storage capacitance in a 3T cell is usually smaller than that in 1T1C cells.

In a 3T memory cell (Figure 2.17) [212] the capacitance C_S on the gate of a transistor M1 stores an amount of charge that represents a binary datum, another transistor M2 is devoted to write access only, and a third MOS device M3 facilitates the read access. At write, device M2 is turned on and a write buffer charges or discharges C_S through a write-bitline BW. Depending on what amount of charge and corresponding voltage is stored on C_S transistor M1 stays either turned on or off when M2 is deactivated. At read M3 is activated, M2 remains in high-resistance state, and the read bitline BR is precharged to voltage V_{PR}. V_{PR} changes significantly if M1 is highly conductive, and changes very little if M1 is in low conductance state. During refresh operation first M2 is turned off and M3 is turned on to provide a signal for read-out. Upon amplification, M2 is turned on, and allows for rewrite a datum on the storage node of the memory cell. In the implementations of this memory cell, the write-bitline capacitance C_{BW}, the read-bitline capacitance C_{BR} and the wordline capacitance C_{WL} are usually significant and restrict speed and power parameters.

Memory Cells   111

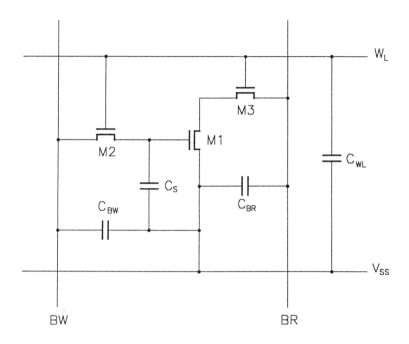

Figure 2.17. A three-transistor memory cell circuit.

2.3.2 Brief Analysis

Write and read signals, which are generated in and by a 3T memory cell, may be characterized by exploiting the analogy between the 3T memory-cell operation and the charge and discharge of a capacitor C through a resistor R. The time functions of charge $v_c(t)$ and discharge $v_d(t)$ signals on C through R are well known as

$$v_c(t) = V_o(1 - e^{-\frac{t}{\tau_c}}) \text{ and } v_d(t) = V_a e^{-\frac{t}{\tau_d}},$$

where V_o is the amplitude of the generator step-signal $V_o 1(t)$ and $\tau = RC$.

In a 3T memory cell, for write operations

$$V_o = V_{DD} - V_T(V_{BG}) \text{ and } \tau = [R_{BW} + r_{d2}(t)]\, c_s(t)$$

may be applied in the equations of $v_c(t)$ and $v_d(t)$, where V_{DD}, V_T and V_{BG} are the supply, threshold and backgate-bias voltages, R_{BW} is the resistance of the write-bitline BW, r_{d2} is the time-dependent drain-source resistance of transistor M2, and $c_s(t)$ is the time-variant storage capacitance. Similarly, for read operations

$$V_o = V_{PR} + \Delta v \text{ and } \tau = [R_{BR} + r_{d1}(t) + r_{d3}(t)]c_{BR}(t)$$

may be used. Here, V_{PR} and Δv are the precharge and the memory-cell generated voltages, R_{BR} is the read-bitline resistance, $r_{d1}(t)$ and $r_{d3}(t)$ are the time-dependent drain-source resistances of transistors M1 and M3, and $c_{BR}(t)$ is the time-variant read-bitline capacitance, may be applied to approach $v_c(t)$ and $v_d(t)$.

For plausibility studies, in the examinations of both write and read signals all parameters $r_{d1}(t)$, $r_{d2}(t)$, $r_{d3}(t)$, $c_s(t)$ and $c_{BR}(t)$ may be replaced by their time-invariant counterparts r_{d1}, r_{d2}, r_{d3}, C_s and C_{BR}. With time-invariant parameters in the equations of $v_c(t)$ and $v_d(t)$, a crude approximation formula for the rise- and fall-times of the charge and discharge signals $t_r = t_f = t = \tau \ln k$ can be obtained, where k is a constant.

The expression of $t = t_r = t_f$ indicates that both fast write and read operations are obtainable at the use of minimum size devices in a 3T dynamic memory cell. Minimum size write and read access transistors M2 and M3 lessen the parasitic capacitance of the write-bitline C_{BW}, of the read bitline C_{BR} and of the wordline C_{WL}. The reduction of the storage capacitance C_S is limited mainly by the required single event upset rate maximum, i.e., the required immunity for atomic particle impact of the memory (Section 5.3). In the memory cell, C_S may be increased without the size-expansion of device M1. Applications of minimum-size transistor devices within a 3T memory cell is important in improving both operational speed and packing density. Operational speed increases by the reduced capacitances C_{BW}, C_{BR} and C_{WL} despite the effects of the larger R_W, R_P, R_1, R_2 and R_3, while packing density increases by the reduced surface area required for the constituent transistors. To further reduce surface area, 3T cells may be designed in stacked configurations, where the transistors are placed each above the other one (Section 2.2.4.4).

2.4 STATIC 6-TRANSISTOR RANDOM ACCESS MEMORY CELL

2.4.1 Static Full-Complementary Storage

Static 6-transistor (6T) full-complementary memory cells are applied in memory designs to satisfy requirements for short access- and cycle-times, high frequency data rates, low power dissipation, radiation hardness, operation in space, high-temperature, noisy and other extreme environments. For the benefits in performance and environmental tolerance 6T cells compromise cell-sizes and, thereby, packing densities.

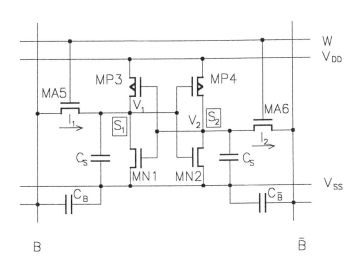

Figure 2.18. A static six-transistor full-complementary memory cell.

Yet, packing densities obtainable by the use of 6T cells are considerably higher than packing densities achieved by other memory cells which provide the same performance and environmental tolerance as 6T complementary memory cells do.

A static 6T complementary memory cell (Figure 2.18) latches a digital datum in a pair of cross-coupled CMOS inverters, which are formed of 4 transistors, MN1, MN2, MP3 and MP4, and uses a pair of access devices MA5 and MA6 to couple and decouple the storage latch to and from other circuits. Both inverters in the latch as well as both access transistors are identical, and the layout design of the cell is mirror-symmetrical.

The data storage capability of this memory cell rests on the well known fact that a pair of complementary inverters in positive-feedback, or as also called in latch, cross coupled, or Ecless-Jordan configuration, has two stable states. Positive feedback exist in the operation region of this circuit where both the low-frequency small-signal loop gain A_L and the total phase-shift in the loop ρ_L fulfills the Barkhausen criteria [213] (Section 3.3.4.2)

$$A_L = A_1 \cdot A_2 = A^2 > 1,$$
$$\rho_L = \rho_1 + \rho_2 = 2\rho = 2\pi,$$

where A_1 and A_2 are the gains and ρ_1 and ρ_2 are the phase angles of the first and second complementary inverters respectively. In a 6T memory cell the two inverters have approximately the same gain $A \approx A_1 \approx A_2$ and the same phase shift $\rho \approx \rho_1 \approx \rho_2$, therefore

$$A = g_{mN} \frac{r_{dN} r_{dP}}{r_{dN} + r_{dP}} \; ; \; \rho = \pi.$$

Here, N and P subscripts indicate n- and p-channel devices respectively; g_m is the transconductance and r_d is the drain-source resistance of the devices when the circuit operates in the vicinity of the metastable state or of the flipping point. Voltages representing the flipping point V_F and both stable states V_1 and V_2 can conveniently be determined by the use of the normal $v_1 = f(v_2)$ and mirrored $v_2 = g(v_1)$ voltage transfer characteristics of the inverters (Figure 2.19). The voltage V_F may also be defined by a variety of different methods comprising the application of the (1) closed loop unity gain, (2) zero Jacobian determinant of the Kirchoff equations, (3) coincidence of roots in flip-flop equations, and the (4) inverters' transfer characteristics.

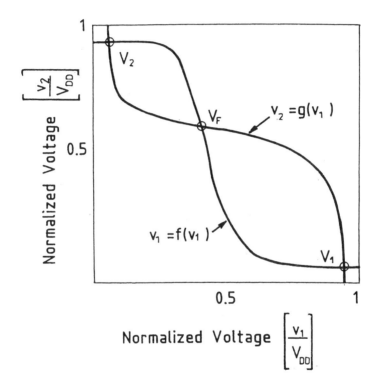

Figure 2.19. Normal and mirror input-output voltage characteristics of the inverters in a six-transistor cell.

In a pair of crosscoupled inverters the pair of the stored voltage levels are able to return to and stay V_1 and V_2 as long as the amplitude of a latch-external signal ΔV on either one or both of the storage nodes $\overline{S_1}$ and $\overline{S_2}$

does not exceed V_F i.e., $\Delta V < V_F - V_{SS}$ or $\Delta V > V_{DD} - V_F$. V_{DD} is the positive supply and V_{SS} is the ground potential. In practice, each of the voltages V_1, V_2, V_F and V_{DD} has a spread of values with determinable minima and maxima $\hat{V}_1$, $\hat{V}_2$, $\hat{V}_F$, $\hat{V}_{DD}$ and $\check{V}_1$, $\check{V}_2$, $\check{V}_F$, $\check{V}_{DD}$. Voltage differences $\hat{V}_F - \check{V}_1$ and $\check{V}_2 - \hat{V}_F$ are widely accepted as noise margins in 6T memory cells. Noise margins, however, may also be defined by the maximum squares between the normal and mirrored transfer characteristics of the inverters [214].

2.4.2 Write and Read Analysis

To write a datum into the cross-coupled inverter circuit, the access transistors MA5 and MA6 and the driver write-amplifier have to be able to provide sufficient write currents I_1 and I_2 to change $V_1 = 0$ to $V_1 > \hat{V}_F$ and $V_2 = V_{DD}$ to $V_2 < \check{V}_F$ when a common gate voltage $V_G < \check{V}_{DD}$ is applied for both MA5 and MA6 and $V_1 < V_2$ (Figure 2.20). Here, V_1 and V_2 are the voltages on storage nodes $\overline{S_1}$ and $\overline{S_2}$, and V_F the flipping voltage in metastable state. In most of the designs the applicable V_G is the same for both read and write. At read, however, currents I_1 and I_2 have to be small enough to disallow a state change by keeping $V_1 < \hat{V}_F$ and $V_2 > \check{V}_F$. Thus, the requirement for state-retention at read, or nondestructive read-out, opposes the requirement for safe and quick write, and sets compromises in the design of I_1 and I_2.

In the write equivalent circuit (Figure 2.21) a pair of high-current write-buffers drives the data signal through the strongly conductive write-enable devices MP9, MN10, MP11 and MN12, column select transistors MN7 and MN8 and cell-access devices MA5 and MA6, into the storage latch of the memory cell. Before devices MA5 and MA6 are turned on in most of the designs, equivalent bitline capacitances C_B and $C_{\bar{B}}$ are brought to bitline voltages $V_B = V_{SS}$ and $V_{\bar{B}} = V_{DD} - V_{TN} (V_{BG})$ respectively. Here, V_T is the threshold voltage of transistors MN7 and MN8, and V_{BG} is the back-gate bias voltage on the effected devices. Because voltages V_B and $V_{\bar{B}}$ are preset and the write transient is short, a voltage generator with step-functions $v_g(t) = \pm 1(t) [V_{DD} - V_{TN}(V_{BG})]$ may be used to substitute the write-amplifier in rough approximations. To approximate the waveforms of the write-signal transients on the storage nodes $\overline{S_1}$ and $\overline{S_2}$ a latch design

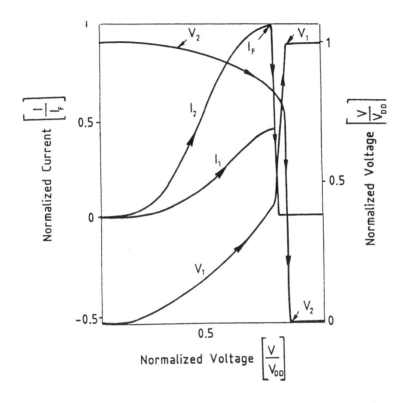

Figure 2.20. Access current and storage voltage changes versus access gate voltage variations.

with a flipping voltage $V_F=[V_{DD}-V_{TN}(V_{BG})]/2$ may be assumed. This V_F divides the development of the signal amplitude on either storage node $\boxed{S_1}$ or $\boxed{S_2}$ into two regions. In the first region the generator or write-data signal works against the effects of the positive feedback, in the second region the positive feedback accelerates the signal development. Presuming that the two regions are nearly equal, then the signal acceleration and deceleration in the two regions can cancel each other, and the presence of positive feed-

118 CMOS Memory Circuits

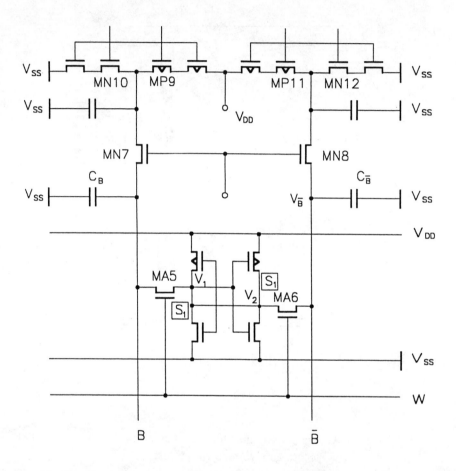

Figure 2.21. Write equivalent circuit for an SRAM designed with 6T memory cells.

back may be disregarded in a first order waveform-estimate model (Figure 2.22). A transient analysis on this model yields the well-known exponential waveform on node S_1 or S_2 (Section 2.2.2), and from that the write-

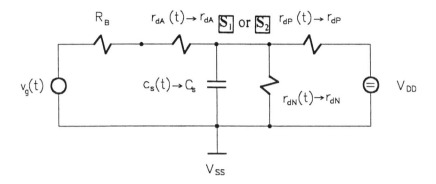

Figure 2.22. Model circuit for first order estimation of write waveforms.

switching time t_W, i.e., the fall time t_{fW} and the rise time t_{rW} of the signals on nodes $\boxed{S_1}$ and $\boxed{S_2}$ may be approximated as

$$t_W = t_{fW} \approx t_{rW} \approx \tau_{W1} \ln \frac{V_{DD} - V_{TA}(V_{BG})}{V_1} + \tau_{W2} \ln \frac{V_2}{V_{DD} - V_{TA}(V_{BG})},$$

$$\tau_{W1} = [(r_{dA} + R_B) \| (r_{dP} \| r_{dN})]C_S \quad \text{for } V_1 \leq V \leq V_{DD} - V_{TA}(V_{BG}),$$
$$\tau_{W2} = (r_{dA} + R_B)C_S \quad \text{for } V_{DD} - V_{TA}(V_{BG}) < V \leq V_2.$$

Here, r_{dA}, r_{dP} and r_{dN} are the respective time-invariant equivalents of the time-dependent $r_{dA}(t)$, $r_{dP}(t)$ and $r_{dN}(t)$ drain-source resistances of the access devices MA5 and MA6, p-channel devices MP3 and MP4, and n-channel devices MN1 and MN2; R_B is the bitline resistance; and C_S is the time-invariant equivalent for the time dependent $c_S(t)$ storage node capacitance.

At read, first the capacitance of the bitline $C_B \approx C_B$ is precharged to V_{PR}, and then the bitline is disconnected from the other circuits except from the accessed memory cell. After that $C_B \approx C_B$ is charged or discharged through the bitline resistance R_B and through the generator resistance R_G of the memory cell. Most of the 6T cells are designed so that access transistors MA5 and MA6 operate in the saturation region, and in each inverter one device is turned on and operates in the triode region during the entire read operation. Thus, the read-signal generated on the bitline B or $\overline{B}$ may conveniently be approximated by using a simple model (Figure 2.23). As discussed previously the read-switching time t_R, fall time t_{fR} and rise time t_{rR} can be obtained as

$$t_R = t_{fR} \approx t_{rR} \approx \tau_R \ln\frac{V_{PR}}{V_{PR} \pm \Delta V_R}, \quad V_{PR} \approx \frac{V_{DD} - V_{TN}(V_{BG})}{2},$$

$$\tau_R = (r_{dsA} + r_{dtP} + R)C_B \quad \text{or} \quad \tau_R = (r_{dsA} + r_{dtN} + R_B)C_B,$$

where, ΔV_R is the read-signal swing on the bitline, r_{dsA} is the drain-source on-resistance of the access device in the saturation region, r_{dtP} and r_{dtN} are the drain-source on-resistances of the p- and n-channel transistors of the inverters in the triode region.

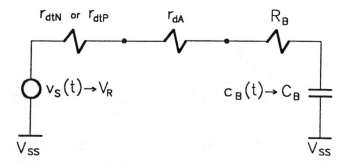

Figure 2.23. Model circuit for read-signal approximation.

The equations for switching times t_W and t_R clearly indicate that the drain-source on-resistance of the access devices should be small for both fast write and quick read operations. Nevertheless, a fast write requires small latch transistors which provide rather high drain-source resistances, while a quick read calls for wide latch-transistors with little drain-source on-resistances. Wide latch transistors improve operation and noise margins, radiation hardness and tolerance of other environmental affects also, up to the limitation imposed by increased drain-source leakage currents. Yet, any increase in any transistor size can expand the silicon surface area of the memory cell, may oppose the conditions for nondestructive read-out, and a number of other design objectives.

2.4.3 Design Objectives and Concerns

The design of the constituent transistors of a 6T static memory cell should approach objectives and satisfy contradictory requirements (Table 2.4) which are similar to those of the 1T1C cell design (Section 2.2.3). To approach the objectives for a 6T cell the design can vary both the width W and length L and, thereby, the aspect ratio W/L in the gain factor β of the individual transistors. Principally, the quotient of the gain factors $\beta_q = \beta_A/\beta_N \approx \beta_A/\beta_P$ where indices A, N and P mark access, pull-down n-channel and pull-up p-channel transistors, have to be designed to assure safe write and nondestructive read operations. Usually, a 6T cell design with $\beta_q \approx 0.35$ allows for the use of a single gate voltage $V_G \approx V_{DD}$ on the access transistors for both write and read functions. Facilitating conditions for safe and quick writes by $V_G > V_{DD}$ is not recommended because of increased hot carrier emission, device-to-device leakage currents, eventual transistor punch-through, instability and breakdown phenomena. Minimum transistor sizes with $\beta_q = 1$ may be designed at the application of midlevel precharge, low-current sense amplifier and high-current write amplifiers. Application of a particular threshold voltage to the access transistors, that is higher than the threshold voltages of the other transistors in the memory cell, is also a widely used method to circumvent the write-read paradox. Higher threshold voltages in the access transistors result in decreased subthreshold leakage currents and, thereupon, in increased noise margins, and allow for higher number of memory cells connectable to a single bitline. The static noise margins in 6T memory cells can be designed by altering the transistor aspect ratios W/L-s and the device size ratios β_q-s and, occasionally, by varying threshold voltages and other device parameters.

Objectives	Access Devices	Latch Devices
(1) Small Surface Area	Small	Small
(2) Many Cells on a Bitline	Small	Small
(3) Large Operation Margins	Small	Large
(4) Nondestructive Read	Small	Large
(5) Speedy Read	Large	Large
(6) Fast Write	Large	Small
(7) Low Power Consumption	Small	Small
(8) Particle Impact Insensitivity	Small	Large
(9) Radiation Hardness	Small	Large
(10) Environmental Tolerance	Small	Large

Table 2.4. Objectives and requirements in transistor sizes.

2.4.4 Implementations

The silicon surface area of 6T memory cells, in planar designs, may be overly large to meet objectives in packing density and performance for a prospective CMOS memory. To improve packing density and speed performance, numerous CMOS processing technologies feature trench-isolation between p- and n-wells e.g., [215] and stacked transistors e.g., [216]. Stack transistor technology, in its most widely used form, places p-channel polysilicon thin-film transistors over n-channel transistors which are implemented in silicon crystals (Figure 2.24). In many thin-film implementations the channel lengths of the transistors have to be extended, and channel offsets have to be introduced, to decrease the subthreshold leakage currents to a required limit, e.g., to 10^{-13} A/m.

Furthermore, the nonlinear current-voltage characteristics of the parasitic diode-like devices (Section 6.3.4), which may occur as a result of combining P⁺ and N⁺ doped materials at junctions of p- and n-channel transistors, have to be considered in the circuit design. This and other three-dimensional designs of 6T memory cells, of course, require increased complexity in the CMOS fabrication technology. Nevertheless, the potential to provide large bit-capacity static memory chips, fast memory operations and high manufacturing yields, exceedingly outweighs the augmentation of the fabrication complexity.

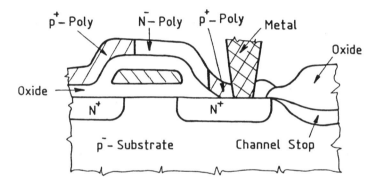

Figure 2.24. Stack-transistor implementation. (Derived from [216].)

An increase in memory circuit complexity is also required for the application of static memory cells. Namely, in memory cells selected by an activated wordline and connected to unselected bitlines, the stored data may be altered by the combined effects of the precharge voltage, coupled noises, and leakage currents of the memory cells connected to the same bitline. To avoid data scrambling in cells connected to the unselected bitlines, the application of bitline loads are needed. These loads are coupled to the bitlines when the precharge is completed (Figure 2.25). In this exemplary bitline-terminating circuit, the serially connected transistor pairs MN9-MP11 and MN10-MP12 act as bitline load devices. MN13 and MN14 are the bitline-select or column-select transistors. Transistors MN16 and MN17, in parallel configuration, determine the precharge voltage $V_{PR} = V_{DD}-V_{TN}$ (V_{BG}), and during precharge MP20 assists to equal-

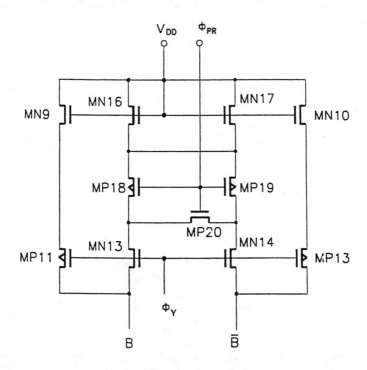

Figure 2.25. A bitline terminating circuit.

ize the voltages on the bitlines B and $\overline{B}$. A precharge of B and $\overline{B}$ occurs when devices MP18, MP19 and MP20 are turned on by impulse ϕ_{PR} simultaneously with the activation of MN13 and MN14 by bitline-select impulse ϕ_Y. At the same time, in the unselected columns ϕ_Y disallows the precharge and data transfer, and connects load devices MN9, MN10,

MP11 and MP12 to the unselected bitlines. The separation of the loads from the selected bitline improves the sensing speed and the operation margins, while the bitline selective precharge greatly reduces the power dissipation of the memory and decreases the substrate currents and the emission of hot-carriers. Less hot-carrier emission results higher reliability in memory operations.

To provide fast read operations not only the sensing and read circuits, but also the precharge of the bitlines must be quick. Speedy precharge requires high β with W/L >10 for devices MN16, MN17, MP18, MP19 and MP20. Yet, minimum size load devices MN9, MN10, MP11 and MP12 may be sufficient to prohibit data alteration in the unselected bitlines. Since in this circuit bitline-select transistors MN13 and MN14 are coupled to the precharge devices and to the bitlines in series configuration, the β of the device pair MN13-MN14 should be about as large as the β of transistors MP18 and MP19.

2.5 STATIC FOUR-TRANSISTOR-TWO-RESISTOR RANDOM ACCESS MEMORY CELLS

2.5.1 Static Noncomplementary Storage

Static noncomplementary four-transistor-two-resistor (4T2R) memory cells are used in memory designs, typically in RAMs, to combine high packing density with short access and cycle times and with simple application in systems. Applications of 4T2R cells allow for obtaining memory packing densities which are between those attainable by designs based on dynamic 1T1C and static 6T memory cells, and which are comparable with designs using dynamic 3T cells. The access and cycle times of memories employing 4T2R cells, however, are much shorter than those obtainable with 1T1C cells and somewhat longer than those performed by memories designed with 6T cells. For long-term data storage 4T2R cells do not need refresh, but in the cells the stored data can be upset by charged atomic particle impacts and by other various environmental effects with much higher probabilities than the data stored in 6T memory cells.

126 CMOS Memory Circuits

The static 4T2R memory cell (Figure 2.26) is a noncomplementary variation of the elementary circuits which use a symmetrical pair of inverters in positive feedback configuration for storage, and a symmetrical pair of transmission devices for access. All active devices MN1, MN2, MA3 and MA4 are uniformly either n- or p-channel devices, and the inverters are implemented as transistor-resistor compounds MN1-R5 and MN2-R6.

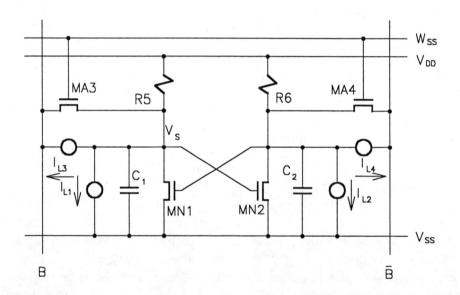

Figure 2.26. A static four-transistor-two-resistor memory cell circuit with leakage currents.

Resistors R5 and R6 are applied, primarily, to avoid loss of data, i.e., compensate the effect of leakage currents I_{L1} and I_{L3} on node potential V_S, when both access devices MA3 and MA4 and one of the driver transistors MN1 are turned off, and to balance I_{L2} and I_{L3} when MA3, MA4 and MN2 are turned off. Here, $I_{L1} + I_{L3} = I_{L2} + I_{L4} = I_L$ is assumed. With I_L and with an allowable V_S the maximum resistance R=R5=R6 can easily be determined. In practice, the maximum applicable resistance R, however, is greatly influenced by ΔR which amounts the variations of R;

$$\Delta R = \Delta R_{PT} + \Delta R_T + \Delta R_E,$$

where resistance-variations ΔR_{PT}, ΔR_T and ΔR_E are functions of the processing technology, temperature and environments including the effects

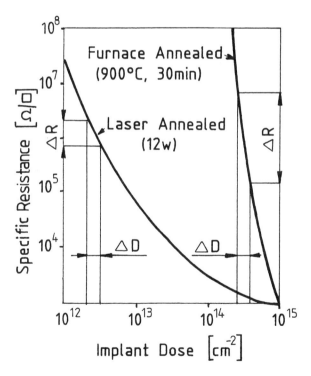

Figure 2.27. Specific resistance versus implant dose for furnace and laser annealed polysilicon.

of radioactive radiation, humidity, and others. Usually resistor R is implemented in polysilicon, and the influence of the CMOS processing technology, specifically the annealing-method, dominates the magnitude of ΔR (Figure 2.27). A large ΔR, as it appears at furnace annealing, may render 4T2R cells unusable due to high standby currents and excessive memory power dissipation. High R, low power dissipation and large bit capacity-per-chip can be obtained by processing improvements, e.g., by the exploitation of laser technology, thermal cycling, etc., in the fabrication of 4T2R memory cells.

2.5.2 Design and Implementation

By all means, CMOS processing technologies should minimize drain-source leakage currents in all the transistors of a 4T2R memory cell. Yet, when either transistor pair MN1-MA3 or MN2-MA4 are off, the leakage currents must have certain ratios I_{L1}/I_{L2} and I_{L2}/I_{L4} to keep the storage nodes on potentials which provide adequate noise margins. I_{L1}/I_{L2} and I_{L2}/I_{L4} may be altered by the variation of device sizes and, thereby, by the gain factor quotients $\beta_q = \beta_1/\beta_3 = \beta_2/\beta_4$. In usual designs, $\beta_q < 0.4$ and a resistor current $I_R \leq 0.1 I_L$ are needed to keep the storage node of $0.1 I_L$ the deactivated access and driver transistors on the required potential. These requirements in β_q and I_R contradict the conditions for sufficient noise margins on the storage node which is connected to a driver transistor either MN1 or MN2 that is turned on (Figure 2.28). The diagram shows that a $\beta_q > 2$ is required for an acceptable noise margin, and that at higher cell-supply voltage V_{CC} larger static noise margin V_{NM} is obtainable. To provide acceptable noise margins in both the on- and off-side of the 4T2R cell and to avoid paradoxical β_q requisites, the threshold voltage of each driver transistor MN1 and MN2 are often set higher than the threshold voltage of each access device MA3 and MA4. Subthreshold currents in MA3 and MA4, nonetheless, must be small to assure sufficient operation margins in the sense circuit (Section 3.1.3) and to allow the connection of a high number of memory cells to the same bitline.

In 4T2R memory cells R5 and R6 provide the functions of the p-channel devices MP5 and MP6 employed in 6T memory cells (Section 2.4), thus $R = R_5 = R_6 = r_{dP} = r_{dtP} = r_{dsP}$ may be used in the equa-

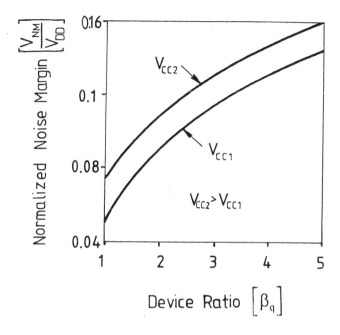

Figure 2.28. Noise margin versus device-size ratio.

tions for A_L, t_{rw}, t_{rR} and t_{fR}. Because R is large in comparison to the drain-source on-resistance of devices MN1 and MN2, and because the inverters are mirror-symmetrical, the Barkhausen-criteria for positive feedback can easily be satisfied. Furthermore, because the R-s are high, above the flipping voltage V_F the effect of the positive feedback is little; and a small write current can rapidly change the information content of the 4T2R memory-cell. During a write operation the low resistances of the write-buffer output and the access device connects parallel with R, and one of the drive transistor MN1 or MN2 is turned off. At a read operation the on-resistance of either transistor pair MN1-MA3 or MN2-MA4 alters the precharge voltage and the current on one bitline significantly, and insignificantly on the other bitline. The more significant change is rapidly sensed and amplified by the sense amplifier, and the datum stored in the memory cell remains unchanged.

Most of the 4T2R memory-cell designs use polysilicon load resistors R5 and R6 and place them over the transistors [217] similarly to the three dimensional design of the 6T memory cell (Section 2.4.4). This three-dimensional placement of constituent elements requires at least two, but most often three, polysilicon layers. Conveniently, the third layer may be applied to extend the storage node capacitances C_1 and C_2. High C_1 and C_2 decreases the inherent susceptibility of 4T2R cells for the effects of charged atomic particle impacts (Section 5.3).

The application of 4T2R memory cells in arrays necessitates the use of load devices to prevent data-loss in the accessed cells which are tied to unselected bitlines. In the write-read equivalent circuit (Figure 2.29) bitline B and $\overline{B}$ are coupled to load devices MN9 and MN10 and to precharge devices MN16, MN17, MP18, MP19 and MP20. Transistors MN13 and MN14 provide bitline selection, MN7 and MN8 are applied for write enable and the other transistors and the two resistors represent the 4T2R cell.

Traditionally, 4T2R memory cells can be designed to occupy considerably smaller silicon surface area and to perform faster write operations than 6T memory cells do. Nevertheless, three issues, (1) stacked device configuration, (2) single event upset rates, and (3) noise margin considerations, may dwindle the size advantage and, somewhat, the write-speed benefits of the 4T2R cell over the 6T cell. In stacked device configuration, both the 4T2R and 6T cells may be designed to take approximately the same area (Section 2.4.4). Furthermore, the area requirements for the 4T2R memory cells may be increased to satisfy requirements in single event upset rates by enlarging storage-node capacitances (Section 5.3.3), and to provide 6T-cell-equivalent noise margins by augmenting the device-size ratios. 4T2R memory cells are prone to the effects of atomic particle impacts, because of their inherently small capacitances and large load resistances. Large load resistances reduce the achievable highest voltage on the cell node, and the reduction makes also the noise margin for log.1 smaller.

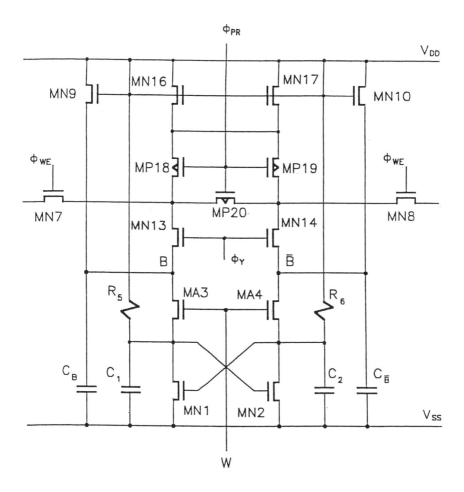

Figure 2.29. Write-read equivalent circuit in an SRAM designed with 4T2R memory cells.

2.6 READ-ONLY MEMORY CELLS

2.6.1 Read-Only Storage

In a read-only memory (ROM) cell, data can be written only one time, the one-time writing or programming is a part of the fabrication process, the cell holds a datum for its life-time and can be read arbitrary times during its life-time. (As contrasts, in so-called programmable read only memory PROM cells, the stored data can be determined by the end-user, and with the exception of fuse and antifuse cells, PROM cells may be programmed more than once.) ROM cells may be applied in all random access, sequential and content addressable memory designs. Yet, in the common use, the denomination ROM implies random access operation, and eventual other access modes are expressed by added attributes, e.g., sequential ROM, content addressable ROM. Commonly, ROM-cells are employed in control and process program stores, look-up tables, function generators, templates, knowledge bases, etc., and in general implementations of Boolean and sequential logic circuits. Most ROM cells used in CMOS memories operate in binary logic system, but an increasing number of CMOS ROMs exploits the benefits multiple-valued nonbinary memory cells.

A ROM cell [218] in CMOS technology is as simple as a single n- or p-channel transistor. In the one-transistor (1T) ROM cell the gate of the transistor serves as the control electrode of an access device, and the single transistor combines both access and storage functions. To binary applications, the transistor can be programmed to provide either a permanently open circuit in NOR configuration (Figure 2.30), or a permanently low-resistance circuit in NAND configuration (Figure 2.31). In both configurations, transistors MPi and MPj are applied for pre-charging and for compensating leakage currents, or for forming resistive loads in the NOR and NAND arrays. A NOR array of n-channel ROM cells requires a high voltage, e.g., the supply voltage V_{DD}, to select a wordline, e.g. W_i, while all other wordlines are kept at ground potential $V_{SS}=0V$. The selected W_i turns all effected unprogrammed transistors on, but the programmed ones provide very high resistances between the bitline, e.g. B_i, and V_{SS}. As results, the bitline B_i with the programmed cell

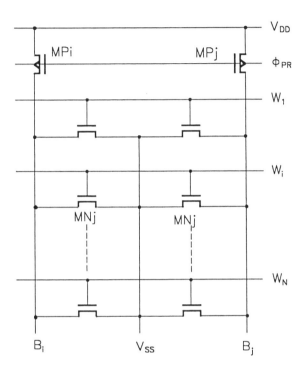

Figure 2.30. Read-only-memory cells in NOR configuration.

remains on high voltage, e.g., $V_{PR} \approx V_{DD}$, and bitlines with unprogrammed cells get discharged toward V_{SS}. In NAND arrays of n-channel ROM cells the selected wordline W_i is brought to a low potential, e.g., V_{SS}, and all the unselected wordlines remain on a high potential, e.g., V_{PR}. The selected W_i turns the unprogrammed transistors off, while all other cells have low drain-source resistances. Thus, the potential of the bitline B_i, that is coupled to a programmed cell, switches to V_{SS} from V_{PR}, while bitlines with unprogrammed cells hold a high potential. This high potential may considerably be less than V_{PR} due to charge redistributions which may plague precharged NAND configurations (Section 2.7.2).

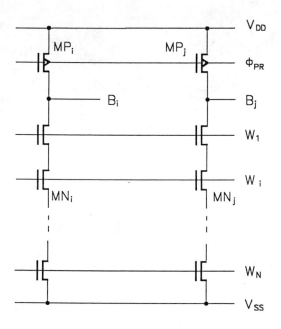

Figure 2.31. Read-only-memory cells in NAND arrangement.

2.6.2 Programming and Design

In both NOR and NAND type of arrays the ROM cells may be programmed by varying the threshold voltage V_T of the transistor in the cell. Conveniently, V_T may be programmed by implanting different ion doses into the channel regions, or by altering the oxide thickness above the channels. Moreover, oxide isolation between the bitline and the transistor-drain may also be used to program open circuits in NOR type of arrays. The choice among the various programming methods is based on the importance of such requirements as highly reliable operation, high yield, and the possibility of programming in the final phases of the CMOS fabrication process. Programming near the end of the processing allows delivery of ROMs in short turn-around times.

The 1T ROM cell can also be programmed to store data in multi-level forms by using a multiplicity of threshold voltages. Multiple-level circuits have much smaller operating and noise margins than binary circuits do. Nevertheless, the development of sense circuits to detect signals with very small amplitudes (Sections 3.3-3.5) and improvements in processing technologies to decrease variations in threshold voltages and in other parameters, make designs with multi-level ROM-cells feasible. A four-level ROM storage e.g., needs three threshold voltages V_{T1}, V_{T2} and V_{T3} which can be programmed in three ranges ΔV_{T1}, ΔV_{T2} and ΔV_{T3} in practice (Figure 2.32). ΔV_{T1} divides the voltage region $V_R = V_{DD} - V_{SS}$ into an upper and a lower part, while ΔV_{T2} and ΔV_{T3} subdivide each of both parts into upper and lower parts too. At read, the stored datum is compared to ΔV_{T1} first, and then either to ΔV_{T2} or to ΔV_{T3} depending on the outcome of the first comparison. For the level comparison, either three sense amplifiers or one sense amplifier with three switchable reference levels, or three precharge voltages, or a combination of these techniques, can be used. A variety of level-detection techniques may be borrowed from analog-to-digital A-D converter circuits [219].

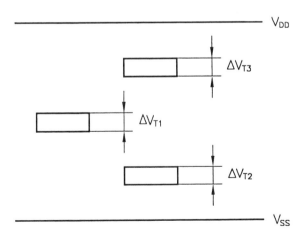

Figure 2.32. Threshold voltage ranges for four-level ROM storage.

1T ROM cell based circuits may be analyzed and designed by the apparatus and methods used for random access memory cells (Section 2.2) and sense circuits (Section 3.1). ROM circuits designed in NOR configurations result in faster read operations than ROMS designed in NAND configurations, because the bitline capacitance is charged and discharged through single transistors which have zero back gate bias. Furthermore, the lack of charge redistribution in NOR arrays provide large operation margins. This, in addition to operation capabilities in extreme environments, makes the 1T ROM cells in NOR-arrays more amenable to multi-level storage than the 1T ROM cells in NAND configurations. Implementations of NAND arrays, however, require only one contact on a bitline and, in turn, may require less semiconductor surface area than NOR arrays formed of 1T ROM cells do. Because in CMOS applications 1T ROM cells can be designed to occupy smaller size and to provide faster operations than any other ROM cells do, most of the CMOS ROM designs use 1T ROM cells.

2.7 SHIFT-REGISTER CELLS

2.7.1 Data Shifting

Shift-register cells are applied in sequentially accessed memories to provide very fast write and read data rates. Fast shift register operations, however, are obtained at the expense of high power dissipations and low memory packing densities. Shift registers move the entire data content of the activated array from the inputs to the outputs, and each addressed datum or data set has a unique time delay from a reference time point. To store and transfer data, a shift register cell (Section 1.3.3) requires control signal with two distinct phases, two access or transmission devices, two storage elements and, eventually, two amplifying components in a single shift register cell (Figure 2.33).

During a first phase, transmission device TR1 is brought to conductive state by impulse ϕ_1 and TR1 transfers a binary datum into the first storage element SE1. The signal that represents a datum may be attenuated by the transfer, and a signal amplifier A1 can be applied to recover the datum for a consequent shift. Furthermore, A1 ensures the direction of the data

Memory Cells 137

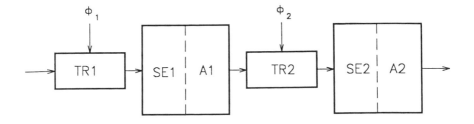

Figure 2.33. Generic shift-register cell composition.

shift by providing a forward amplification $A_f \gg 1$ and a reverse amplification $A_r \ll 1$. For the time of a first data transfer, storage and amplification, transmission device TR2 is turned off. In a second phase, impulse ϕ_2 turns TR2 on, and impulse ϕ_1 deactivates TR1. Thus, TR1 separates storage and amplifying elements SE1 and A1 from the preceding stage, and the datum stored in SE1 is transferred through A1 and TR2 to storage element SE2 and amplifier A2. Circuit complexes TR1-SE1-A1 and TR2-SE2-A2 are considered as the two halves of a single shift register stage or shift register cell.

The structure and the operation of a shift register indicate three important issues. (1) The load on the output of each half-cell is minimum (Fanout = 1) resulting in very quick data shifts. (2) One half of all shift-register cells are activated at each phase of the operation, which results in very high power dissipation. (3) Shift operations in an N-bit array need 2N access and 2N storage elements and, thereby, limit packing densities. Although amplifying elements are used in each half of a shift register cell, their applications are necessary only in those cells in which the signal degradations reduce the noise margins to magnitudes unacceptable for memory operations.

Fast sequential, buffer and specialty memories, e.g., FIFOs, LIFOs, register-files and nonparallel associative memories, benefit mostly from designs with shift-register cells. Random access and specialty memory designs also employ shift registers as supplemental circuits, e.g., in signal multiplexers, parallel-serial and serial-parallel converters and timing generators. In serially accessed memories the use of shift-register cells is limited to rather moderate bit-capacities by the excessive power

138   CMOS Memory Circuits

dissipation P. P increases with the operating frequency f, total charged and discharged capacitance C, and voltage swing V, i.e., $P = fCV^2$; and imposes restrictions in operation speed and in packing density. Nevertheless, extremely low power dissipation and very high operational speed at high packing densities can be combined by the use of shift-register cells in shuffle memories (Section 1.3.4). Shuffle memory designs allow for charge and discharge only of diminutive fractions of the total array capacitance, and apply only a half shift-register cell, or a single memory cell per bit for both data-move and data-storage.

CMOS implementations of shift-register cells may exploit the availability of both n- and p-channel devices, and the very large input and out-put-off resistances of the MOS transistors. The alternating use of n- and p-channel transistors as transmission gates, eliminates the need for two separate wires to deliver two separate clock impulses to each shift register cell. Signal changes in the cells can be very fast, due to the full-complementary operation of the inverters and due to the small load capacitances. In many cases, the parasitic capacitances can be used as temporary data storage elements. Depending on the data storage mechanism CMOS shift-register cells, like the random access memory cells, may be of dynamic or static types.

2.7.2 Dynamic Shift-Register Cells.

In the widely applied CMOS dynamic six-transistor shift-register (6T SR) cell (Figure 2.34), the transmission devices are formed of a single n- and a single p-channel transistor, the data are stored in the input capacitances C_{S1} and C_{S2} of the CMOS of inverters, and the inverters amplify the signal amplitude to $V=V_{DD}-V_{SS}$. For convenience the amplitude of the clock signal V_ϕ is set $V_\phi=V_{DD}$. By using this V_ϕ, the data-signal amplitude V gets reduced by the threshold voltage $V_{TN}(V_{BG})$ of the n-channel device MN1 or by the threshold voltage $V_{TP}(V_{BG})$ of the p-channel transistor MP4. Here, V_{DD}, V_{SS} and V_{BG} are the supply, ground and backgate bias voltages. The reduced V results in slower operation. Furthermore, both the operational speed and the power dissipation are unfavorably effected by the one-clock two-phase operation that allows direct-current paths between V_{DD} and V_{SS} during the transient times of clock signal ϕ.

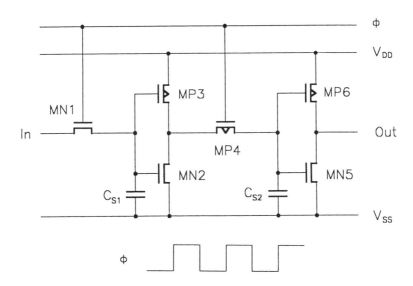

Figure 2.34. A dynamic six-transistor shift-register cell.

By the application of two clocks ϕ_1 and ϕ_2 with a voltage amplitude $V_\phi = V_{\phi 1} = V_{\phi 2} > V_{DD} + V_{TN}$ (V_{BG}), and by the use of uniform n-channel transmission devices MN1 and MN4 (Figure 2.35), both the operational speed and power consumption can be improved. The control of the transmission devices by two distinct clock signals reduces significantly the time when a direct-current path may appear between V_{DD} and V_{SS}, while the elevated clock signal amplitude $V_{DD} + V_{TN}$ (V_{BG}) allows to drive the inverters by a full voltage swing $V_\phi = V_{DD} - V_{SS}$.

To obtain full voltage swings on the inverter inputs at the use of p-channel transmission devices, a more negative than V_{SS} is needed, i.e., $V_\phi < V_{SS} - V_{TP}(V_{BG})$. The generation of such negative V_ϕ requires the application of an extra isolated n-type well in the substrate, and that is usually cost prohibitive for shift register implementations.

140   CMOS Memory Circuits

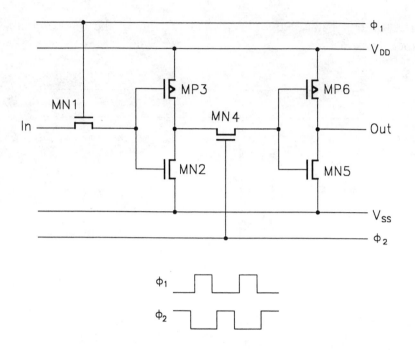

Figure 2.35. A two-phase dynamic shift-register cell.

The dynamic eight-transistor shift-register 8T SR cell (Figure 2.36) provides very high speed operation without the need for elevated voltages for clock impulses. No threshold voltage drop appears in this circuit, since both the transmission devices MN1-MP2 and MN5-MP6 and the inverter transistors MN3-MP4 and MN7-MP8 are complementary pairs. Furthermore, the full voltage swings which occur on storage capacitors C_{S1} and C_{S2}, degrade very little, because in this circuit configuration the effect of the charge redistribution is minimal.

Charge distributions may significantly reduce data signal amplitudes in a formally different but functionally equivalent half shift-register cell (Figure 2.37) where the transmission gates MN1 and MP2 are connected to the power supply poles V_{CC} and V_{SS}. At the time t_o, devices MN1 and MP2 are turned on by impulse ϕ and $\bar{\phi}$, and the voltage $v_{C1}(t)$ on capacitor C_{S1} is $v_{C1}(t_o) = V_{SS}$. At t_1, the voltage $v_{C2}(t)$ on capacitor C_{S2} raises to $v_{C2}(t_1) = V_{DD}$. At t_2, voltage $v_{c1}(t)$ is switched to $v_{c1}(t_2) = V_{DD}$, device MN3 is turned on, and all other transistors are in high impedance state. Between

Memory Cells   141

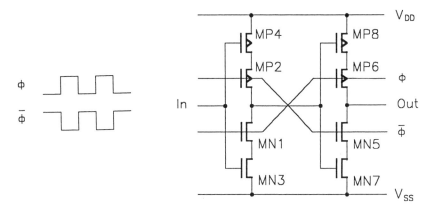

Figure 2.36. A high-speed eight-transistor shift-register cell.

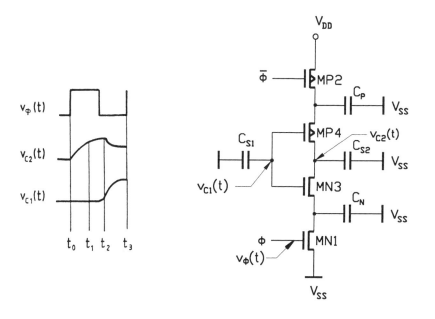

Figure 2.37. Charge distribution effect in a half shift-register cell circuit.

t_2 and t_3, the highly conductive MN3 allows a charge redistribution on capacitors C_{S2} and C_N, and at t_3 the voltage $v_{C2}(t)$ reduces to

$$v_{C2}(t_3) \approx \frac{C_{S2}}{C_{S2}+C_N} V_{CC} .$$

Similar voltage swing degradation may occur due to the charge redistribution on capacitor C_{S2} and C_p.

In the dynamic four-transistor-two-diode shift-register (4T 2D) SR cell [220] (Figure 2.38) both the charge redistribution and the direct conductance between supply poles are avoided by the application of two pairs of control impulses ϕ_1 - ϕ_2 and ϕ_3 - ϕ_4. All the four control signals ϕ_1, ϕ_2, ϕ_3 and ϕ_4 can easily be generated from a common system clock ϕ. At a data shift operation, ϕ_1 charges C_{S2} to $V_\phi = V_{DD}$ through diode D3 if C_{S1} stores a low voltage $V_{S1} < V_{TN}$, or through devices MN1, MN2 and D3 if C_{S1} is on a high potential $V_{S1} \gg V_{TN}$. Simultaneously, parasitic capacitance C_N is also charged to $V_\phi = V_{CC}$, because clock ϕ_2 turns transistor MN2 on. Both transistors MN1 and MN2 are conductive after the appearance of impulse ϕ_1 if $V_{S1} > V_{TN}$, thus MN1 and MN2 discharges both C_{S2} and C_N. If $V_{S1} < V_{TN}$, then both capacitances C_{S2} and C_N remain charged to $V_\phi = V_{DD}$, because transistor MN1 is turned off and diode D3 is reverse biased. The other half of the cell MN4, MN5 and D6 is controlled by clocks ϕ_3 and ϕ_4, and operates as described for the half-cell MN1, MN2 and D3. For the implementation of diodes D3 and D6, CMOS processing technology is required. At some compromise in shifting speed, MOS devices may replace the diodes.

All dynamically operating CMOS shift-register cells store data on capacitors in forms of charge pockets. Due to leakage currents the amount of stored charges change, and the voltages which represent data, get corrupted (Section 2.2.1). To prevent data degradation or loss, dynamic shift-registers must operate faster than a minimum shifting rate, and during a long term data storage the information stored in the cells must be periodically refreshed. The required time period for refresh may be calculated with the method described for dynamic random-access memory cells. Refresh operations in shift-register based memories are very simple, by connecting the input and the output of a shift register the stored data can completely be recycled during the storage time and also at write and read operations.

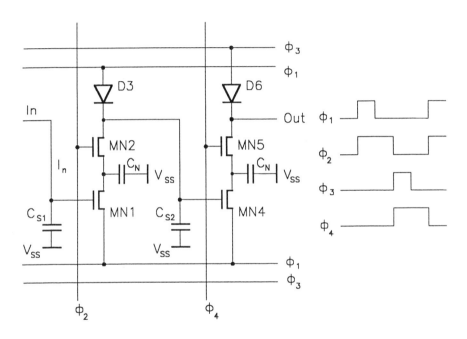

Figure 2.38. High-speed dynamic shift-register cell with diodes. (After [220].)

2.7.3 Static Shift-Register Cells

CMOS static shift register cells can combine permanent data storage, without requirements for data-recycling, with very high speed data-shift operations. Seven-transistor static shift register (7T SR) cells (Figure 2.39)

144 CMOS Memory Circuits

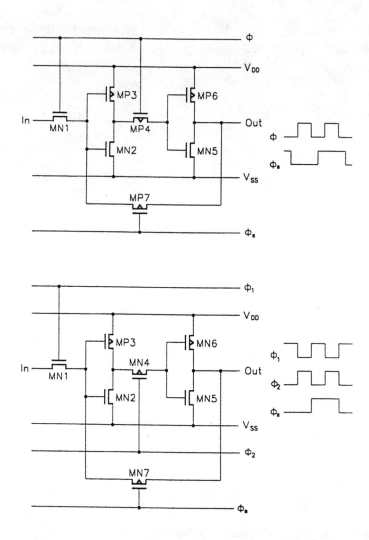

Figure 2.39. Static seven-transistor shift-register cells.

employ a single feedback transistor MP7 or MN7 to form a data latch with devices MN2, MP3, MP4/MN4, MN5 and MP6. When the feedback device MP7 or MN7 is turned off, the 7T SR cells operate as dynamic shift-register stages do. When MP7 or MN7 is brought to a highly conductive state by clock ϕ or ϕ_S, 7T SR cells store the data in the cross-coupled inverters permanently.

If permanent data storage in each half-cell is required, the number of transistors per shift-register stage has to be increased. A twelve-transistor shift register 12T SR cell (Figure 2.40) corresponds to two cascaded 6T random access memory cells. In this 12T SR cell, transmission gates MP6

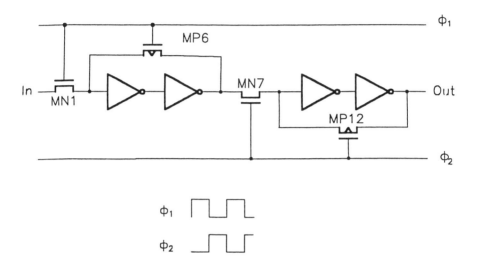

Figure 2.40. A static twelve-transistor shift-register cell.

and MP12 make the link for positive feedback when devices MN1 and MN7 are turned off by clock signals ϕ_1 and ϕ_2. Feedback devices MP6 and MP12 may be replaced by resistors R6 and R12, as it is exemplified in the ten-transistor-two-resistor shift register 10T2R SR cell (Figure 2.41). Here, the higher the resistances R6 and R12 are, the faster the data shift operation is. Yet, in this shift-register cell the maximum resistance for R6 and R12 is limited by the required immunity against the effects of atomic particle impacts (Section 5.3) and by the leakage currents flowing through devices MN1 and MN7. Here, the alternating application of n- and p-channel transistors, rather than n-channel transistors only, as transmission gates MN1 and MN7, makes possible to use a single clock ϕ for two-phase control. The use of a single control clock and resistive feedbacks allows for 10T2R SR cell designs which require considerably smaller silicon surface area than 12T SR cell designs need.

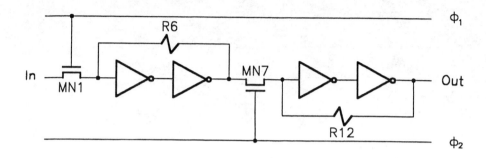

Figure 2.41. Static shift-register cell with feedback resistors.

For the analyses, designs and implementations of the dynamic and static shift register cells described here, the techniques which are presented in the discussion of the dynamic and static random-access memory cells (Section 2.2-2.5) can be adopted.

2.8 CONTENT ADDRESSABLE MEMORY CELLS

2.8.1 Associative Access

Content addressable memory (CAM) or associative access cells are the fundamental elements of all-parallel associative memories, and of many in cache and other data-associative memory devices. Memories using CAM cells feature very fast operation in comparing an arbitrary data-word, or a data-set in a word, to a multiplicity of data-words or sets, in finding identical or similar data-words or sets, in determining their address information, and in establishing the degree of similarity. Random address memory (RAM) cells, however, occupy much less semiconductor surface area than CAM cell implementations do. Thus, the application of RAM cells are preferred in designs where the requirements for the time of the associative data search is noncritical, e.g., in word-parallel-bit-serial and bit-serial-word-parallel CAMs.

A CMOS CAM-cell combines a complete RAM cell with a one-bit digital comparator stage (Figure 2.42). The RAM cell includes a storage and one or more access elements, and the digital comparator comprises logic gates

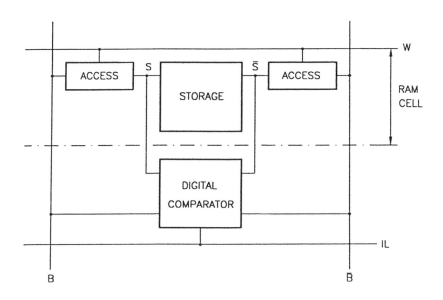

Figure 2.42. Content-addressable-memory cell structure.

that can provide an XAND, XOR, XNOR or XNAND function. To the logic gate, the input variables are the true and the complement values of an argument datum B and $\overline{B}$ and the true and the complement values of the stored bit S and S. In the comparator gate, B and $\overline{B}$ are compared with S and S, and the resulting match or mismatch is coupled to an interrogator-line IL. An interrogation of a set of CAM cells indicates whether the data content of an argument word or bit-set, is identical with the data content of one or more words or bit-sets of CAM cells. In the case of a data match or a mismatch, a flag-signal appears on one or more IL-s. If more than one flag signals occur, then a multiple response resolver selects one of the flagged lines. The selected IL activates the wordline W of a flagged set of cells, and allows for write and read operations in the effected storage elements. During write and read the storage and access devices, the pair of bitline and the wordline in CAM cells have the same functions as those do in RAM cells. Although any RAM cell and any comparator logic circuit may be used to compose a CMOS CAM cell, packing density, operational speed and power considerations have reduced the practically applied CMOS CAM cells to those which include 6T, 4T2R and 1T1C RAM cells and precharged NOR and NAND gate combinations.

2.8.2 Circuit Implementations

In the ten-transistor content addressable memory 10T CAM cell (Figure 2.43), transistors MN1, MN2, MP3, MN4, MP5 and MN6 form a static six-transistor random access memory 6T SRAM cell, and devices MN7, MN8, MN9 and MN10 with a cell-external precharge transistor MP11 constitute a one-bit comparator circuit. A set of CAM cells are coupled to an interrogation line IL, and when in a CAM cell either transistor pair MN7-MN8 or MN9-MN10 is in low-resistance state; the voltage on line IL $V_{IL} = V_{PR} \approx V_{DD}$ changes to $V_{IL} \approx V_{SS}$. When wordline W selects a set of CAM cells and V_{IL} remains $V_{IL} \approx V_{DD}$, then a match occurs between the CAM-cell set and the argument data. At a match, array-external inverters I1 and I2 generate a flag signal that activates the access devices MN1 and MP6, and permits write and read operations to all 10T CAM cells which are coupled to wordline W.

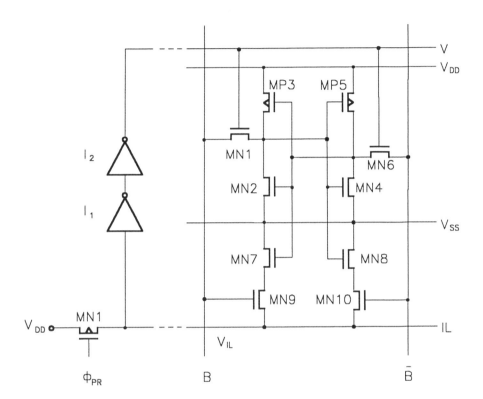

Figure 2.43. A ten-transistor content-addressable memory cell circuit.

In an eight-transistor-two-resistor 8T2R CAM cell [221] the p-channel devices MP3 and MP5 of the cross-coupled inverters are replaced by a pair of resistors R3 and R5 (Section 2.5). Resistive loads, rather than active loads, may allow for cost reduction in manufacturing but degrades read performance and environmental tolerance including the immunity against the effects of high temperature, atomic particle impacts and radioactive radiation.

150 CMOS Memory Circuits

Operational speed and environmental tolerance are traded for benefits in packing density and manufacturing costs in the four-transistor-two-capacitor 4T2C CAM cell (Figure 2.44) [222]. In this CAM cell transistor-capacitor pairs MN1-C2 and MN3-C4 form two dynamic one-transistor-one-capacitor 1T1C RAM cells (Section 2.3), and devices MN5 and MN6 in combination with the cell- and array-external precharge transistor MP7 and drive inverters I1 and I2 make up a one-bit comparator. Cell- and array-external inverters shape the comparator signals, and provide the flag-signal that indicates a match between the data content of the argument register and the data stored in a set of 4T2R CAM cells.

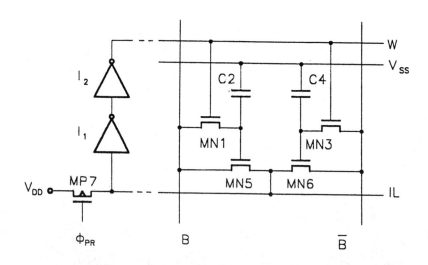

Figure 2.44. A dynamic content-addressable-memory cell. (After [213].)

The detailed operation, analysis, design and implementations of CAM cells are obtainable from those discussed previously for random access memory cells (Section 2.2-2.5) and from the theory and design of CMOS logic circuits.

CMOS 10T CAM cell based circuits are applied mostly in space and military equipment to implement other than traditional Von Neumann type of computing systems, which have to operate with high computational speed in extreme environments. For operations in less severe environments, designs using 8T2R CAM cells are cost-effective alternatives. In the future, the 4T2C CAM cells and similar small CAM cells have great potential to satisfy the need for high-packing-density low-cost CMOS components in the increasing number of nontraditional computing systems. For special computations, of course, a diversity of special CAM cells may be devised by combining a RAM cell with a simple logic circuit, e.g., [223].

2.9 OTHER MEMORY CELLS

2.9.1 Considerations for Uses

An abundance of memory cells may be devised to optimize designs for specific combinations of requirements in speed, power and environmental tolerance and to overcome limitations dictated by an available CMOS processing technology. Yet, CMOS compatible memory cells, which can be implemented only at deviations from the main-stream fabrication technology, are very seldom or never used, regardless of their potentially outstanding features. Furthermore, memory cells, which include nonstandard design features, but can be fabricated without any change in the processing technology, are also infrequently applied to CMOS memory designs. Namely, a divergence from either of both the main-stream CMOS processings and the memory cell designs, may result in expansion of manufacturing costs, risks, and time-to-market of a memory product. For stimulation of creative memory designs, however, a rather arbitrary selection of memory cells, which are compatible with CMOS technologies, are very briefly introduced in this section.

Innovations and research works in memory-cell technology are directed to improve memory packing density, performance and environmental tolerance. Revolutionary memory cells may result in broad changes in the semiconductor technology. Most likely, however, the main-stream CMOS memory technology will apply predominantly 1T1C and 6T

152 CMOS Memory Circuits

memory cells also in the foreseeable future, while an increasing share of memory designs will use nonvolatile memory elements.

2.9.2 Tunnel-Diode Based Memory Cells

Tunneling or delta-doped diodes applied as storage elements [224] can combine very fast operation, capability to function in high-temperature and radiation-hardened environments, very high packing density and static data storage. Tunnel diode based CMOS memory cells (Figure 2.45) store data by using stable quiescent operation points, e.g., S_1 and S_2, in the two

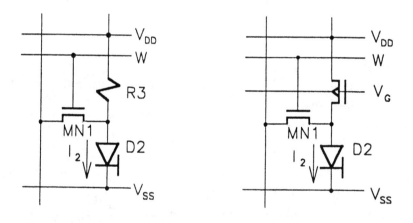

Figure 2.45. Tunnel-diode based memory cells. (After [224].)

positive-resistance regions of the current-voltage I-V characteristics of the diode D2 (Figure 2.46). A quiescent operating point, that occurs in the negative-resistance region of the I-V curve, e.g., (S_3), is unstable, because an incremental raise of voltage across the diode reduces the current through the diode, and a voltage decrease enlarges the current. To provide bistable storage in diode D2, the resistance of the load device R3 or MP3 has to be fitted into a rather small domain. Access device MN1 must be able to provide sufficient diode currents, i.e., $I_2 > I_p$ at V_p and $I_2 < I_v$ at V_v, to flip the circuit from one stable state to the other one. Here, I_p and I_v are peak and valley currents, V_p and V_v are peak and valley voltages in the diode I-V characteristics. Resonant tunnel diodes [225] may exhibit more

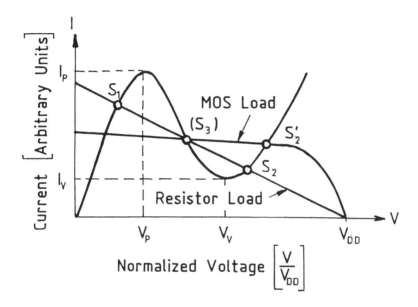

Figure 2.46. Current-voltage characteristics and load curves for a tunnel-diode.

than one negative resistance regions (Figure 2.47) which make possible for multi level data storage. In memory cells which apply tunnel diodes, data can be switched very quickly because the negative resistance greatly reduces the time constant that appears by the operation of the memory cell τ_c (Section 2.2.2) in the memory cell. Moreover, these memory cells are inherently amenable to operate in high temperature and radiation environments, since tunnel diodes are implemented in highly doped material and the percentage change in the doping concentration of the tunnel diodes is little in extreme environments. Sizes of tunnel-diode based cells may be competitive with the sizes of dynamic one-transistor-one-capacitor cells. Namely, the load and diode devices can readily be placed under and above the access transistor by the exploitation of the technologies which are available for dynamic and static RAM-cell fabrications. Memory cells using tunnel diodes do not need data refresh, but may require considerable stand-by currents for static data storage. Eventual cell current reductions are limited by the diode-parameters I_p, I_v, V_p and V_v, as well as by the required noise margins requirements, load-device characteristics and sense circuit operations.

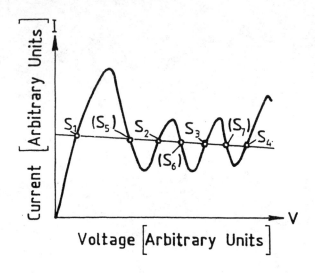

Figure 2.47. Current-voltage characteristics of a resonant tunnel-diode. (After [225].)

2.9.3 Charge Coupled Device

Charge couple device (CCD) based CMOS sequential memories can be implemented in very high packing densities and perform high input and output data rates. CCDs, e.g., [226], store data as minority-carrier charge pockets in potential wells under the gate electrodes of MOS devices in the semiconductor material (Figure 2.48). The potential wells are created by the electric fields of the gates, and the strength and contour of the electric fields can be controlled by clock impulses brought on the gates. A minimum of two clocks ϕ_1 and ϕ_2 makes possible to move the charge pockets along the semiconductor surface, similarly to the operation of shift-registers (Section 1.3.3). A CCD cell needs very little semiconductor area because only two overlapping gates and no drain, source and contact are required for its implementation. The lack of drain and source electrodes reduces parasitic capacitances and provides fast shifting operation. To facilitate shifting operation, nevertheless, all gates in the entire CCD memory have to be charged and discharged by impulses ϕ_1 and ϕ_2 within a single shift period, which results high operating power consump-

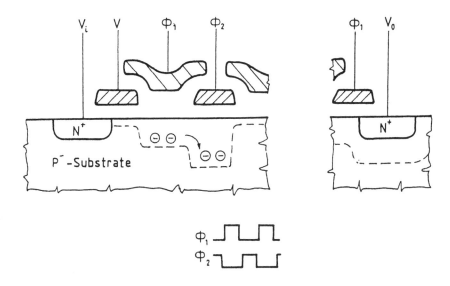

Figure 2.48. A charge-coupled-device structure.

tion. In CCDs data have to be periodically refreshed, and downscaling of feature sizes increase the susceptibility of the stored data for the effects of atomic particle impacts. Furthermore, the effects of charge trapping at interface states, parasitic diffusion and drift currents result in incomplete transfer of minority carriers from well to well. The transfer efficiency η is the fraction of charge transferred from one potential well to the next one

$$\eta = \frac{Q_i}{Q_{i+1}} = 1 + \varepsilon,$$

where Q_i and Q_{i+1} are the charge amounts in well i and in well i+1, and ε is the transfer inefficiency. Transfer efficiencies set the limit for the number of CCD stages that may be cascaded without amplification, and they tend to decrease with increasing clock frequencies. Operations at higher frequencies [227] may be facilitated by injecting a trickle charge (fat zero) to keep fast interface traps filled, by using a substrate with <100> crystal orientation, low doping concentration and buried channels to reduce the number of interface traps, by constructing potential slope inside each well to propagate charge by built-in fields rather than by thermal diffusion. Thermally generated background charges cause spurious channel currents (dark currents) and, thereby, further restrains in speed and power performances. Nonetheless, very high speed CCD operations can be combined with extremely low power dissipations and with utmost packing densities by applying CCD cells in a shuffle like-memory schema (Section 1.3.4).

2.9.4 Multiport Memory Cells

Multiport memory-cells are employed to accommodate simultaneous or parallel write and read operations. Parallelism in computing and data processing system greatly increases the data throughput rate. Multiport memory cells [228] which stores a single datum, may include two or more access devices MA1-MA4. The access devices may be connected to separate write-bitlines BW1 and BW2, read-bitlines BR1 and BR2, write-wordlines WW1 and WW2 and read-wordlines WR1 and WR2 (Figure 2.49a). Furthermore, multiport cells may comprise also a write-enable device ME1 that is controlled through a write-enable line WE1 (Figure 2.49b), and may feature a cell-internal read amplifier A_R in addition to a storage element S_E (Figure 2.49c). In most of the multiport memory-cell designs, traditional dynamic or static storage elements and transmission gate type of access devices are employed (Section 2.3-2.5). Implementations of multiple access memory cells result in large cell sizes, but many systems, e.g., multiprocessor, superscalar, and graphic systems, do not require the use of very high bit-capacity multiport memories.

Memory Cells 157

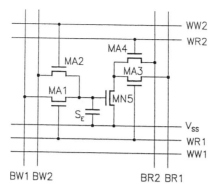

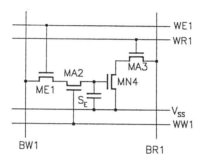

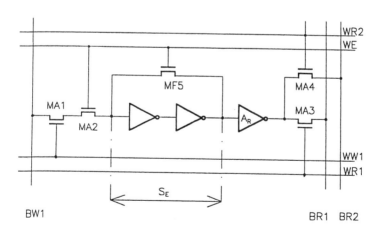

Figure 2.49. Multiport memory cells.

2.9.5 Derivative Memory Cells

Clearly, the dynamic 1T1C, the static 6T and 4T2R memory cells are the mostly applied elements to all random, sequential and special access memory designs. The 1T read-only memory cell looks to be the exclusive choice for fixed program memories. High-speed sequential memory designs prefer the use of dynamic 6T or 8T shift-register cells. Apart from the main-stream demands, memory designs may need to satisfy some specific circuit and process technological requirements. Specific design objectives may justify the use of derivative or innovative memory cells. From the huge variety of memory cells which could be derived from the concepts of the heretofore discussed memory cells, the dynamic 3T and 4T (Figure 2.50), static 5T and 6T (Figure 2.51) memory cells, and dynamic 6T and static 9T shift-register cells (Figure 2.52) can most likely be put to practical use.

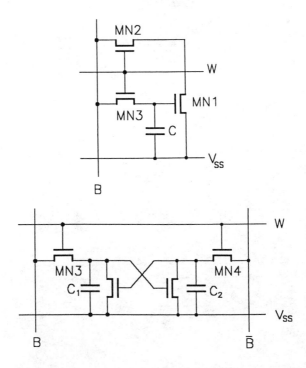

Figure 2.50. Derivative dynamic memory cells.

Memory Cells 159

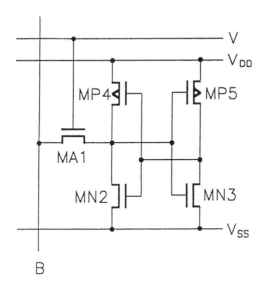

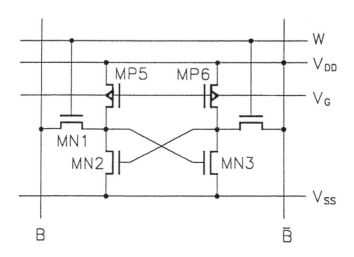

Figure 2.51. Alternative static memory cells.

160 CMOS Memory Circuits

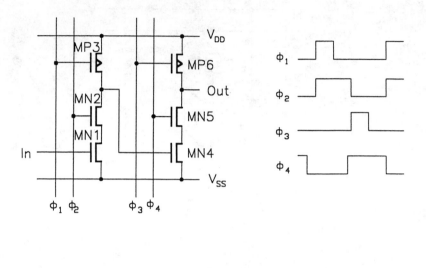

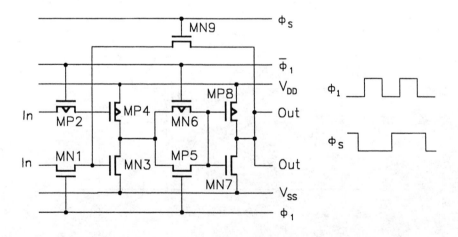

Figure 2.52. Dynamic and static shift-register cell variations.

The cell circuits, shown here, represent only a few variations which have gained applications in memory designs. Future requirements in CMOS memory technology may necessitate the use of different and novel memory cells.

3

Sense Amplifiers

Sense amplifiers, in association with memory cells, are key elements in defining the performance and environmental tolerance of CMOS memories. Because of their great importance in memory designs, sense amplifiers became a very large circuit-class. In this chapter, for the first time in publications, the sense amplifier circuits studied systematically and comprehensively from the basics to the advanced current-sensing circuits. The study includes circuit and operation descriptions, direct current, alternative current and transient signal analyses, design guides and performance-enhancement methods.

3.1 Sense Circuits

3.2 Sense Amplifiers in General

3.3 Differential Voltage Sense Amplifiers

3.4 Current Sense Amplifiers

3.5 Offset Reduction

3.6 Nondifferential Sense Amplifiers

3.1 SENSE CIRCUITS

3.1.1 Data Sensing

In an integrated memory circuit "sensing" means the detection and determination of the data content of a selected memory cell. The sensing may be "nondestructive," when the data content of the selected memory cell is unchanged (e.g., in SRAMs, ROMs, PROMs, etc.), and "destructive," when the data content of the selected memory cell may be altered (e.g., in DRAMs, etc.) by the sense operation.

Sensing is performed in a sense circuit. Typical sense circuits (Figure 3.1) are mirror-symetrical in structure and comprise (1) a sense amplifier, (2) circuits which support the sense operation such as precharge, reference and load circuits, (3) bitline decoupler/selector devices, (4) an accessed memory cell, and (5) parasitic elements including the distributed capacitances and resistances of the bitline, and the impedances of the unselected memory cells connected to the bit line.

The combined impedance, which is introduced by the supporting circuits and parasitic elements coupled to a bitline, effects significantly the operation of random access memories and of many sequential access and associative memories. Because the effective bitline capacitances and resistances are large, and because a memory cell's energy output is small at read operations, an accessed memory cell can generate only small current and voltage signals. These signals have long switching and propagation times and insufficient amplitudes to provide the minimum log.0 and maximum log.1 levels which are required to drive the peripheral logic circuits of the memory (Figure 3.2).

To improve the speed performance of a memory, and to provide signals which conform with the requirements of driving peripheral circuits within the memory, sense amplifiers are applied. Sense amplifiers must work in the ambience of the other sense circuit elements. Fundamental conditions for sense circuit and sense amplifier operations can most conveniently be obtained from the operation margins of the prospective sense circuits.

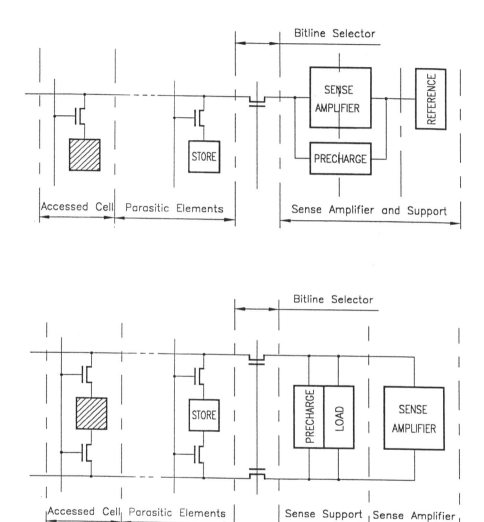

Figure 3.1. Typical Sense Circuits.

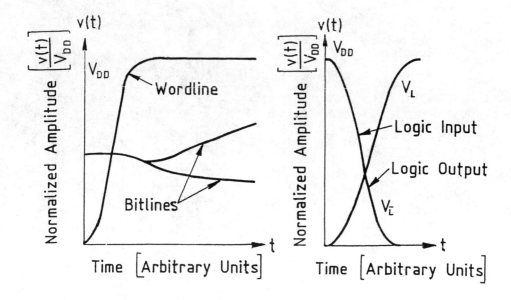

Figure 3.2. Unamplified data-signals on the bitlines (a) and standard data-signals in the peripheral logic circuits (b).

The following sections provide understanding of the terms determining operation margins, analyze the circuit design of the most important sense amplifier types and of the other substantial elements of sense circuits.

3.1.2 Operation Margins

Operation margins in a digital circuit are those domains of voltages, current, and charges which domains unambiguously represent data throughout the entire operation range of the circuit. In a specific circuit, the extent of an operation margin depends on the (1) circuit design, (2) processing technology and (3) environmental conditions. The particular issues, which effect operation margins, may include circuit configurations, sizes of individual transistors, input and output loads, parasitic elements, active and passive device characteristics, furthermore the variations of these parameters which may result from the vicissitudes in semiconductor processing, supply voltage, temperature and radioactive irradiation.

Generally, the operation margins at the input of a sense amplifier differ from those required to drive the peripheral Boolean-logic circuits in positions and in sizes (Figure 3.3). In a sense circuit the sense amplifier has to amplify the small log.0 and log.1 levels which appear on the bitline and on the input of the sense amplifier, to the larger levels which are required for the operation of the logic circuit coupled to the sense amplifier.

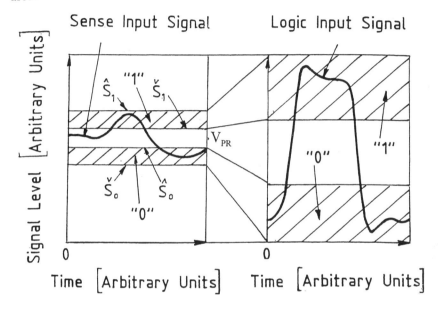

Figure 3.3. Operation margins in the sense-amplifier inputs and in the peripheral logic circuits.

The relationship between the operation margins of the sense circuits and that of the logic circuits indicates

(1) Minimum log.0 and log.1 signal amplitudes $\check{S}_0$ and $\check{S}_1$ and maximum log.0 and log.1 signal amplitudes $\hat{S}_0$ and $\hat{S}_1$ which are detectable by the sense circuit and which must be generated by the selected memory cell on the bitline,

(2) Target precharge voltage V_{PR} and the initial quiescent operation voltage $v_i(0) = V_{PR}$, e.g., for symmetrical "0" or "1" margins V_{PR}

that is

$$V_{PR} = v_i(0) = \frac{\check{S}_1 - \hat{S}_0}{2},$$

(3) Required minimum gain $\check{A}$ for the sense amplifier

$$\check{A} = \frac{\check{L}_1 - |\hat{L}_0|}{\check{S}_1 - |\hat{S}_0|}.$$

where $\check{L}_1$ and $\hat{L}_0$ are the respective minimum log.1 and maximum log.1 logic levels required for the peripheral circuit inputs.

In the design of a sense amplifier, the operation margins are of fundamental importance, and these in combination with the requirements for speed, power and reliability, determine the complexity and layout area of the sense circuit. In a sense circuit, the principal terms, which demarcate the internal operation margins, include

(1) Supply voltage,

(2) Threshold voltage drops,

(3) Leakage currents,

(4) Charge couplings

(5) Imbalances,

(6) Other specific effects,

(7) Precharge level variations.

Moreover, through the variations of these terms the operation margins are also functions of parameter fluctuations caused by

(A) Semiconductor processing,

(B) Temperature changes,

(C) Voltage biasing conditions,

(D) Radioactive radiations

Resulting from the fluctuations caused by A, B, C and D each terms (1) through (7) have a maximum. These maxima of term-variations have to be considered in obtaining the worst-case "0" and "1" operation margins (Figure 3.4). Here and in the following margin analyses the levels are expressed in voltages, but the concepts introduced with voltage levels can well be used to operation margins formulated by current or charge levels.

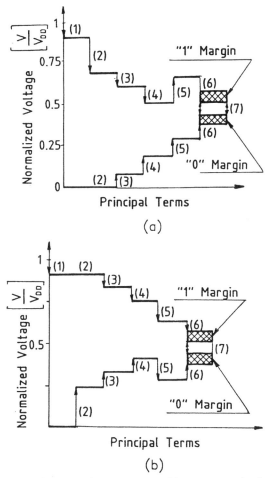

Figure 3.4. Operation margins determined by the principal terms where n-channel (a) and where p-channel (b) access devices are used.

The "0" or "1" or both operation margins may disappear at a certain temperature T (Figure 3.5a) or radiation dose D (Figure 3.5b). A disappearance of the operation margins can effectively be avoided by applying the worst-case operation margins, which can occur under predetermined processing, voltage-bias and environmental conditions, to the design of the sense circuit. Moreover, certain design techniques can substantially increase the operation margins, e.g., by bootstrapping the output voltage of the wordline drivers to eliminate the term resulting from the threshold voltage drop; and can also balance differing "0" and "1" operation margins, e.g., by setting the cell-plate voltage to the center of the effective log.0 and log.1 levels occurring in the sense circuit.

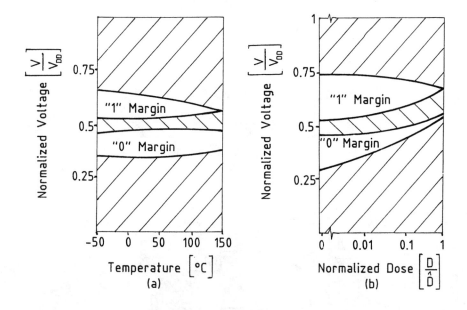

Figure 3.5. Reduction and disappearance of the operation margins.

A sense amplifier has to operate under the conditions that are dictated by the operation margins. The principal terms determining the operation margins are discussed in the next sections.

3.1.3 Terms Determining Operation Margins

The computation of the principal terms determining the internal operation margins of a sense circuit should be based on worst-case electrical parameters. Here, worst-case parameters are those which result in a maximum reduction in the "0" or "1" operation margin of a sense circuit. To the determination of the worst-case operation margins, the principal voltage terms may be obtained as follows.

3.1.3.1 Supply Voltage

In the determination of operation margins, the initial voltage level V_I is the minimum supply voltage $V_I = V_{DD}$ (min.) $= V_{DD}$. V_I can be obtained from the supply voltage range, e.g., $V_{DD} \pm 10\%$, $V_{DD} \pm 5\%$, that is specified for the memory.

3.1.3.2 Threshold Voltage Drop

The maximum threshold voltage drop V_{TA} across the access transistor of a selected memory cell MA1 reduces either the "0" or the "1" operation margin in the sense circuit (Figure 3.6). On the bitline, the "1" margin decreases when the access transistor MA1 is an n-channel device, and the "0" margin recedes when MA1 is a p-channel device, if a positive supply voltage V_{DD} and a positive logic convention is assumed. By the use of an n-channel access device MA1, the maximum available bitline voltage V_B can abate from $V_B = V_{DD}$ to $V_B = V_1 - V_{TA}$. Here, V_1 is the minimum log.1 output level allowed in the circuit, and V_{TA} is the maximum threshold voltage drop through MA1. Similarly, by the application of a p-channel device as MA1 the minimum obtainable bitline voltage V_B increases from $V_B = V_{SS}$ to $V_B = V_o + |V_{TA}|$. Here, V_o is the maximum log.0 output level allowed in the circuit, and operation in the saturation region is presumed for MA1.

Generally, the maximum threshold voltage drop V_{TA} across MA1 is a function of the backgate bias V_{BG}, temperature T and radioactive radiation dose D. The cumulative effect of V_{BG}, T and D on V_{TA} may be expressed

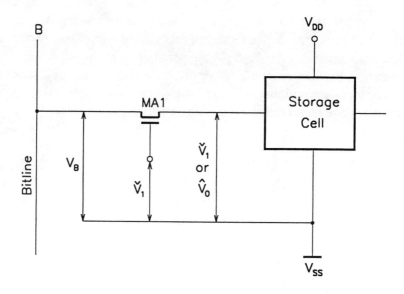

Figure 3.6. Threshold voltage drop through the access transistor.

through the maximum of the threshold voltage change $\Delta \hat{V}_T(V_{BG},T,D)$ as $\hat{V}_{TA}(V_{BG},T,D) = \hat{V}_{TO} + \Delta \hat{V}_T(V_{BG},T,D)$, where $\hat{V}_{TO}$ is the maximum threshold voltage at $V_{BG}=0$ and at $T=25°C$. For most of the approximate computations, $\hat{V}_{TA}(V_{TO},V_{BG},T,D)$ may be considered as the linear superposition of the individual maximum threshold voltage shifts $\hat{V}_{TO}$, $\Delta \hat{V}_T(V_{BG})$, $\Delta \hat{V}_T(T)$, and $\Delta \hat{V}_T(D)$ so that

$$\hat{V}_{TA}(V_{TO}, V_{BG}, T, D) = \hat{V}_{TO} + \Delta \hat{V}_T(V_{BG}) + \Delta \hat{V}_T(T) + \Delta \hat{V}_T(D).$$

Usually, all the components of $V_{TD}(V_{TO},V_{BG},T,D)$ are measured and provided by the processing technology in the list of the electric device parameters, yet in lack of measured results, $\Delta V_T(V_{BG})$, $\Delta V_T(T)$ and $\Delta V_T(D)$ may also be approximated as follows:

$\Delta V_T(V_{BG})$ may be computed by using the Fermi function ϕ_F and the material constants K_1 and K_1' in simple empirical expressions [31] such as

$$\Delta V_T(V_{BG}) \approx K_1 (2\phi_F + V_{BG})^{½}$$

for long-channel transistors, and

$$\Delta V_T (V_{BG}) \approx K_1' V_{BG}$$

for short-channel devices.

ΔV_T (T) for CMOS devices is often approachable by a linear function [32] in the traditional ranges of the operation temperatures as

$$\Delta V_T (T) \approx (|\phi_F| - K_M) \Delta T/T \approx K_T/T,$$

where K_M is a material dependent term, ΔT is a temperature increment and K_T is the linear temperature coefficient, e.g., $K_T = 2.4$ mV/°C.

Greatly nonlinear and voltage bias dependent is the variation of V_{TD} as a function of the total absorbed radioactive-radiation dose D [rad(Si)] [33]. Radiation induced threshold voltage changes $\Delta V_T(D)$s are experimentally obtained data. These changes are large, and can cause the disappearance of operation margins (Section 6.2.2) at a low total dose [rad (Si)]. In addition to radiation total dose effects, transient radiation induced dose rate D [rad(Si)/Sec] imprints may also substantially expand $\Delta V_T(D)$ (Sections 6.1.3 and 6.1.5).

If, in a design V_{TA} (V_{TO}, V_{BG},T,D) appears to be prohibitively large, $\Delta V_T(V_{BG}$,T,D) can be eliminated or greatly reduced by increasing V_1 on the gate of an n-channel MA1 or by decreasing V_o on the gate of a p-channel MA1 by $\Delta V_T(V_{BG}$,T,D).

3.1.3.3 Leakage Currents

Leakage currents I_L-s reduce both "0" and "1" operation margins, since they decrease the signal amplitudes on the bitline by $V_{BL} \approx I_L R_B$, and degrade the levels of data stored in memory cells by $V_{CL} \approx I_L R_D$. Here, I_L is the maximum leakage current, R_B comprises the maximum resistances of the access transistor and the bitline, R_D is the maximum resistance between the data-storage node and the supply or ground node in the memory cell, and V_{BL} and V_{CL} are the maximum I_L-induced voltages which degrade the operation margins on the bitline and in the cell, respectively.

In the computation of the maximum margin-degradation $\hat{V}_L = \hat{V}_{BL} + \hat{V}_{CL}$, $\hat{I}_L$ represents the highest leakage current which may appear in the operational, processing and environmental worst cases.

The operational worst-case occurs on the bitline of N memory cells when the accessed memory contains a log.1 and all other N-1 memory cells store log.0 (Figure 3.7), or vice versa. When such "all-but-one" data pattern is stored, the cell-current I_C, that is generated by the datum of the accessed memory cell, flows against the accumulated leakage currents of the N-1 unselected memory cell I_L.

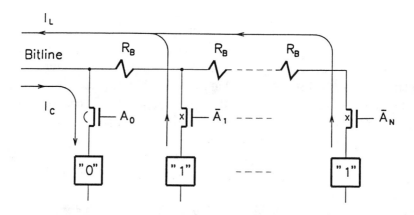

Figure 3.7. A memory-cell current opposes the cumulative leakage current on a bitline.

Inside of a memory cell the degradation of log.0 and log.1 levels depends mainly on those leakage currents which percolate through the access and the eventual load devices. Cell interim load devices, as in an SRAM cell, compensate the level degrading effects of leakage currents, while the lack of load device, as in a one-transistor-one-capacitor DRAM cell, causes such a significant storage-charge loss by leakage currents that the cell's storage capacitor has to be recharged periodically (Section 2.2.1).

From the various types of leakage currents, which may appear in a memory, the subthreshold current I_{ST}, the junction leakage current I_j, and the eventual radiation-induced leakage current I_r influence the operation margins most significantly. For obtaining the maximum of the cumulative leakage current

$$I_L \approx (I_{ST} + I_j + I_r)(N-1),$$

the maximum currents I_{ST}, I_j and I_r as well as their environmental dependency data are usually available in the list of the electrical parameters that is provided by the processing technology prior to the design start.

If at the design start measured data are unavailable the following approximation [34] to I_{ST} may be applied:

$$\hat{I}_{ST} \approx I_0 \frac{\hat{W}}{L} e^{(\hat{V}_{GS} - \check{V}_T)/\check{V}_C},$$

where

$$I_0 = \frac{L}{W} I_{DS}^x , \quad V_C = \frac{kT}{q}(1 + \alpha n^{\frac{1}{2}} t_{ox}),$$

I_{DS}^x is the maximum specific drain-source current at $V_{GS} = V_T$, $\hat{W}$ is the maximum channel width, L is the maximum channel length, V_{GS} is the maximum gate-source voltage, V_T is the maximum threshold voltage, V_C is the minimum V_C, k is the Boltzmann constant, T is the temperature in °K, q is the charge of an electron, α is an adjustment factor, n is the doping concentration and t_{ox} is the oxide thickness.

Usually I_j is dominated by the Sah-Noyce-Schockley generation-recombination current I_{gr} that results from the presence of defects and impurities in the semiconductor crystals, and I_j may be approached [35] by

176   CMOS Memory Circuits

where
$$I_j \approx I_{gr} = \frac{Aqn_i w}{2\tau},$$

$$n_i = \frac{kT^{\frac{3}{2}}e^{-\frac{1.21}{kT}}}{V_n},$$

A is the maximum junction area, $\hat{W}_d$ is the maximum depletion width, τ is the minority carrier lifetime, n_i is the maximum number of electrons residing in the conduction band at maximum operating T, and V_2 is the minimum volume. In addition to I_{ST} and I_j the occurrence of other leakage currents [36] may also be momentous.

The maximum leakage current I_L can mightily be aggrandized by radiation-induced leakage currents I_r-s (Section 6.1). Great I_r-s occur in memories which operate in radioactive environments (Sections 6.1 and 6.3). The effects of radioactive radiations on leakage currents are analyzed and calculated in the literature, e.g., [36], but an actual memory design can only rely on the experimental data obtainable on that specific CMOS processing technology which is to be used for the fabrication of the memory. In some memory designs, the effects of other type of leakage currents [37] may also be important.

3.1.3.4 Charge-Couplings

Charge-couplings $v_c(t)$-s occur mainly through the gate-source or gate-drain and gate-channel capacitances C_{GS}, C_{GD} and C_{GC} of the memory cells' access devices and, eventually, of the bitline decoupling devices (Figure 3.8), and alter temporarily the signal levels in the memory cells as well as on the bitline and sense-amplifier inputs. The maximum amount of charge-coupling V_{CC} causes either a voltage level increase or a decrease in both log.0 and log.1 levels, and modifies both "0" and "1" operation margins unidirectionally. Magnitude of level- and margin-changes are effected not only by the capacitances C_{GS}, C_{GD} and C_{GC}, but also by the wordline resistance R_W and capacitance C_W, sense-enable-line resistance R_E and capacitance C_E, bitline resistance R_B and capacitance C_B, and by the waveforms of the wordline and sense-enable control signals.

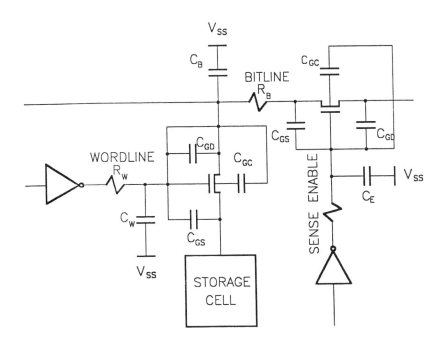

Figure 3.8. Charge-coupling through parasitic capacitances.

The signal shape, as a function of time in the bitline or on the sense-amplifier inputs $v_c(t)$, can conveniently be obtained by computer simulations of the sense circuit. Without the use of a simulation program, an approximate $v_c(t)$ function, the maximum charge coupling induced voltage shift V_c and, thereby, its influence on the operation margins, may crudely be estimated by a linear model (Figure 3.9). In this model, r_o is the equivalent output resistance of the wordline driver or of the sense-enable driver circuit, C_C is combined of C_{GS}, C_{GD} and C_{GC} and it is the coupling capacitance between the gate of the access transistor and the data storage node of the memory cell, or between the gate of the sense-enable device and the bitline node, and r_c is the resistance between the storage node of the memory cell and the ground V_{SS} or supply voltage V_{DD} or, for the sense-enable device, between the bitline node and the V_{SS} or V_{DD} node. Here, the effects of all other resistances, capacitances and eventual active elements as well as of all nonlinearities are neglected.

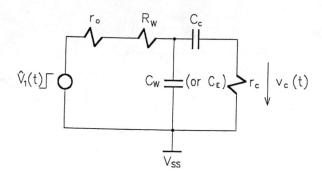

Figure 3.9. Simple charge-coupling model.

A linear analysis of the model circuit, using operator impedances and Laplace transforms, results $v_c(t)$ in the bitline or on the sense amplifier input node as

$$v_c(t) = \frac{\tau_c}{\tau_w - \tau_c} \hat{V}_1 (e^{-\frac{t}{\tau_w}} - e^{-\frac{t}{\tau_c}}) ,$$

$$\tau_w = (r_o + R_w)C_w , \quad \tau_c = r_c C_c ,$$

where $\hat{V}_1$ is the maximum log.1 level that is allowed to drive the access device. The maximum charge-coupling induced voltage change $\hat{V}_c$ that degrades the operation margins can easily be obtained by plotting the $v_c(t)$ functions or by deriving it from the $v_c(t)$ equations. As the expression for $v_c(t)$ clearly shows, the amount of changes in the operation margins depends strongly on both time constant τ_w and τ_c in addition to the amplitude of the driver signal V_1. Usually, the parameters for τ_w and τ_c are such that the charge coupling induced margin degradations are more significant in the access transistors of the memory cells than in the sense-enable devices.

3.1.3.5 Imbalances

Imbalances caused operation margin degradations V_{IB}-s are specific to those sense circuits which use differential sense amplifiers, and the imbalances reduce both "0" and "1" margins. The phrase imbalance indicates the nonuniform topological distribution of parameters in the transistor-devices and in the interconnects which constitute a differential sense circuit.

Ideally, a differential sense circuit is designed to be electrically and topologically symmetrical. Symmetrical means, here, that the two half circuits of the sense amplifier, the pair of bitlines, bitline loads, memory cells coupled to the bitlines, precharge devices, parasitic and eventual other elements, are mirror images of each other (e.g., Figure 3.10). Despite

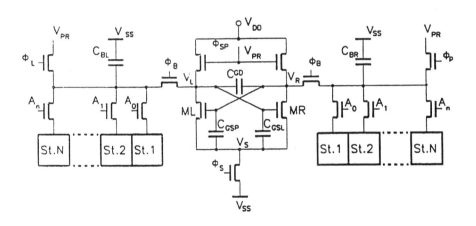

Figure 3.10. Differential sense circuit.

the most careful efforts to mirror the two half-circuits the effects of semiconductor processing, voltage biases, substrate currents, temperature changes and radioactive radiations, can cause small and nonuniform variations in the threshold voltage V_T, gain factor β, bitline capacitances C_B, gate-source capacitances C_{GS}, gain-drain capacitances C_{GD}, and in other parameters. These nonuniform parameter variations manifest themselves in offsets [38], i.e., in voltage and current differences between the two outputs of the sense amplifier, when identical precharge voltages appear on its input pair (Section 3.5.1). The sense amplifier offset may act against either the log.0 or the log.1 level. Therefore, imbalances may reduce both "0" and "1" operation margins in a differential sense circuit.

Approximations of the maximum imbalance-caused margin reductions require computer aid, because the effects of the parameter variations are time-dependent, nonlinear, and interacting. Nevertheless, the qualitative effects of variations in a few parameters V_T, β, C_B and C_{GS} may be illustrated by differentiating the Kirchoff equations, which describe the differential sense circuit in matrix form [39], when devices MPL and MPR are used for precharge only and $V_{PR} > V_{DD} - V_T$ (Section 3.3.2);

$$\frac{d}{dt}\begin{bmatrix} V_L \\ V_R \end{bmatrix} = -\frac{1}{[C]}\frac{1}{2}\begin{bmatrix} \beta_L (V_R - V_S - V_{TL})^2 \\ \beta_R (V_L - V_S - V_{TR})^2 \end{bmatrix} + \begin{bmatrix} C_{GSL} \\ C_{GSR} \end{bmatrix}\frac{dV_S}{dT} ,$$

$$[C] = \begin{bmatrix} C_{GSL} + C_{BR} + C_{GD} & -C_{GD} \\ -C_{GD} & C_{GSR} + C_{BL} + C_{GD} \end{bmatrix}.$$

Here, indices L and R designate left and right symmetrical elements of the circuit, V_L and V_R are the voltages on the input and output of e sense amplifier, and V_S is the common source voltage of transitors ML and MR. From the matrices the estimate of the output signal differential $d(V_L-V_R)/dt$ may be summarized in three terms

$$\frac{d}{dt}(V_L - V_R) \approx \Delta V_T + \frac{K_1}{C_\Delta}\beta(C_{GS} + C_B) + K_2\frac{d}{dt}V_S ,$$

$$C_\Delta = (C_{GSL} + C_{BR} + C_{GD})(C_{GSR} + C_{BL} + C_{GD}), \quad C_B = \frac{1}{2}(C_{BL} + C_{BR}),$$

where K_1 and K_2 are adjustment factors which are affected by the device parameters of ML and MR. In the equation of $d(V_L-V_R)/dt$, the first term is the threshold voltage difference for transistors ML and MR, the second term includes the effects of capacitance and gain-factor fluctuations, while the third term indicates the dependency on the speed of the sense signal transient. Larger parameter nonuniformity and faster transient signals cause larger imbalance, and, by that, larger margin degradations at a given operating temperature T. An imbalance-caused margin degradation voltage V_{IB} may be calculated as

$$V_{IB} \approx d/d_t(V_L-V_R) \Delta t,$$

where Δt is the sense amplifier setup time.

In practice, the voltage imbalance V_{IB} may be considered by the anticipated DC offset V_{off} of the sense amplifier, so that $V_{IB} = V_{off}$. Both the offset voltage and offset current may change slightly with the variations of temperature and greatly with the amount of radioactive radiations.

3.1.3.6 Other Specific Effects

A variety of circuit-specific effects may also considerably degrade the operation margins. For estimation of circuit-specific degradations one must thoroughly understand the operation of the effected circuit and the characteristics of the effect. From the variety of the margin reducing effects, the following are just a few examples for possible consideration in sense circuit designs.

Operation margins may significantly be reduced by the effects of high-amplitude fast-changing electromagnetic noises which may be coupled from memory external sources into the sense circuit (Sections 5.2.1 and 5.2.4). Furthermore, in high-density memory circuits the array-internal crosstalk noises (Section 5.2.2), may also decrease the operation margins substantially. Both "0" and "1" margins may be reduced by the appearance of either one or both internal and external noise-signals in a sense circuit.

Operation margins in a sense circuit may not merely be decreased, but may disappear due to the effects of ionizing atomic-particle impacts (Section 5.3) and of various radioactive radiation events on the transmitter and circuit-paramaters (Section 6.1) on the transistor- and circuit-parameters.

Dynamic sense circuits may be plagued by both incomplete bitline restore V_{BR} and bitline droop V_{BD} [39]. Both V_{BR} and V_{BD} may counteract the data signal development on the bitline, and may increase the imbalance in a symmetrical differential sense circuit. Thus, both "0" and "1" operation margins may be reduced by V_{BR} and V_{BD} during sense operation.

3.1.3.7 Precharge Level Variations

Precharge level variations ΔV_{PR}-s due to the effects of semiconductor processing and environmental variations may reduce both or either one of the "0" and "1" operation margins. Both margins are decreased when "midlevel" precharge, and either the "0" or the "1" margin may be degraded when "low" or "high" level precharge is applied.

A precharge is applied on the bitlines as well as on the inputs and in many designs, also on the outputs of a differential sense amplifier. In many sense circuit designs the precharge voltage serves as a temporary reference level for the discrimination of log.0 and log.1 information, and it may define the initial quiescent operation point for the sense amplifier as well.

The desirable reference level can be determined by the use of operation margin diagrams, and can be generated from the supply voltage V_{DD} by a precharge circuit. A precharge circuit consists of MOS transistors, as well as passive capacitive and resistive components. The supply voltage, transistor and passive device parameters vary as results of processing, temperature, and radioactive radiation effects, and these variations cause deviations from the desired precharge level. Although precharge level changes can be caused by the changes of a wide variety of circuit parameters, yet the maximum fluctuation of a precharge level $\Delta \hat{V}_{PR}$ is influenced predominantly by the percentage variations of the divider impedances, or of the threshold voltages of the voltage references (Section 4.2.2) (Figure 3.11).

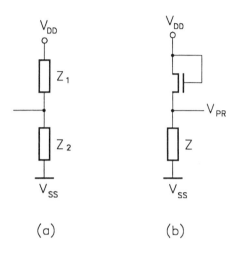

Figure 3.11. Simplified impedance reference circuit (a) and threshold voltage reference circuit (b).

In the divider circuits the precharge voltage V_{PR} fluctuations ΔV_{PR} may be expressed as

$$\Delta V_{PR} = \frac{\Delta Z_1 + \Delta Z_2}{Z_1 + Z_2} V_{DD} K ,$$

while in threshold voltage reference circuits the ΔV_{PR} may be approximated by

$$\Delta V_{PR} = \frac{\Delta V_T}{V_T} K .$$

Here, ΔZ_1, ΔZ_2, and ΔV_T indicate the total amount of parameter change in impedances Z_1 and Z_2 and in threshold voltage V_T, and K is an attenuation factor. K=1 in unattenuated circuits, and K<<1 can be provided by the use of voltage stabilizer circuits.

In precharge generator circuits V_{PR} can track the fluctuations of the supply voltage V_{DD} and often of other device parameters, e.g., V_T, β, C_{GS}, etc. Parameter tracking (Section 6.2.2) implemented in sense circuits increase both "0" and "1" operation margins.

3.2 SENSE AMPLIFIERS IN GENERAL

3.2.1 Basics

A sense amplifier is an active circuit that reduces the time of signal propagation from an accessed memory cell to the logic circuit located at the periphery of the memory cell array, and converts the arbitrary logic levels occurring on a bitline to the digital logic levels of the peripheral Boolean circuits (Figure 3.2). The sense amplifier circuit has to operate within the conditions that are set by the operation margins (Section 3.1).

The operation margins constrain (1) the minimum and maximum input signal amplitudes V_i and V_j, (2) the initial quiescent operation voltage $V_i(0) = V_{PR}$, and (3) the minimum gain A for the sense amplifier. The gain A, however, is a function of the initial voltage level or precharge voltage V_{PR} and of the input signal swing $\Delta V_i = V_j - V_i$ (Figure 3.12). Thus, the combination of V_i, V_j, V_{PR} and A determines basic conditions for sense amplifier designs. In most sense circuits A influences the sense delay t_{PS}, but a high A does not necessarily reduce t_{PS}. Usually, t_{PS} has to be compromised for reduced power consumption and layout area, and for better tolerance of environmental effects.

Layout area restrictions for sense amplifiers are specific for memory designs. In memories sense amplifier layouts should fit either in the bitline pitch when each bitline requires individual data sensing as in DRAMs, or in the decoder pitch when a multiplicity of bitlines are connected to a single sense amplifier as in SRAMs and ROMs. The bitline pitch is determined by the size of a memory cell, and the decoder pitch is limited by the number of parallel running decoder wires in the layout design.

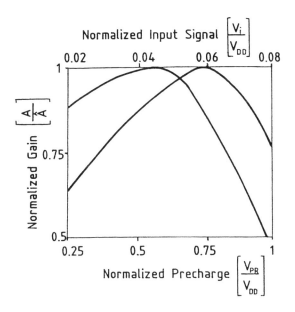

Figure 3.12. Sense amplifier gain as a function of precharge voltage and input-signal swing.

The circuit design of a sense amplifier need not to aim linearity in amplification; in fact, sense amplifiers operate in both linear (small) and nonlinear (large) signal-gain domains of their I_{DS} (V_{DS}, V_{GS}) and $V_o(V_i)$ characteristics (Figure 3.13). Here, I_{DS} is the drain-source current, V_{DS}, V_{GS}, V_o and V_i are the drain-source, gate-drain, output and input voltages in the sense amplifier. Linear amplification appears in the vicinity of the assumed quiescent operation point Q where both n- and p-channel MOS transistors operate in their saturation regions, and where the saturation characteristics of MOS devices are nearly linear. Signals outside of the linear region of the transfer characteristics result in distorted nonlinear amplification. This dual, linear and nonlinear, property of MOS sense amplifiers indicates the application of DC, AC and transient analyses in designs, and the mixed nature of sense amplifier characterization. Parameters characterizing a sense amplifier include amplification A, sensitivity S, offsets V_{off} and I_{off} and common mode rejection ratio CMRR, rise time t_r, fall time t_f and sense delay t_{PS}.

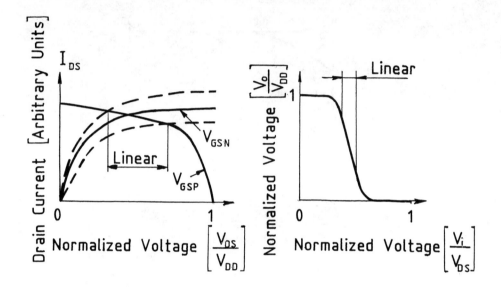

Figure 3.13. Linear gain regions in the I_{DS} (V_{DS}, V_{GS}) and $V_o(V_i)$ characteristics.

In sense circuits, S is the amplitude of the minimum detectable signal, A is the ratio of the output signal amplitude to the input signal amplitude, and specifically in differential sense amplifiers V_{off} and I_{off} are the signal differences between the output pairs when a common mode signal pair appear on the input pairs, CMRR is the ratio of amplifications for differential and common mode signals in the vicinity of the precharge or initiation levels, t_r and t_f means the time from the 10% to the 90% amplitude of the transient signal, and t_{PS} is measured between the 50% amplitude of the word select signal's leading transient and the 50% amplitude of the sense amplifier's output-signal transient.

3.2.2 Designing Sense Amplifiers

Sense amplifier design objectives combining

- minimum sense delay,
- required amplification,
- minimum power consumption,
- restricted layout area,
- high reliability,
- specified environmental tolerance

are difficult to meet due to the contradictory effects of circuit complexity and transistor sizes on the individual design goals.

To optimize the combination of design goals the circuit designer may manipulate only the circuit configuration, and the sizes and shapes of the circuit-constituent transistors, capacitors, resistors and interconnects. The other transistor and passive-element parameters and their variations are determined by the effects of the processing technology, power supply, temperature and radioactive radiations. Because the design parameters have to satisfy a number of contradictory requirements, and because less number of equations than the number of unknowns are at the designer's disposal, the sense circuit design is a highly iterative procedure.

Sense amplifier design procedures aim to reduce the number of iterations by partitioning the design into four major phases: (1) preliminary design, (2) circuit analysis, (3) reliability and environmental tolerance analysis, and (4) final design (Table 3.1).

The Preliminary Design phase makes possible to adopt or devise a basic sense amplifier circuit that most likely satisfies the design goals. Investigations on experimental designs of candidate sense-circuit architectures and schematics provide operation margins and precharge levels, as well as speed, gain and layout data, which are approximate. These approximations are, however, sufficient to select a basic sense amplifier circuit that performs acceptably within the limitations dictated by the processing technology and environmental effects.

The results of the Circuit Analysis assists to approximate the final circuit diagram, MOS transistor sizes, operation, timing and layout of the basic sense amplifier. Initially, a direct current (DC) analysis of the sense

1.	**Preliminary Design**
	Sense Circuit Architecture and Schematic
	Operation Margin and Precharge Analysis
	Sense Amplifier Circuit
2.	**Circuit Analysis**
	Direct Current
	Small Signal Alternative Current
	Large Signal Transient
	Timing
	Layout
3.	**Reliability and Environmental Analysis**
	Operation Stability
	Hot Carrier Suppression
	Temperature Effects
	Atomic Particle Impact
	Radiation Hardness
4.	**Final Design**
	Layout Integration
	Timing Integration
	Complete Analysis

Table 3.1. Design phases.

amplifier circuit establishes the quiescent operation point and voltage biases and, furthermore, the crude aspect ratios of the MOS transistors. Next, the aspect ratios, and the circuit itself, may be modified to provide the desirable alternative current (AC) characteristics, such as gain as a function of the precharge voltage and input voltage swing, common mode rejection ratio, offset, etc. Following the AC analysis an examination of large signal transient or time (t) behavior reveals the switching signal forms as functions of time. To approach the objectives in switching times and delays, usually further changes in MOS device sizes and capacitors and the inclusion of additional circuit elements are needed. The operation of the circuit elements cause ripples, spikes, delays, and other anomalies in the sense operation. Often, these anomalies can be minimized by proper timing of the part circuits. Moreover, the timing is of fundamental importance to provide the conditions that are assumed for the analyses. Furthermore, the analysis must consider the layout limitations, and investigate whether the circuit can be placed into the available place and how the physical implementation of the circuit effects the operation of the memory circuits.

Reliability and Environmental Analysis may effect the memory circuits considerably, but usually it does not impose substantial changes in the basic sense amplifier circuit. The sense amplifier circuit may need additional circuit elements, and sometimes complete circuits, to provide stable operation throughout the specified temperature range, to avoid hot carrier emission induced reliability degradations, to reduce soft-error-rates resulting from impacts of alpha or of a variety of cosmic particles in space, and to harden against eventual ionizing radiations in nuclear environments.

A Final Design integrates the layout and timing of the memory cell array, decoders, write and read circuits with the layout and timing of the sense circuit. The unification of layout and timing may effect the sense amplifier operation and design. To finalize the design a complete reanalysis that includes all changes and effects has to be performed.

The analysis of the sense amplifier circuit heavily relies on computer aid, and requires the use of the most sophisticated MOS device models in circuit analysis programs. Particularly, the MOS device model for a submicrometer sense circuit simulation should comprise velocity saturation, substrate currents, subthreshold characteristics, drain induced barrier lowering, drain-source capacitance, threshold voltage dependency from channel length and width in addition to the parameters of the traditional MOS device models.

3.2.3 Classification

Sense amplifiers may be classified by circuit types such as differential and nondifferential, and by operation modes such as voltage, current and charge sense amplifiers (Table 3.2)

Sense Amplifiers	
Circuit Types	Operation Mode
Differential	Voltage
Nondifferential	Current
	Charge

Table 3.2. Sense amplifier classification.

Differential sense amplifiers are applied in the vast majority of CMOS memories including all SRAM, DRAM, many ROM, and other memory designs. In such designs the sense amplifiers are coupled to a pair of identical bitlines. Nevertheless, the bitline pairs do not necessarily carry a pair of complementary signals (log.0 and log.1), but on one of the bitlines a reference level may be provided, while the other one supplies the data signal. The minimum signal amplitudes, which are distinguishable from noises on bitline pairs, are significantly smaller than those on single bitlines. Since a differential sense amplifier can distinguish smaller signals from noise than its nondifferential counterpart, the signal detection can start sooner than that in a nondifferential sense amplifier. Although differential sensing compromises some silicon area, yet in most of the designs the use of differential amplifiers allow to combine very high packing density with reasonable access time and low power consumption.

Nondifferential sense amplifiers find application in those nonvolatile and sequential memories, where the memory cells are capable to generate significantly larger and faster signals on a bitline than DRAM and SRAM memory cells do. Nonetheless, the evolution of the nonvolatile and sequential memory technology toward higher density and performance, places stringent requirements on sense amplifiers which can hardly be satisfied by nondifferential approaches.

In conventional memories both differential and nondifferential sense amplifiers operate in voltage amplification mode, because the very large input resistance of the MOS transistors allow for obtaining high voltage gain and voltage swings by application of simple circuits. However, the speed of sense circuits, which sense voltage differences, is limited by the charge and discharge times of the circuit-inherent capacitive elements.

In advanced memories the speed limitation resulting from high bitline capacitances, may be encountered by the application of current-mode sense amplifiers. Current-mode sense amplifiers greatly can reduce the charge and discharge times of the capacitances by their low input and output resistances. Furthermore, current-mode sense amplifier designs can provide speed-power products, which are superior to other approaches.

Some memories, as an alternative approach to improve sensing speed, apply charge-transfer or other type of preamplifiers placed before a voltage-mode sense amplifier. Nonetheless, the performance of the pre-amplifier-plus-voltage amplifier compound is inferior to that of purely current-mode sense amplifiers, in most cases.

This chapter, therefore, focuses on differential voltage-mode and differential current-mode sense amplifiers, and discusses nondifferential and charge transfer sense amplifiers in accordance with their declining importance in memory designs.

3.3 DIFFERENTIAL VOLTAGE SENSE AMPLIFIERS

3.3.1 Basic Differential Voltage Amplifier

3.3.1.1 Description and Operation

Differential amplifiers have been known for a long time [310], and the differential sense amplifiers have been derived from the basic MOS differential voltage amplifier (Figure 3.14). Importantly, this basic circuit contains all elements required for differential sensing, easy to analyze, and the results, principles and tradeoffs are readily applicable in the design of all other differential voltage sense amplifier circuits. Moreover, the design tradeoffs in this basic circuit demonstrate the necessity and justify the use of more complex sense amplifier circuits.

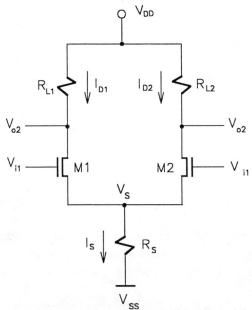

Figure 3.14. Basic differential voltage amplifier circuit.

The basic voltage differential amplifier consists of two enhancement-mode MOS devices M1 and M2 and three resistive elements R_{L1}, R_{L2} and R_S. Transistors M1 and M2 are assumed to be identical and to operate in their saturation region, and load resistors R_{L1} and R_{L2} are presumed to be the same.

The operation of the basic differential amplifier as a sense amplifier begins with precharging both inputs to an identical voltage level $V_{PR}= v_{i1}= v_{i2}$. After v_{i1} and v_{i2} reach V_{PR} the precharge generator is disconnected from the sense amplifier, and V_{PR} is temporarily stored on the parasitic input capacitances C_{i1} and C_{i2}. As long as the input voltages are the same $v_{i1} = v_{i2}$, both output voltages are assumed to be identical $v_{01} = v_{02}$. Next, when a memory cell is accessed, the datum stored in the cell generates a potential difference $\Delta v_i = |v_{i1} - v_{i2}|$ between the inputs. Subsequently, this Δv_i is amplified by the circuit to an output voltage difference $\Delta v_0 = |v_{01} - v_{02}|$.

3.3.1.2 DC Analysis

Source resistor R_S provides an approximately constant bias current I_S to M1 and M2 during the time that the input voltages v_{i1} and v_{i2} are nearly equal and Δv_i is sufficiently small for linear operation. At $v_{i1} = v_{i2}$, I_S is split evenly between the two matched transistors M1 and M2.

$$\frac{I_S}{2} = I_{D1} = I_{D2} = \frac{v_S - V_{SS}}{2R_S} = I_D$$

where I_{D1}, I_{D2} and I_D are the drain currents of devices M1, M2 and M respectively and R_S is the source resistance. This current-split allows for a theoretical bisection of the basic differential amplifier into two equivalent circuits (Figure 3.15) provided that the analysis takes place in the vicinity of $v_{i1} = v_{i2}$. Thus, each of the half-circuits can be construed of R_{L1}, M and $2R_S$, where for the load resistor $R_L = R_{L1} = R_{L2}$ is assumed.

The DC load line of M (Figure 3.16) may be determined by the node-potential method

$$V_{DS}(M) = v_{01} - v_s = (V_{DD} - V_{SS}) - I_D(R_L + 2R_S),$$

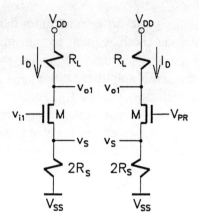

Figure 3.15. Bisected basic differential amplifier circuit.

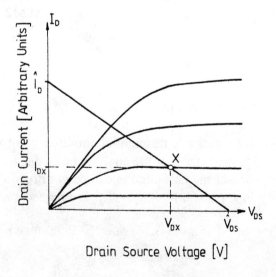

Figure 3.16. Determination of the DC load line.

where v_{o1} is the output voltage and v_s is the source potential of transistor M. Since in most of the memories $V_{SS} = 0$, then the two extreme points of the DC load line in the characteristic field of M1 or M2 may be determined as $I_D = V_{DD}/(R_D+2R_S)$ at $V_{DS}=0$ and $V_{DS}=V_{DD}$ at $I_D=0$. By means of the DC load line and the gate voltage of transistor M, e.g., $V_{GM}=v_{il}-v_s$, for any quiescent operation point X; the drain current I_{DX} and voltage V_{DX} and, in turn, the channel width W and channel length L of transistor M and the resistances R_L and R_S, can be approximated. Namely, the saturation current of a submicrometer MOS transistor I_{DSAT} may roughly be estimated [311] by

$$I_{DSAT} = WC_{OX}v_\infty[(V_{PR}-v_S)-V_T(V_{BG})] \text{ and } C_{OX} = \frac{\varepsilon_{OX}}{t_{OX}},$$

where ε_{ox} is the specific permittivity of the gate oxide, t_{ox} is the thickness of the gate oxide, v_∞ is the saturation velocity of the electrons or holes, V_{PR} is the precharge voltage, V_T is the threshold voltage, and V_{BS} is the substrate bias of device M. This estimation is rough because the above equation for I_{DSAT} treats the carrier transport problem within the channel inaccurately, and disregards the two-dimensional current flow. For description of the saturation current I_{DSAT} of a long channel MOS device the traditional first-order approach [312] may be used;

$$I_{DSAT} = \frac{W}{L}C_{OX}\mu[(V_{PR}-v_S)-V_T(V_{BG})]^2,$$

where μ is the mobility and $\mu = f(V_{PR} - v_s)$.

Since the same amount of current flows through MOS device M and resistors R_L and $2R_S$ current I_D can be expressed as

$$I_D = I_{DSAT} = \frac{V_{DD}-v_o}{R_L} = \frac{v_s}{2R_S}.$$

In the introduced equations, L, C_{ox} v_∞ and μ are MOS device parameters which are determined by the processing technology, the supply voltage V_{DD} is available from the design specification, precharge, source and output voltages V_{PR}, V_s and v_o are predetermined by the internal operation and noise margins of the sense circuit and of the circuit driven by the sense amplifier. With these device and circuit parameters, W or W/L as well as R_L and R_S can be obtained from the above equations to any practical I_D.

The drain current I_D can well be approximated, because it has upper and lower boundaries I_{DU} and I_{DL}. I_{DU} is constrained by the allowable substrate current to avoid hot-carrier emission caused reliability problems, by the available layout area to fit into a cell or decoder size dictated pitch, and by the maximum permissible power dissipation of the memory predetermined as one of the design goals. I_{DL} is restrained by the amplification required to provide adequate signal amplitudes for the inputs of the peripheral logic circuits, by the CMRR sufficient to suppress common mode signals below the permissible noise level, and by the maximum propagation delay allowed for the sensing. With the maximum and minimum permissible drain currents $\hat{I}_D$ and $\check{I}_D$ the gain factor β as well as the channel width W and length L for the transistor M can be approached.

3.3.1.3 AC Analysis

The small-signal, low-frequency Norton and Thevenin equivalents (Figure 3.17) of the basic differential voltage amplifier circuit (Figure 3.44) can be used to estimate the differential gain A_d, common mode gain A_c and common mode rejection ratio CMRR [313].

Applying Kirchoff's current-loop law for the Thevenin equivalent circuit, the differential-mode gain A_d, the common-mode gain A_c, and the differential output voltage Δv_o may be expressed as

$$A_d = -g_m(r_d \| R_L) \quad , \quad A_c = \frac{-g_m(r_d \| R_L)}{1 + 2g_m R_S} \quad ,$$

$$\Delta v_0 = -2(r_d \| R_L)[(V_{PR} - v_S) - V_T(V_{BS})]\Delta v_i .$$

Sense Amplifiers 197

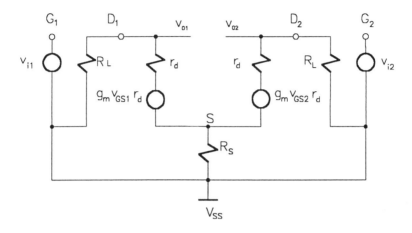

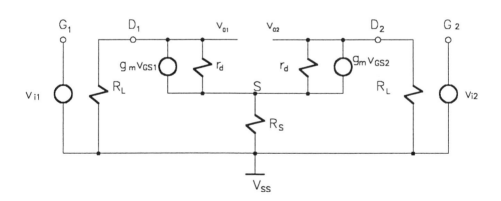

Figure 3.17. Norton and Thevenin equivalents of the basic differential amplifier circuit.

Here, g_m is the transconductance and r_d is the dynamic drain-source resistance for driver devices M1 and M2, R_L it the load resistance, and R_S is the common source resistance. If $r_d >> R_L$, then the differential and common mode gains A_d and A_c are

$$A_d \approx -g_m R_L \quad , \quad A_C \approx \frac{-g_m R_L}{1 + g_m R_S} \quad ,$$

and the common mode rejection ratio, CMRR is

$$\mathrm{CMRR} = \frac{|A_d|}{|A_c|} = \frac{1 + 2 g_m R_S}{2} \quad .$$

Parameters g_m and r_d may most conveniently be varied by the channel-width W and the channel-length L of the identical driver devices M1 ad M2. The aspect ratio W/L of M1 and M2 and resistances R_L and R_S should be designed so that the position of the precharge voltage V_{PR} approximates the center of the high-gain linear portion of the input-output transfer curve $v_o = f(v_i)$ (Figure 3.18) to assure high initial amplification A_d. To match the

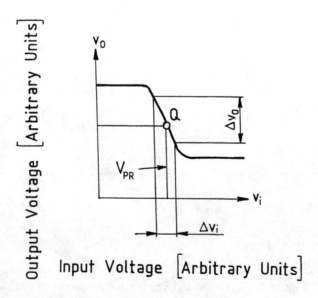

Figure 3.18. Precharge voltage position for high initial gain.

high-gain linear portion of the transfer curve with the predetermined V_{PR} by varying W, L, R_D and R_S is an untrivial task, because A_d is a function also of the gate-source voltage V_{GS} of transistors M1 and M2 and thereby of V_{PR} through the carrier mobility $\mu(V_{GS},...)$, $g_m(V_{GS},...)$, and $r_d(\mu,V_{GS},...)$ besides the dependence of A_d from other parameters.

In practice, increased g_m is obtained by increased device aspect ratio W/L and by lowered V_T, and larger r_d is acquired by decreased bias voltage V_{PR}-V_S on transistors M1 and M2. Moreover, for M1 and M2 some sense amplifier designs use depletion devices to improve g_m and r_{ds}.

Because g_m/area of n-channel MOS devices has been larger than that of p-channel devices, traditional designs apply NMOS enhancement devices for M1 and M2. However, when $L_{eff} < 0.15\mu m$, p-channel devices may provide about the same g_m/area as n-channel devices do due to the effects of the carrier velocity saturation. Since p-channel devices have somewhat larger r_d than their n-channel counterparts do on the same layout area, p-channel devices may also be used for M1 and M2.

Improvements in g_m and r_d of transistors M1 and M2 alone, nonetheless, would result in little increase in A_d and CMRR if resistances R_D and R_S are small. A direct application of large R_D and R_S in the basic differential voltage differential amplifier, however, would result in unacceptable slow output signal changes when the outputs drive the rather large capacitive loads C_{L1} and C_{L2}. Moreover, the imperfect symmetrical nature of M1-M2, R_{L1}-R_{L2} and C_{L1}-C_{L2} pairs result in high offset voltages, which limits the detectable signal amplitudes and delays the start of the sense operation.

Because of the rather slow operational speed provided at considerable power dissipation and because of the inherently high offsets, the basic differential voltage amplifier is not applied in memories in the discussed primitive form. Nevertheless, the DC and AC analyses of the basic differential voltage amplifier have fundamental importance in designing other type of sense amplifiers.

3.3.2 Simple Differential Voltage Sense Amplifier

3.3.2.1 All-Transistor Sense Amplifier Circuit

The simple differential sense amplifier (Figure 3.19) applies transistors MN1 and MN2 as driver and MN3 as source devices, transistors MP4 and MP5 as open-circuit, i.e., very high resistance, loads at small signal amplification and as medium-resistance load-transistors at large signal switching. At the start of the amplification the open-circuit load allows for

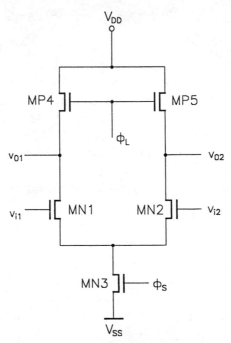

Figure 3.19. Simple differential voltage sense amplifier circuit.

high initial amplification A_d and low offset voltage V_{off}. A_d becomes larger with increasing load-resistances, and V_{off} gets smaller because the effects of nonsymmetricity in the load-device pair MP4-MP5 are eliminated. Furthermore, source transistor MN3 acts as a nearly constant current source and, thus, increases CMRR during small-signal operation. When the output-signal swing is large enough, i.e., it can be discriminated from the noises, then requirements for great A_d, large CMRR and small V_{off}

become unimportant, and load-transistors MP4 and MP5 can be turned on. The activated MP4 and MP5 provide fast large-signal pull-ups through their rather small drain-source resistance. The drain-source resistance of MN3 may also be increased by temporary gate voltage increase. In this differential voltage sense amplifier devices MN1, MN2, MN3, MP4 and MP5 have the same respective functions as elements M1, M2, R_S, R_{L1} and R_{L2} in the basic differential amplifier (Section 3.3.1) do.

The operation of the voltage sense amplifier commences at a turned-off MN3 and turned-on MP4 and MP5. Devices MP4 and MP5 precharge the outputs to $V_{Po} = v_{o1} = v_{o2}$, and other devices (not shown) precharge the inputs to $V_{Pi} = v_{i1} = v_{i2}$. An identical precharge voltage $V_{PR} = V_{Po} = V_{Pi}$ for both the inputs and outputs are applied in many sense circuits for reductions of transistor counts. After the precharge, V_{PR} is temporarily stored on the input capacitances, and both devices MP4 and MP5 are turned off. Next, source device MN3 is turned on slightly, and a memory cell connected to the bitlines is accessed. The accessed memory cell generates a small voltage difference Δv_i between the two input nodes, and Δv_i is amplified to an output signal $\Delta v_o = A_d \Delta v_i$ which appear between the two output nodes, while any common mode signal change is reduced by the factor of CMRR. A_d and CMRR reduce gradually as Δv_o exceeds the linear region of the sense amplifier operation and when Δv_o becomes larger than the noise levels, load devices MP4 and MP5 and source device MN3 are turned on again to enlarge output currents.

3.3.2.2 AC Analysis

During the linear amplification the load resistances are very large and source current I_{Sc}, is approximately constant, because transistors MP4 and MP5 are turned off and MN3 operates in its current saturation region (Figure 3.20). The nearly constant current and the large resistive loads promote high A_d and high CMRR.

When load transistors MP4 and MP5 are turned off, their drain-source resistances $r_{d4} = r_{d5} = r_{dL}$ are determined by the output leakage currents I_{oL4} and I_{oL5}

$$r_{dL} = \frac{V_{DD} - V_{PR}}{I_L} \quad \text{and} \quad I_L = I_{oL4} \approx I_{oL5} ,$$

while the initial drain-source resistance r_{d3} of MN3 may be viewed as the output impedance of a current source [314].

$$r_{d3} = \frac{1}{\lambda I_{DS}} .$$

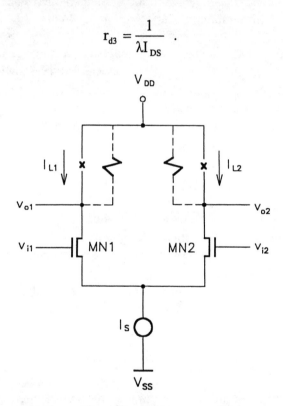

Figure 3.20. Rudimentary model for small signal amplification.

Here, λ is an empirical saturation coefficient and I_S is the drain-source current of device MN3. If the drain-source resistances r_{d1} and r_{d2} of the driver transistors MN1 and MN2 and the resistance r_{d3} are much larger than r_{dL}, then both A_d and CMRR are large. However, neither high A_d nor high CMRR are needed after the amplitude of Δv_o clearly exceeds the noise levels, e.g., $v_o = 0.1 V_{DD}$, but then speedy pull-ups and pull-downs of the individual output voltages v_{o1} and v_{o2} toward the potentials V_{DD} or V_{SS}, are required. The large-signal transients of Δv_o are fast even when the amplifier drives large capacitive loads, if the output currents are large. Increases in the output currents i_1 and i_2 are obtainable through decreased output resistances, which decrease may be provided by turning devices MN3, MP4 and MP5 on hard and, thereby, greatly reducing their drain-source resistances r_{d3}, r_{d4} and r_{d5}.

3.3.2.3 Transient Analysis

The transient analysis should include the nonlinearities in the characteristics of all constituent devices MN1, MN2, MN3, MP4 and MP5 as well as of the equivalent load capacitance C_L, which makes the computations of signal rise, fall and propagation delays rather difficult. Therefore, computer programs with high-level complex device models are routinely used in sense amplifier designs. Nonetheless, to reduce design times and to understand the effects of individual parameters on the transient characteristics crude approximations, which disregard nonlinearities, may beneficially be applied.

The linear equivalents of the bisected basic CMOS sense amplifier that charges and discharges a linear capacitive load C_L (Figure 3.21) allow to employ Laplace-transform, and to obtain the operator impedance $Z(p)$ of the circuit

$$Z(p) = \frac{1}{pC_L} + r_{dD} + 2r_{ds} \; .$$

where r_{dD} and $2r_{ds}$ are the drain-source resistances of devices MD and MS respectively.

The Laplace-transformed of the discharge current $I_f(p)$ of C_L, when C_L is precharged to a midlevel V_{PR} that provides approximately the same extents for both "0" and "1" operation margins, is

$$I_f(p) = \frac{V_{PR}}{r_{dD} + 2r_{ds}} \cdot \frac{1}{p + \dfrac{1}{\tau_f}} \quad \text{and} \quad \tau_f = C_L \left(r_{dD} + 2r_{ds} \right) \; ,$$

204 CMOS Memory Circuits

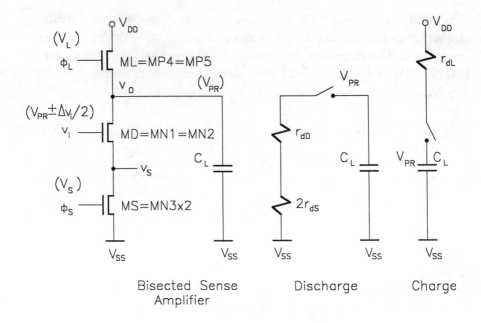

Figure 3.21. Bisected simple CMOS sense amplifier and its linear equivalent circuits.

The reverse Laplace transformation of $I_f(p)$ into the time domain gives the time function of the fall-current $i_f(t)$, and $i_f(t)(r_{dD} + 2r_{dS})$ results the time function of the fall voltage $v_f(t)$;

$$i_f(t) = \frac{V_{PR}}{r_{dD} + 2r_{ds}} e^{-\frac{t}{\tau_f}} \quad \text{and} \quad v_f(t) = V_{PR} e^{-\frac{t}{\tau_f}}.$$

Similarly, the time function of the rise current $i_r(t)$ and voltage $v_r(t)$ during the charge of C_L may be obtained by using time contstant $\tau_r = C_L r_{dL}$, where r_{dL} is the drain-source resistance of ML. Assumptions, here, include that all elements r_{dD}, r_{dS}, r_{dL} and C_L can be characterized by linear concentrated parameters, the effects of parasitic elements on the transient signals are negligible, and the current through ML is very little in comparison to the total discharge and charge currents.

Both the discharge and charge are clearly nonlinear operations, because r_{dD}, r_{dS}, r_{dL} and C_L are functions of $i_f(t)$, $v_f(t)$, $i_r(t)$ and $v_r(t)$. Yet, each of the output signals $v_f(t)$ and $v_r(t)$ (Figure 3.22) may be approximated by using piecewise linear functions in MD's $V_{DS} = f(I_D, V_{GS})$ char-

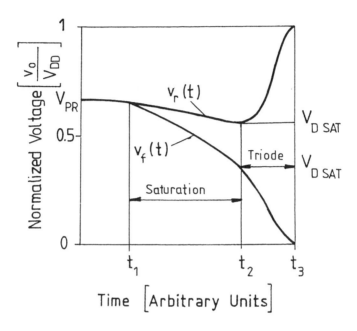

Figure 3.22. Output signal of the simple CMOS sense amplifier.

acteristics for two separate time intervals t_2-t_1 and t_3-t_2. Here, t_2-t_1 is the time interval during MD operates in the saturation region, and t_3-t_2 is the time interval MD operates in the triode region. Drain-source resistances r_d-s in both saturation and triode regions may be approached by linear functions, if it can be assumed that during t_2-t_1 the gate voltages $V_{PR} \pm \Delta v_i/2$ and V_s of devices MD and MS, and during t_3-t_2 the gate voltage V_L of device ML, change very little. Applying the equations of $v_f(t)$ and $v_r(t)$ to intervals t_2-t_1 and t_3-t_2 and using the piecewise linear approximation to r_{dD},

r_{dS} and r_{dL}; the fall-time t_f and the rise time t_r of the output signal can be estimated as

$$t_f = (t_2 - t_1) + (t_3 - t_2) \approx \tau_{fSAT} \ln \frac{0.9 V_{PR}}{V_{DSAT}} + \tau_{fTRI} \ln \frac{V_{DSAT}}{0.1 V_{PR}},$$

$$\tau_{fSAT} = C_L (r_{dDSAT} + 2 r_{dSSAT}) \quad, \quad \tau_{fTRI} = C_L (r_{dDTRI} + 2 r_{dSTRI}),$$

$$t_r = (t_2 - t_1) + (t_3 - t_2) \approx \tau_{rSAT} \ln \frac{V_{DSAT}}{0.1(V_{PR} - V_{DSAT})} + \tau_{rTRI} \ln \frac{0.9(V_{DD} - V_{PR})}{V_{DSAT}},$$

$$\tau_{rSAT} = C_L r_{dLSAT}, \quad \tau_{rTRI} = C_L r_{dLTRI}.$$

In these equations, designations SAT and TRI indicate MOS device operations in the saturation and triode regions of MOS devices, respectively, and V_{DSAT} is the drain-source voltage at which the carrier velocity saturates. V_{DSAT} is not only a function of the critical electrical field strength at which the carrier velocity saturates E_C, but also of the effective channel length of the MOS device L_{eff} and the voltages V_{GS}, V_T and V_{BG} [315] as

$$V_{DSAT} = \frac{E_c L_{eff} [V_{GS} - V_T (V_{BG})]}{E_c L_{eff} + [V_{GS} - V_T (V_{BG})]},$$

where V_{GS} is the gate-source voltage of the MOS device.

To shorten the duration of signal-transients, resistances r_{dD}, r_{dS}, r_{dL} and the voltage V_{DSAT} can be manipulated in the design by varying the effective channel width W_{eff} and length L_{eff}, and by adjusting the gate-source voltage V_{GS} in each individual MOS device MD, MS and ML, but only within the boundaries imposed by the DC and AC conditions of operation. For large A_d and CMRR the AC analysis suggests large r_{dD}, r_{dS} and r_{dL}, but the expressions of transient times t_f and t_r indicate that all r_{dD}, r_{dS} and r_{dL} should be small for short sensing time. This tradeoff, A_d and CMRR versus t_f and t_r, is alleviated in most practical designs by the previously described switching from initial large r_{dL} and r_{dS} to small r_{dL} and

r_{ds} by changing the gate voltages of transistors ML and MS at a certain small output signal amplitude.

Reductions in t_f and t_r, by increasing W_{eff} in devices MD, ML and MS, are limited by restrictions in layout area, power dissipation, substrate current and input capacitance of the sense amplifier, while the minimum L_{eff} is determined by the processing technology. The magnitude of the output-load capacitance C_L of the sense amplifier depends mainly on the architecture of the memory array through parasitic capacitances of long interconnects and by the input capacitances of the circuits driven by the sense amplifier. By placing a buffer amplifier in the immediate vicinity of the sense amplifier output the capacitance C_L can greatly be reduced.

Apart from decreasing r_{dL}, r_{dS}, r_{dD} and C_L a widely applied method to reduce t_f and t_r is the limitation of the output signal swing Δv_o to an small optimized voltage (Section 3.3.6.5).

In both full-swing and optimized-amplitude operation modes only one of both appearing output signals, the rising one, can be accelerated by the simultaneous switching of both load devices MP4 and MP5 from high to low drain-source resistances. This simultaneous switching of load devices is a fundamental drawback of the simple differential voltage sense amplifier in obtaining fast sensing operation.

3.3.3 Full-Complementary Differential Voltage Sense Amplifier

3.3.3.1 Active Load Application

The full-complementary sense amplifier (Figure 3.23) reduces the duration of the signal-transients by using active loads in large-signal switching, and improves the small-signal amplification A_d and common mode rejection ratio CMRR by providing virtually infinite load resistances and approximately constant source current of the inception of signal sensing. In these sense amplifier the active load is implemented in transistors MP4 and MP5, and transistors MN3 and MP6 serve as switchable source devices. When devices MN3, MP6 and MN7 are activated transistor triad MP4-MP5-MP6 operates with triad MN1-MN2-

MN3 in synergy, and together they form a high-speed complementary push-pull amplifier.

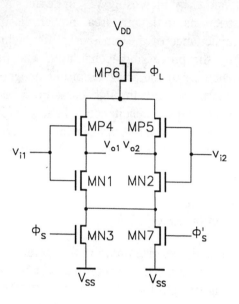

Figure 3.23. Full-complementary differential voltage sense amplifier.

The operation of the active-load full-complementary differential voltage sense amplifier is similar to that of the previously described simple differential voltage sense amplifier. Prior to the sense amplifier activation source transistors MN3, and MP6 and MN7 are turned off, and all input and output nodes are precharged, through devices which are not shown here, to $v_{i1} = v_{i2} = v_{o1} = v_{o2} = V_{PR}$. First, device MN3 is turned on, this activates the differential amplifier combined of devices MN1, MN2, MN3 and of open-circuit-loads, and the circuits amplifies $\Delta v_i = |v_{i1} - v_{i2}|$ to $\Delta v_o = |v_{o1} - v_{o2}|$. When Δv_o is large enough to approach the cutoff voltage of either MN1 or MN2, source device MP6 is turned on and MP6 connects load devices MP4 and MP5 to V_{DD}. When either one MP4 or MP5 gets cut off, the other one pulls the positive-going signal toward the supply voltage rapidly through the decreasing drain-source resistance of MP4 or MP5. Similarly, low resistance path occurs also for the negative-going signal through device MN1 or MN2. The activation of high-current device MN7 shunts source transistors MN3 and further accelerates the output-signal development. For fast signal pull-ups the drain-source current of device MP6 is also high.

3.3.3.2 Analysis and Design Considerations

All methods and results of DC, AC and transient analyses used previously in the examination of the basic and simple differential voltage sense amplifiers (Sections 3.3.1 and 3.3.2), can also be applied to the analysis and design of the full-complementary differential sense amplifier. The operation of the full-complementary sense amplifier may be divided into three segments (1) small signal amplification, (2) signal pull-down and (3) signal pull-up, and the circuits made up by the devices participating in these three operational segments (Figure 3.24) demonstrate the affinity between the full-complementary and the basic and simple differential sense amplifiers.

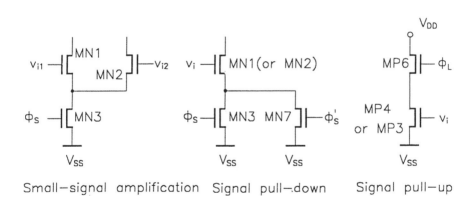

Figure 3.24. Devices participating in small-signal amplification, large signal pull-down and pull-up.

The full-complementary differential sense amplifiers' speed-power product can be enhanced by designing its operation so that at an optimum output voltage swing (Section 3.3.6.5) $\Delta v_{opt} = \Delta v_o \ll V_{DD}\text{-}V_{SS}$ one of both devices MN1 or MN2 and simultaneously one of both devices MP5 or MP4 cut off in a complementary-differential fashion. This selective cut off interrupts the direct current flow between V_{DD} and V_{SS}, decreases the signal transient times and the power dissipation. Moreover, a signal amplitude limitation by turning off devices MN3, MP6 and MP7 at Δv_{opt} reduces the propagation delay and power of the signal transmission between the sense amplifier and the read circuits.

The operational speed of the full-complementary differential amplifier circuit is often hiked also by implementing transistors MN1, MN2, MP4 and MP5 as zero-threshold-voltage devices in separate p⁻ and n⁻ wells [316] (Figure 3.25). A threshold voltage that is set to zero increases the effective gate voltage $V_{Geff} = V_{GS} - V_T(V_{BG})$ of the MOS device, and the higher V_{Geff} results in higher drain-source current and faster signal detection. The well separation allows for minimizing the $V_T(V_{BG})$ fluctuation by controlling V_{BG} through V_{BP} and V_{PN}, and the reduced $V_T(V_{BG})$ fluctuation decreases the offset and the time to develop a valid output signal, in addition to the improvement in the signal detection sensitivity.

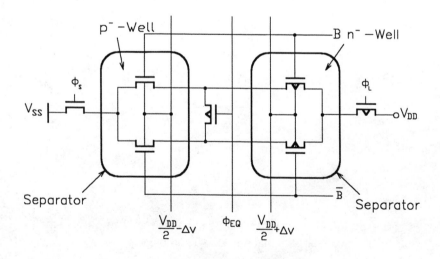

Figure 3.25. Reduction of back-gate bias effects by well separation.

The full-complementary differential sense amplifier is able to combine high initial gain, common mode rejection ratio and fast operation, and has a large input and a small output impedance. The operation can be made even faster by using positive feedback (Sections 3.3.4 and 3.3.5) which provides an enhanced initial differential amplification and an instant data rewrite into the memory cells at destructive read-out.

3.3.4 Positive Feedback Differential Voltage Sense Amplifier

3.3.4.1 Circuit Operation

The positive feedback in differential sense amplifiers (1) makes possible to restore data in DRAM cells simply, (2) increases differential gain in the amplifier, and (3) reduces switching times and delays in the sense circuit.

The positive feedback differential amplifier (Figure 3.26) has two data-terminals ⓛ and ②, and each of both terminals acts as a common input and

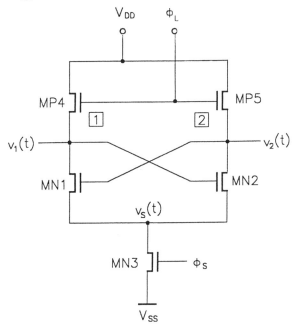

Figure 3.26. Positive feedback differential voltage sense amplifier circuit.

output for the circuit [317]. In the circuit, a simple crosscoupling between the drains and gates of devices MN1 and MN2 implements the positive feedback.

Before the start of the positive-feedback sense operation, the accessed memory cell generates a small voltage difference $v_1(0)-v_2(0)$ on the bitlines and on the sense amplifier nodes [1] and [2], and the source device MN3 and load devices MP4 and MP5 are turned off and feedback devices are biased to operate in the saturation region. The positive feedback operation, and the sense signal development (Figure 3.27) start, when clock ϕ_s turns MN3 on, and input/output nodes [1] and [2] are decoupled from the bitlines (Section

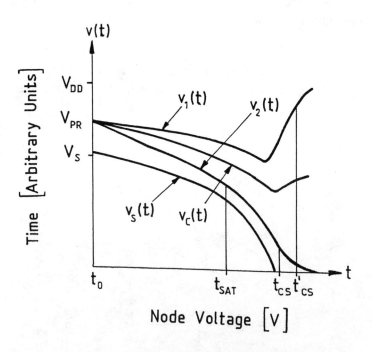

Figure 3.27. Input/output, common mode and source signals.

3.3.6.2.) at the time $t=t_o$. After $t=t_o$ both potentials $v_1(t)$ and $v_2(t)$ of nodes [1] and [2] fall simultaneously toward V_{ss} through the dynamic drain-source resistances r_{d1}, r_{d2} and r_{d3} of devices MN1, MN2 and MN3 and, concurrently, potential difference $v_1(t)-v_2(t)$ increases. As $v_1(t)$ and $v_2(t)$ fall and $v_1(t)-v_2(t)$ grows simultaneously, one of both devices MN1 or MN2, e.g., MN1, is cut off and the other device, e.g., MN2 begins to operate in the triade region at the time $t=t_{SAT}$. About at $t=t_{SAT}$, clock ϕ_L turns load devices MP4 and MP5 on, and the rising signal, e.g., $v_1(t)$, is pulled toward V_{DD} rapidly. The pull-down of the descending signal, e.g., $v_2(t)$, is also quick, because MN2 receives high gate voltage from the rising $v_1(t)$, and because MN3 can be turned on hard or shunted by an additional high current device (Section 3.3.3.1).

3.3.4.2 Feedback Analysis

Positive feedback effects can exist exclusively in the operating region where the complex loop gain $A_1 A_2$ satisfies the Barkhausen criteria

$$\dot{A}_1 \dot{A}_2 = A_1 e^{-jp_1} \cdot A_2 e^{-jp_2} > 1 \ .$$

Here, without the numerical subscripts, $\dot{A}$ is the complex amplification, A is the low-frequency small-signal gain, and p is the phase angle for one half of the bisected symmetrical differential sense amplifier. In mirror-symmetrical feedback amplifiers the Barkhausen criteria defines two requirements

$$A^2 > 1 \text{ and } p_1 + p_2 \pm \Delta p = 2\pi n \ ,$$

where $n = 0, 1, 2, \ldots$ and $\Delta p = |p_1 - p_2| \to 0$. Both requirements should be fulfilled for a possible wide range of $v_1(t)$ and $v_2(t)$ rather than in a small vicinity of V_{PR} to benefit from the effects of positive feedback. The differential gain A'_d of a differential positive feedback sense amplifier (Section 3.3.1.3), that is mirror symmetrical, can be expressed as

$$A'_d = A^2{}_d \approx -g_m{}^2 r_d{}^2 \ ,$$

where A_d is the differential amplification without feedback, g_m and r_d are the transconductance and the drain-source resistance of the driver transistors MN1 and MN2.

214 CMOS Memory Circuits

Transient analysis of positive feedback amplifiers, even in rudimentary approaches are cumbersome, because of the interdependency of the parameters which determine the sense signal $v_1(t)$ and $v_2(t)$. To simplify the analysis of the large signal model (Figure 3.28), however, two important observations can be used: (1) the time function of the common mode voltage $v_C(t) = [v_1(t)+v_2(t)]/2$ follows the source signal $v_S(t)$ closely until the time either one of the devices MN1 or MN2 enter into its saturation region t_{SAT} and somewhat loosely between t_{SAT} and the time either MN1 or MN2 cuts off t_{cut} and (2) during most of the switching time until $t = t_{SAT}$ both MN1 and MN2 operate in their saturation region [318]. Thus, until the time when MN1 or MN2 cuts off, the voltage difference $v_C(t) - v_S(t) = v_{GS}(0) = V_{PR} - v_S(0) \approx$ Constant, where $v_{GS}(0)$ and $v_S(0)$ are the initial gate-source and source potentials of MN1 and MN2 just before $t = t_o$.

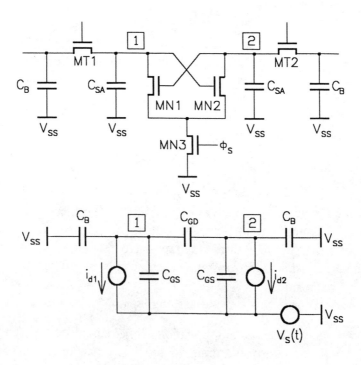

Figure 3.28. Large signal model.

If the sense amplifier starts to operate at $t = t_o$, and the load devices MP4 and MP5 are turned on at $t = t_{SAT}$, then the time of the differential signal development t_d can be approached as a sum of two terms

$$t_d = (t_{SAT} - t_o) + (t_{CS} - t_{SAT}) ,$$

where t_{CS} is the time when the signal $v_1(t)$ or $v_2(t)$ reaches the amplitude of $0.1 V_{DD}$ or $0.9 V_{DD}$, whichever appears earlier.

Term $(t_{SAT} - t_o)$ may first be approached by assuming that the source voltage $v_S(t) = v_S(0) =$ Constant, and by modeling the crosscoupled circuit of MN1 and MN2 as a primitive flip-flop or static memory cell (Sections 2.4 and 2.5) in which the load devices are of extremely high resistances. When a perfectly symmetrical flip-flop brought to its equilibrium voltage $v_C = v_1(0) = v_2(0) = v_{GS}(0)$ an infinitely small initial voltage jump $\Delta V_o 1(t)$ causes exponential changes on both node voltages $v_1(t)$ and $v_2(t)$;

$$v_1(t) = v_{GS}(0) - \Delta V_0 e^{\frac{t}{\tau}} \quad \text{and} \quad v_2(t) = v_{GS}(0) + \Delta V_0 e^{\frac{t}{\tau}} ,$$

where v_{GS} is the drain-source voltage of the driver device, MN1 or MN2, and τ is a constant, as it is known from the feedback theory. In this positive feedback voltage sense amplifier, however, the source voltage $v_S(t)$ changes with time, and $v_S(t) = v_S(0)$ is valid only at t_o. For an arbitrary $v_S(t)$, with $v_C(t) - v_S(t) = V_{PR} - v_S(0)$, may be written

$$v_1(t) = V_{PR} - v_s(0) - V_T(V_{BG}) + v_s(t) - \Delta V_0 e^{\frac{t}{\tau}} ,$$

$$v_2(t) = V_{PR} - v_s(0) - V_T(V_{BG}) + v_s(t) - \Delta V_0 e^{\frac{t}{\tau}} .$$

where V_T is the threshold voltage, and V_{BG} is the substrate bias. A subtraction of $v_2(t)$ from $v_1(t)$, at the assumption of constant and identical a threshold voltage V_T for the transistors MN1 and MN2, results the differential voltage change $v_d(t)$ from t_o to t_{SAT}

$$v_d(t) = v_2(t) - v_1(t) = 2\Delta V_0 e^{\frac{t}{\tau_{SAT}}} ,$$

and the time until either MN1 or MN2 enters to its triode region

$$t_{SAT} - t_0 = \tau_{SAT} \ln \frac{v_d(t_{SAT})}{2\Delta V_0}.$$

At the assumptions that devices MN1, MN2, MP4 and MP5 are of identical sizes, and both devices MN1 and MN2 operate in the saturation region, the time constant τ_{SAT} may be approximated by

$$\tau_{SAT} \approx \frac{C_B + C_{GS} + 4C_{GD}}{\beta[V_{PR} - v_s(0) - V_T(V_{BG})]},$$

where C_B is the bitline capacitance, C_{GS} and C_{GD} are the gate-source and gate-drain capacitances for devices MN1, MN2, MP4 and MP5 and β is the gain factor for devices MN1 and MN2. By applying this τ_{SAT} and setting $v_d(t_{SAT}) = V_T(V_B)$ the duration of t_{SAT}-t_o can be estimated.

In the estimation of the time-period t_{CS}-t_{SAT}, the presumption that t_{CS} is the time point $t_{0.9}$ when the rising output signal $v_1(t)$ achieves $0.9(V_{DD}-V_{PR})$, rather than $0.9V_{DD}$, can often be used. Here, V_{DD} is the supply voltage, and V_{PR} is the precharge voltage. The falling output signal $v_1(t)$ is usually shorter than $v_2(t)$, because MN1 and MN2 operate in positive feedback configuration while MP4 and MP5 are in nonfeedback configuration, and because the mobility in n-channel devices MN1, MN2 and MN3 is larger than in p-channel devices MP4 and MP5 as long as the effective channel length $L_{eff} < 0.12\mu m$. In case, when the signal pull-up by MP4 and MP5 determines in the time-period t_{CS}-t_{SAT}, then

$$t_{CS}\text{-}t_{SAT} \leq t_{0.9}\text{-}t_{SAT} = \tau_{TRI}(\ln 0.9(V_{DD}\text{-}V_{PR})/V_{DSAT}), \tau_{TRI} = C_L r_{dTRIg}$$

where C_L is the load capacitance on node [1] or [2], r_{dTRI} is the effective drain-source resistance of MP4 and MP5 in the triode region, on V_{DSAT} is the velocity saturation. Since, in numerous designs clock ϕ_L drives a multiplicity of MP4s and MP5s, the switching time $t_{f\phi}$ of ϕ_L can be longer than $t_{0.9}$-t_{SAT}. In such a design, simply

$$t_{CS} - t_{SAT} = t_f$$

can be used.

The presented equations indicate that the differential signal development time t_d can be reduced by increased ΔV_o, and decreased time constants τ_{SAT} and τ_{TRI}. Furthermore, the observation that the time function of the average differential output voltage $v_C(t)$ follows the time function of $v_S(t)$ allows to reduce t_d by finding the optimum waveform for $v_S(t)$. A quicker fall time of $v_S(t)$ results in shorter t_d. The optimization of the waveform of $v_S(t)$ may be approached as a time optimum control problem [319], but in practice the DC and AC design conditions and the operation margins limit the shaping of $v_S(t)$ to a certain fall-time.

3.3.5 Full-Complementary Positive-Feedback Differential Voltage Sense Amplifier

The full-complementary positive feedback sense amplifier (Figure 3.29) improves the performance of the previously analyzed simple positive feedback amplifier (Figure 3.26) by using an active load circuit constructed of devices MP4, MP5 and MP6 in positive feedback configuration [320].

In practice, device pairs MP4-MP5 and MN1-MN2 can not be completely matched despite carefully symmetrical design. Usually the nonsymmetricity between the p-channel MP4 and MP5 is more substantial than that between the n-channel MN1 and MN2, because most of the CMOS processes optimize n-channel device characteristics. To avoid a large initial offset resulting from the added effects of imbalances in the n- and p-channel device pair, source devices MN3 and MP6 are not turned on simultaneously, but first the n-channel and later the p-channel complex is activated by impulses ϕ_S and ϕ_L respectively.

The delayed activation of transistor triad MP4-MP5-MP6 by clock ϕ_L results that until the time MP6 is turned on, device triad MN1-MN2-MN3 operates alone and can be analyzed the same way as shown previously for the simple positive feedback sense amplifier (Section 3.3.4).

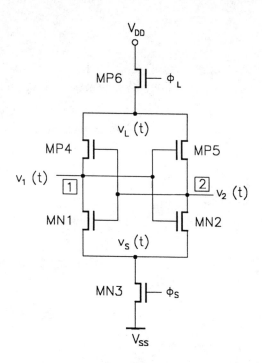

Figure 3.29. Full-complementary positive-feedback sense amplifier circuit. (Source [320].)

When the sense signal on the bitline is large enough, e.g., when the drain-source voltage of either MN1 or MN2 reaches the saturation voltage V_{DSAT}, clock ϕ_L activates triad MP4-MP5-MP6. The activated feedback in MP4-MP5-MP6 introduces a pair of time dependent load resistances $r_{L1}(t) = r_{d4}(t) + 2r_{d6}(t)$ and $r_{L2}(t) = r_{d5}(t) + 2r_{d6}(t)$. Here, $r_d(t)$ is the time-dependent drain-source resistance, and indices 4, 5 and 6 represent devices MP4, MP5 and MP6. The resistances of these devices may be considered as time invariant parameters during the activation of MP6 t_{SAT}, so that $r_L = r_{L1} = r_{L2}$ may be used. With this modified r_L, the formerly introduced

Sense Amplifiers

DC and AC formulas (Sections 3.1-3.4) can be reapplied in the DC and AC analyses of this circuit also.

In the transient analysis, the differential signal development time t_d during the presence of impulse ϕ_S until the appearance of clock ϕ_L is determined by the switching time of the n-channel triad t_{dN}, and thereafter t_d is dominated by the transient time of the p-channel triad t_{dP} (Figure 3.30). With the assumptions used in the transient analysis of the previously discussed positive feedback differential voltage sense amplifier (Section 3.3.4.2) the sense-signal development time in the full-complementary positive feedback differential voltage sense amplifier t_d may be

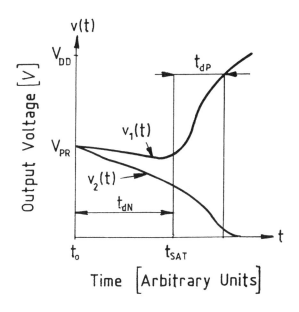

Figure 3.30. Output signal development.

approached as

$$t_d = t_{dN} + t_{dP} = \tau_{dN} \ln \frac{V_{DSAT}}{2\Delta V_0} + \tau_{dP} \ln \frac{0.9(V_{DD} - V_{PR})}{V_{DSAT}},$$

where

$$\tau_{dN} \approx \frac{C_B + C_{GSN} + 4C_{GDN}}{\beta_N[V_{PR} - v_s(0) - V_{TN}(V_{BG})]} \quad , \quad \tau_{dP} \approx \frac{C_B + C_{GSP} + 4_{GDP}}{\beta_N[v_L(0) - V_{PR} - |V_{TP}(V_{BG})|]} \quad ,$$

indices N and P designate n- and p-channel, devices, V_{DSAT} is the saturation voltage, ΔV_o is the amplitude of the initial voltage difference generated by the accessed memory cell on nodes [1] and [2], V_{PR} is the precharge voltage, C_B is the bitline capacitance, C_{GS} and C_{GD} are the gate-source and gate-drain capacitances, and β is the individual gain factor for devices MN1, MN2, MP4 and MP5, $v_S(0)$ and $v_L(0)$ are the initial potentials on the drains of device MN3 and MP6, V_T is the threshold voltage and V_{BG} is the backgate bias.

The equation of t_d demonstrate that in a full-complementary positive-feedback differential sense amplifier quicker operation can be obtained by increasing the gain factors β_N and β_P, by decreasing the parasitic gate-source capacitance C_{GS} and gate-drain capacitance C_{GD} of the n- and p-channel latch devices MN1, MN2, MP4 and MP5, and by decreasing the bitline capacitance C_{BL}. Additionally, reductions in the fall time of $v_S(t)$ and in the rise time of $v_r(t)$ also shorten t_d.

3.3.6 Enhancements to Differential Voltage Sense Amplifiers

3.3.6.1 Approaches

The performance of sense circuits can be improved by adding a few devices to the differential voltage sense amplifier. From the great variety of possible enhancements to the basic amplifier the evolution of the memory technology reduced the number of approaches to a few which can be efficiently implemented in CMOS memories;

(1) Temporary decoupling of the bitlines from the sense amplifiers,

(2) Separating the input and output in feedback sense amplifiers,

(3) Applying switchable constant current sources to the source devices,

(4) Optimizing the output signal amplitude.

Approaches (1) and (2) decrease the capacitive load of the sense amplifier. By approach (3) the sense amplifier's source resistance is virtually increased to achieve high gain, and by approach (4) the amount of switched charges are decreased.

3.3.6.2 Decoupling Bitline Loads

In memories that are designed with positive-feedback differential voltage sense amplifiers, obtainable sensing speeds are greatly reduced by the high load capacitances C_L-s coupled to the sense amplifiers. Generally, C_L is dominated by the capacitance of the memory cells connected to the bitline and by the stray-capacitance of the bitline itself. A significant decrease in capacitance C_L requires major modifications in process technology and in sense circuit design.

By a small design alteration, C_L may be reduced by placing a pair of MOS devices MT1-MT2 (Figure 3.31), or a pair of preamplifies, next to the sense amplifier inputs to decouple the bitline capacitance from the sense amplifier for the time of the initial high-gain amplification.

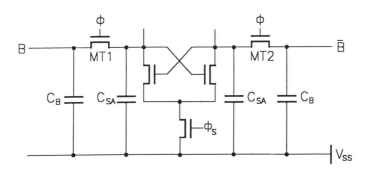

Figure 3.31. Decoupling of bitline capacitances from a sense amplifier.

At the time t_1, decoupler devices MT1 and MT2 are turned on, the accessed memory cell generates a small signal difference $\Delta v_{SA}(t_1)$ on the bitline and on the inputs of the amplifier. During this time, the sense amplifier is inactive and the load on its input-output node is $C_L(t_1) \approx C_B + C_{SA}$, where C_B is the total bitline capacitance, and C_{SA} is the total input-output capacitance of the sense amplifier and $C_B >> C_{SA}$. The sense amplifier is activated at the time t_2 when $\Delta v_{SA}(t)$ achieves the minimum detectable amplitude defined by the operation margins (Section 3.1.2). At the time t_2 devices MT1 and MT2 are turned off. Thus, MT1 and MT2 decouples the C_B from the sense amplifier input and reduces $C_L(t_1)$ to $C_L(t_2) = C_{SA}$, and with the smaller load capacitance $C_L(t_2)$, the sense amplifier can rapidly amplify $\Delta v_{SA}(t)$. At the time t_3, when the amplified signal $\Delta v_{SA}(t)$ is large enough for rewriting the memory cell, decoupler devices MT1 and MT2 are turned on again, therefore $C_L(t_3) = C_{BL} + C_{SA}$ appears for the sense amplifier.

The switching of MT1 and MT2 may be eliminated by the application of depletion mode transistors and by cross-coupling MD1 and MD2 (Figure 3.32). Preamplification in addition to decoupling can be obtained

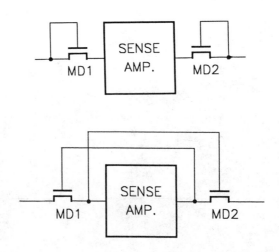

Figure 3.32. Decoupling provided by depletion mode and cross-coupled devices.

if MT1 and MT2 are designed to operate as a charge transfer or as another nondifferential sense amplifier (Section 3.6). The use of preamplifiers to a positive-feedback sense amplifier, nevertheless, may not result in a respectable speed improvement, because the increase in offsets and in parasitic capacitances and the preamplifiers' inherent delay counteract the speed gain obtained by preamplification.

A widely applied sense amplifier (Figure 3.33) incorporates the decoupler transistors MT1 and MT2 by taking advantage of the sequential activation of n- and p-channel transistor triads MN3-MN4-MN5 and MP6-MP7-MP8. Initially, clock ϕ_S activates the n-channel triad and clock ϕ_T

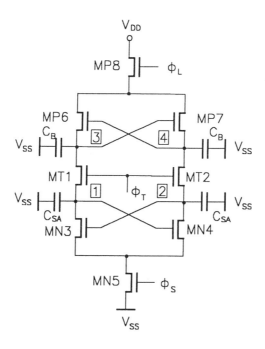

Figure 3.33. Sense amplifier incorporating decoupler devices. (Derived from [320].)

turns devices MT1 and MT2 on. MT1 and MT2 are turned off, however, when the differential signal $v_d(t)$ between nodes [1] and [2] reaches the minimum signal amplitude that is detectable by the sense amplifier. From this time, MN3-MN4-MN5 can amplify $\Delta v_d(t)$ rapidly, because the bitline capacitance C_B is decoupled from nodes [1] and [2]. During this time C_B appears on each nodes [3] and [4] . When $v_d(t)$ exceeds an intermediate amplitude, e.g., $v_d(t) = V_T$, ϕ_T turns on MT1 and MT2 again, and the sense amplifier provides a rapid complementary large signal amplification. The switching of MT1 and MT2 requires certain time, but the overall delay time of the sense amplifier may significantly be reduced by the temporary decoupling of the large bitline capacitances from the transistor triad MN3-MN4-MN5.

3.3.6.3 Feedback Separation

Positive feedback in differential voltage sense amplifiers is implemented mostly by crosscoupling of two simple inverting amplifiers.

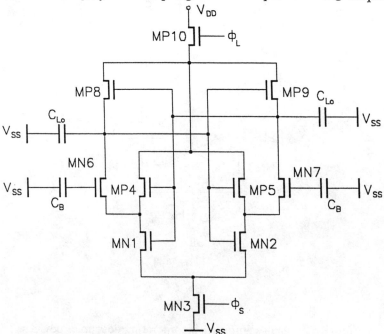

Figure 3.34. Feedback separating in a sense amplifier.

The crosscoupling renders an input and an output to a common node, which makes the input and output load capacitances the same.

A separation of the input from the output at retaining the positive feedback (Figure 3.34) can decrease the load capacitance of the output C_{Lo}. The reduced C_{Lo} shortens the signal transient times t_f, t_r and t_p, while the positive feedback enlarges the amplification A_d at the tradeoff of increased complexity [321]. Without complexity increase, but at the sacrifice of some feedback effects, very fast sensing can be provided by combining the nonfeedback triad MN1-MN2-MN3 with a feedback active-load MP4-MP5-MP6 (Figure 3.35) in single amplifier circuit. Variations of the positive feedback circuits, which feature with separate input and output terminals, are applied generally to memory cells which allow for non-destructive readouts, e.g., in SRAMs, ROMs and PROMs.

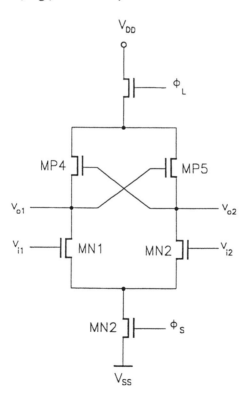

Figure 3.35. Feedback active load with nonfeedback small-signal amplifier.

3.3.6.4 Current Sources

A current source device keeps its output or source current I_S approximately constant and, thereby, provides a very large output resistance r_o in a certain domain of circuit parameters. The large r_o may be used to improve the differential amplification A_d and common mode rejection ratio CMRR of a sense amplifier. Although both A_d and CMRR get higher with larger drain-source resistances r_d, r_{dD}, and r_{dL} (Sections 3.3.1.3, 3.3.2.2, 3.3.4.2); the enlargement of these resistances increases also the sense signal's fall, rise, propagation, and development times t_f, t_r, t_p and t_d (Sections 3.3.2.3, 3.3.4.2, 3.3.5).

To combine short t_f, t_r, t_p and t_d with high initial A_d and CMRR a current source [322] (Figure 3.36), in which the output transistor MS1 can be shunted by a high-current switch transistor MS2, may beneficially be used. At the start of a sense operation, MS2 is turned off, and all other

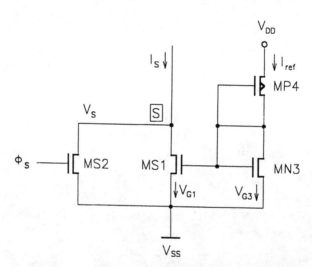

Figure 3.36. Current source for sense amplifier.

devices operate in the saturation region. The source current I_S, in the saturation region, may be estimated by

$$I_s = \beta_1 (V_{S1} - V_{TN})^2 (1 + \frac{\lambda}{L_{S1}} V_s) \ ,$$

and due to the current mirroring

$$I_s = \frac{\beta_1}{\beta_3} I_{ref} \quad , \quad I_{ref} = I_{C3} = I_{C4} \quad ,$$

Since the reference current $I_{ref} = I_3 = I_4$ and the gate voltage of MP4 $V_{G4} = V_{DD} - V_{G3}$, the gate voltage of MN3 V_{G3} may be expressed as

$$V_{G1} = V_{G3} = \frac{\left(\frac{\beta_1}{\beta_3}\right)^{\frac{1}{2}} V_{DD} - V_{TN}(V_{BG}) - V_{TP}}{\left(\frac{\beta_1}{\beta_3}\right)^{\frac{1}{2}} + 1} \quad ,$$

In the equations, subscripts 1, 3 and 4 indicate devices MS1, MN3 and MP4, V_G is the gate-source voltage, β is the gain factor, λ is the channel-length modulation factor, L is the channel length, and V_s is the voltage on node [S], V_{TN} and V_{TP} are the n- and p channel threshold voltages. With V_{G1}, L_1, λ and β_1 the output resistance

$$r_o = r_{d1} = \frac{2}{\lambda \beta_1} \frac{L_1^2}{W_1} (V_{G1} - V_{TN})^{-2} \quad .$$

can be designed by varying MS1's channel width W_1 and, in turn, by altering, its drain-source resistance r_{d1}. Shunt device MS2 is turned on, when the sense signal safely exceeds the noise levels, and its small drain-source resistance r_{d2} in parallel coupling with r_{d1} results in high I_S, low r_o and short t_f, t_r, t_p and t_d.

The output resistance r_o of the previously described current source may be increased by making use of additional transistors MN5 and MS6 (Figure 3.37) in the circuit. An analysis of this circuit's Norton equivalent shows that the output resistance r_o is

$$r_o = r_{d1} + r_{d3} + r_{d1} \, r_{d3} \, (1+d) \, r_{d1},$$

where indices 1 and 3 designate transistors MS1 and MN3, r_d is the drain-source resistance, $d=1/g_m(\partial I_s/\partial V_{BG})$, g_m is the transconductance, and V_{BG} is the backgate bias.

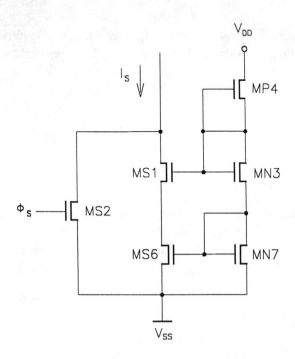

Figure 3.37. Improved current source.

The implementation of current sources in CMOS RAMs seems to require large silicon area. Nevertheless, in most of the CMOS RAM designs, a single current source can be used to a multiplicity of sense amplifiers, which allows for efficient circuit layouts. CMOS sense amplifier designs, yet, apply very seldom current sources to provide large load resistances. During small signal sensing the load resistances are very high anyway, because the load devices are turned off, and appear as open circuits with very little leakage currents to the drive transistors.

3.3.6.5 Optimum Voltage-Swing to Sense Amplifiers

A widely applied method to improve speed and power performances of many sense amplifiers is the limitation of the output, or of the common input/output, signal swing Δv_o to a small optimized voltage swing v_{opt}. A $\Delta v_o = v_{opt}$ exists because an increasing effective gate voltage $V_{Geff} = V_{GS} - V_T(V_{BS}) \approx V_{PR} \pm \Delta v_o/2$ on the input of the circuit, that is driven by the sense amplifier, results in faster output switching times t_s in the driven circuit, but the switching of a greater V_{Geff} and a larger charge package Q_L on the same load capacitance C_L, with the same output current i_d, requires longer t_s. The existence of an optimum can be made plausible by setting the current of a transistor i_d equal to the current of the load capacitance i_c, i.e., $i_d = C_L[dv_o(t)/dt]$ (Figure 3.38), and expressing t_s by a linear approximation of the discharge time t_f yields

$$t_s = t_f \sim \frac{C_L}{\beta V_{Geff}} \ln \frac{V_{Geff} - V_\sigma(0)}{V_\sigma(0)} .$$

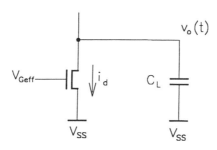

Figure 3.38. Simplified discharge model.

Here, V_o is the output log.0 level required to drive the logic circuit that follows the sense amplifier. The equation of t_s shows that both V_{Geff} decreases and the logarithm of V_{Geff} increases the switching time t_s and, thus, t_s curves a minimum over an optimum output voltage swing v_{opt}

(Figure 3.39). At v_{opt} the power dissipation of the sense amplifier approaches a minimum also.

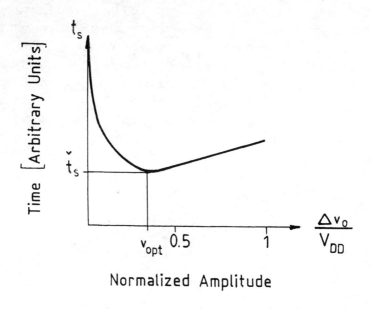

Figure 3.39. Optimum voltage swing.

In addition to substantial improvements in speed and power characteristics, the reduction of voltage swings becomes imperative in designs for deep-submicrometer CMOS technologies. Reduced voltage swings, namely, results in decreased hot-carrier emissions, cross-talkings and noises, and operation margin degradations.

For output voltage swing limitation [323] the two most widely used techniques are the amplitude timing (fixed time) and the voltage clamping (fixed voltage). Amplitude timing can easily be implemented by deactivating the sense amplifier at $v_o(t) = v_{opt}$ at the time point t_x when $\Delta v_o = v_{opt}$ (Figure 3.40a). Applications of the fixed time technique, however, may result in large variations in Δv_o due to device parameter changes. Less prone to device parameter fluctuations is the voltage limitation technique that uses voltage clamping (Figure 3.40b). Both techniques, the amplitude timing and the voltage clamping, can be applied also for limiting over- and under-shots of the signals on the bit- and wordlines.

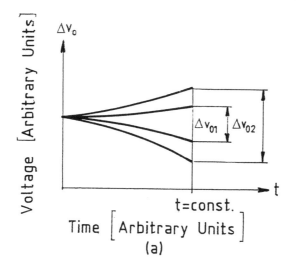

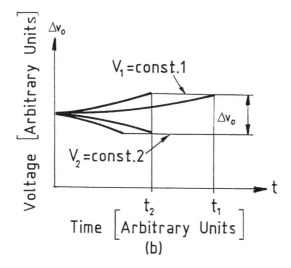

Figure 3.40. Voltage swing limitation by fixed time (a) and fixed voltage (b) methods.

3.4 CURRENT SENSE AMPLIFIERS

3.4.1 Reasons for Current Sensing

The fundamental reason for applying current-mode sense amplifiers in sense circuits is their small input impedances and, in cross-coupled feedback configuration, their small common input/output impedances. Benefits of small input and input/output impedances, which are coupled to a bitline, include significant reductions in sense circuit delays, voltage swings, cross talkings, substrate currents and substrate voltage modulations.

The reduction in sense circuit delays, that results from the use of current amplifier, can be made plausible by comparing the voltage signal v(t) and the current signal i(t) which appear in the simplified Thevenin

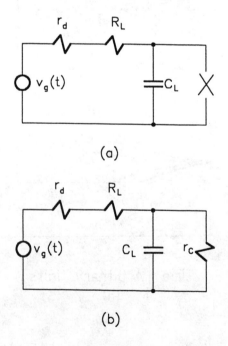

Figure 3.41. Simplified equivalents of a voltage (a) and a current (b) sense circuit.

equivalents of the voltage (Figure 3.41a) and of the current (Figure 3.41b) sense circuits. In these sense circuit equivalents, the accessed memory cell is represented by a voltage generator $v_G(t)$ and a resistor r_d, the bitline load is simplified to a capacitance C_L and a resistance R_L, the voltage amplifier's impedance is modeled by an open circuit, and the current amplifier's impedance is comprised in a small resistance r_c. Operator impedances for each of the equivalent circuits are

$$(a)\ Z(p) = (r_d + R_L)C_L + \frac{1}{pC_L} \quad \text{and} \quad (b)\ Z(p) = (r_d + R_L) + \frac{r_c}{pr_c C_L + 1}.$$

Assuming that the generator voltage is an ideal voltage jump $v_G(t) = V_G 1(t)$ then its Laplace-transformed is V_G/p. The reverse Laplace-transformation of the bitline current $i(t)$ for both equivalent circuit may be written as

$$i(t) = \frac{V_G}{r_d + R_L} e^{-\frac{t}{\tau}},$$

and from $i(t)$ the approximative fall and rise times $t_f = t_r = 2.2\tau$ can be obtained. For the voltage amplifier's equivalent $\tau_v = (r_d + R_L)C_L$ while for the current amplifier's equivalent $\tau_c = [(r_d + R_L)\|r_c]C_L$. Evidently, the sense circuit with the current amplifier has a much smaller τ than the circuit with the voltage amplifier does, i.e., $\tau_c \ll \tau_v$, because of the shunting effect of r_c. The smaller τ results that the current sense circuit provides shorter t_f and t_r than the voltage sense circuit does, and this is manifested clearly by the normalized current and voltage transient signals $i_c(t)$, $i_v(t)$, $v_c(t)$ and $v_v(t)$ (Figure 3.42). Here, the signals are obtained from the equation of $i(t)$, and indices c and v designate current and voltage sense amplifiers respectively, and parameters r_d, R_L and C_L are the same for both the current and the voltage sense circuits.

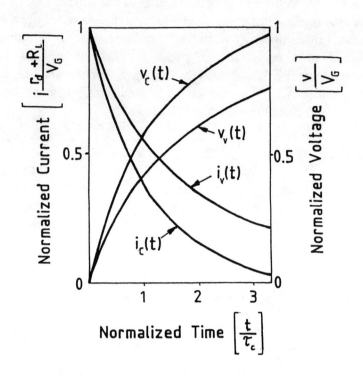

Figure 3.42. Comparison of transient signals appearing in voltage and current sense circuits.

In sense circuits where the bitline has to be modeled as a distributed parameter network, the simple Norton equivalent of the circuit (Figure 3.43) may be used to compare the signal propagation delay of a voltage-mode amplifier t_{pv} to that of a current-mode amplifier t_{pc}. The equivalent for the sense amplifier circuit models the accessed memory cell by a current generator with transconductance g_m and output resistance r_{go}; the bitline by a ladder of n incremental capacitor C and resistor R; and the sense amplifier by its input resistance r_{SA}. The signal delay for both voltage and current signals t_p, in this sense circuit model, if the generator

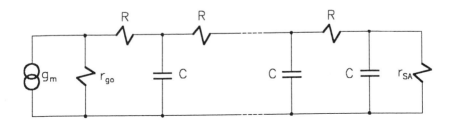

Figure 3.43. Sense circuit equivalent modeling the bitline as a distributed parameter network.

signal is assumed to be a linear ramp signal, may be obtained by applying Laplace-transforms, and the reverse Laplace-transforms for t_p results [324]

$$t_p \approx \frac{n^2 RC}{2} \frac{r_{go} + \frac{nR}{3} + r_{SA}}{r_{go} + nR + r_{SA}} + r_{gr} r_{SA} nC \frac{1}{r_{go} + nR + r_{SA}}.$$

For voltage amplifiers $r_{SA} \to \infty$ while for current amplifiers $r_{SA} \to 0$, thus the signal delay of the voltage amplifier t_{pv} and of the current amplifier t_{pc} may be approximated as

$$t_{pr} \approx \frac{n^2 RC}{2}(1+\frac{2r_{go}}{nR}) \quad \text{and} \quad t_{pc} \approx \frac{n^2 RC}{2} \frac{r_{go} + \frac{nR}{3}}{r_{go} + nR}.$$

Since $nR \ll r_{go}$, then $t_{pv} \gg t_{pc}$, which indicates the superiority of the current sense amplifier.

By the amplification of current signals rather than voltage signals, the voltage swings can be reduced without penalties in signal amplifications.

236 CMOS Memory Circuits

Smaller voltage swings result in reduced crosstalkings, substrate currents and substrate voltage modulations and, in turn, in increased reliability of memory operations.

Current amplification in memories are implemented almost exclusively in feedback circuits. Yet, numerous feedback circuits other than current amplifiers, can also provide small input- or small common input-output impedances. Generally, sense amplifier input and output impedances may be optimized for specific memory cell type, load circuit, amplification and other requirements. Clearly, the design should use that amplifier type, or that combination of various amplifiers, which provides the highest performance at the least costs when combined with the other parts of the sense circuit.

The following brief overview of feedback circuits is an aid to find the feedback type that approaches optimum in the sense circuit design.

3.4.2 Feedback Types and Impedances

Commonly, a feedback system comprises a main amplifier with a gain A and a feedback circuit with a gain or attenuation B (Figure 3.44). The

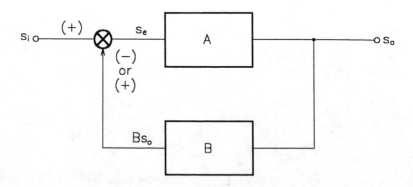

Figure 3.44. General feedback system.

closed-loop gain or amplification in both negative feedback $A^{(-)}$ and positive feedback $A^{(+)}$ are hyperbolic functions of the open-loop gain AB

$$A^{(-)} = \frac{s_o}{s_i} = A\frac{1}{1+AB} \quad , \quad s_i = s_e + Bs_o \quad ,$$

$$A^{(+)} = \frac{s_o}{s_i} = A\frac{1}{1-AB} \quad , \quad s_i = s_e + Bs_o \quad .$$

Since the input signal s_i can be either a voltage, v_i or a current i_i, and the output signal s_o can also be a voltage v_o or a current i_o; four combinations [325] of these parameters results in signal amplifications (Table 3.3). Each

Types Parameters	Voltage	Current	Transfer – Impedance	Transfer – Admittance
Amplification A	$A_V = \frac{v_o}{v_i}$	$A_C = \frac{i_o}{i_i}$	$A_Z = \frac{v_o}{i_i}$	$A_Y = \frac{i_o}{v_i}$
Input – Impedance $Z_i / Z_i (A)$	$(1 \pm AB)$	$\frac{1}{1 \pm AB}$	$\frac{1}{1 \pm AB}$	$(1 \pm AB)$
Output - Impedance $Z_o / Z_o (A)$	$\frac{1}{1 \pm AB}$	$1 \pm AB$	$\frac{1}{1 \pm AB}$	$(1 \pm AB)$
Configuration	Series – Parallel	Parallel – Series	Parallel – Parallel	Series - Series

Z_i – feedback input-impedance
$Z_i(A)$ – nonfeedback input-impedance
Z_o – feedback output-impedance
$Z_o(A)$ – nonfeedback output-impedance

Table 3.3. Feedback types and impedances. (Source: [325].)

of the amplifier types, voltage current, transfer-admittance, have different input and output impedances Z_i and Z_o. Impedances Z_i and Z_o can arbitrarily be set by varying the open-loop gain AB. Thus, by increasing AB→∞ the input-impedance decreases Z_i→0 in both the current- and transfer-impedance amplifiers. Furthermore, at expanding AB→∞ the output-impedance reduces Z_o→0 in both voltage- and transfer-impedance amplifiers.

All four feedback types have importance in designing sense amplifiers to specific memory cells and architectures. Nevertheless, the implementation of the current amplifier results in such a combination of high performance, small layout area and high memory reliability which is difficult to match by other approaches in random access memory circuits. The circuit implementations of current amplifiers may widely vary, and the following sections discuss the basics of those that have gained applications or have good potentials for future use in sense circuits.

3.4.3 Current-Mirror Sense Amplifier

The traditional form of the primitive current amplifier is the current-mirror amplifier (Figure 3.45). In the current-mirror amplifier [326], if devices M1 and M2 are identical, then the bitline or input current i_i is the same as the readline or output current i_o, because the gate-source voltage V_{GS} is common for both devices M1 and M2. If M1 and M2 differ only in

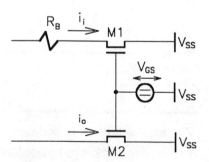

Figure 3.45. Current mirroring and multiplication.

their aspect ratios W/L otherwise they are identical, then the application of MOS current equations yields

$$i_o = \beta_q \frac{(1+\lambda V_{DS2})}{(1+\lambda V_{DS1})} i_i \quad , \quad \text{if} \quad V_{DS1} \neq V_{DS2},$$

$$i_o = \beta_q \frac{\left(\frac{W}{L}\right)_{M2}}{\left(\frac{W}{L}\right)_{M1}} i_i \quad , \quad \text{if} \quad V_{DS1} = V_{DS2},$$

where $\beta_q = \beta_2/\beta_1$, β_1 and β_2 are the respective gain factors of M1 and M2, V_{DS1} and V_{DS2} are the respective drain-source voltages for M1 and M2, and λ is the channel-length modulation factor, $(W/L)_{M1}$ and $(W/L)_{M2}$ are the respective aspect ratios of M1 and M2, W is the channel width, and L is the channel length.

Current mirroring and multiplication by β_q are implemented usually by shorting the drain and gate of device M1 (Figure 3.46), rather than by using an extra bias voltage source V_{GB}. An additional device M3 combines the bitline selection with the current sense circuit. In this circuit, the common gate voltage V_{GS} tends to increase when i_i increases. An increased

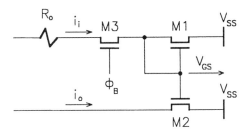

Figure 3.46. Current sensing combined with bitline selection.

V_{GS}, however, reduces the drain-source resistance r_{d1} in M1, and V_{GS} tends to decrease. At little changes in V_{GS}, small variations in current i_i can be detected and amplified. The practical limit of the current amplification A_i is set by the silicon surface area available for the output device M2 which is determined by the column or by the decoder pitch in most of the designs.

Designs for combining high current amplification and small area may apply a small-size linear amplifier A_m between the gates of M1 and M2 (Figure 3.47) or a positive feedback. (Sections 3.4.4, 3.4.6, 3.4.8-3.4.10).

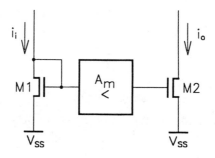

Figure 3.47. Combining high current-amplification and small silicon surface area.

3.4.4 Positive Feedback Current Sense Amplifier

Very small input resistance and some built-in compensation of offsets can be provided by connecting two identical primitive current mirror amplifiers in positive feedback configuration [327] (Figure 3.48) in which the closed loop gain is unity or less. In this configuration when an input voltage Δv_i increases, then the input current i_i rises and the drain-source voltage V_{DS} becomes greater. Since $V_{DS} = V_{GS}$ in MN2, the increased V_{GS} tends to increase the current drive capability of M2 and, thereby, to decrease V_{DS} of M2. The current of MN2 is mirrored to MN3, and the effective gate voltage $|V_G - V_{ref}|$ of MP4 grows. Here, V_G is the gate voltage of MP1 and MP4, and V_{ref} is the reference voltage. Nonetheless, a larger $|V_G - V_{ref}|$ lowers the drain-source resistances r_{ds} of MP4 and MP1, and thus the gain in both $|V_G - V_{ref}|$ and Δv_i get attenuated while a considerable current change occurs. Since the closed loop gain is designed to be less than unity, the circuit is stable. To provide the near unity gain and minimum offsets all four devices MP1, MN2, MN3 and MP4 have the same gain factor β.

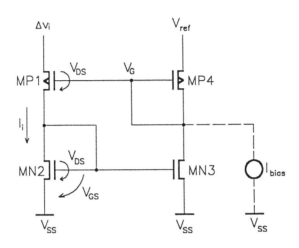

Figure 3.48. Simple positive feedback current sense amplifier. (Source: [327].)

The positive feedback, through devices MP1, MN2, MN3 and MP4, transforms the nonfeedback input impedance $Z_i(A) \approx 1/g_{m1}$ to the feedback-input impedance Z_i;

$$Z_i = \frac{1}{g_{m1}}(1-A).$$

Using the Thevenin equivalent of this circuit the amplification A may be approximated by

$$A \approx \frac{g_{m1}}{g_{m4} + \dfrac{1}{r_{d4}} + \dfrac{1}{r_{d3}}} \cdot \frac{g_{m3}}{g_{m2} + \dfrac{1}{r_{d2}} + \dfrac{1}{r_{d1}}} \leq 1.$$

242 CMOS Memory Circuits

To ensure A<1 over the variation range of transconductances g_m-s and drain-source resistances r_d-s, and to provide initial bias after selection, an additional stabilizer bias current source I_{bias} may be added to the circuit. Here, subscripts 1, 2, 3 and 4 are added to the indices of g_m and r_d to designate transistors MP1, MN2, MN3 and MP4.

The positive feedback in MP1, MN2, MN3 and MN4 results also in an offset compensation effect. Namely, the feedback mechanism keeps not only the memory cell generated input voltage swing Δv_i at very small amplitude, but compensates also the circuit imbalance induced offset voltage V_{off}. Here, V_{off} is the offset voltage without the positive feedback. To demonstrate the effect of positive feedback on V_{off} the total of the

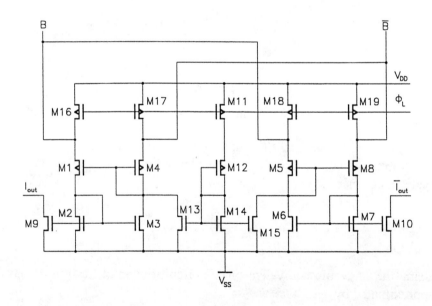

Figure 3.49. A complete positive feedback current sense amplifier circuit.

Sense Amplifiers 243

parameter imbalances may arbitrarily be combined in a single term ΔV_T, and the offset voltage with positive feedback $V_{off}^{(+)}$ can be approximated by

$$V_{off}^{(+)} \approx V_{off} + (1 + \frac{g_{m2}}{g_{m1}} A - \frac{g_{m3}}{g_{m1}} \cdot \frac{1-A}{1 - \frac{g_{m3}}{g_{m2}}}) \Delta V_T .$$

The reduced offset voltage $V_{off}^{(+)} \ll V_{off}$ allows for sensing of smaller input signals imbalances, the sensing can start earlier, and the total sense time becomes shorter.

A complete positive feedback current sense amplifier for static memories (Figure 3.49) includes two positive feedback quads M1-M4 and M5-M8, two output transistors for current multiplication M9 and M10, a current bias circuit M11-M15 and two bitline load pairs M16-M17 and M18-M19. Although this sense amplifier needs 19 transistors, yet, the transistors can be designed to near minimum sizes allowed by the processing technology. Designs with this circuit benefit in short data sense and transmission times, and in insensitivity to a large range of circuit parameter variations.

3.4.5 Current-Voltage Sense Amplifier

At destructive readout a rewrite capability may be obtained by combining a current sense amplifier with a voltage sense amplifier so that the benefits of current sensing may be retained (Figure 3.50). The current-voltage sense amplifier has two terminal pairs; one of both pairs D-$\overline{D}$ is for read data transfer, while the other pair WR-$\overline{WR}$ is for rewrite the sensed voltage or for write a new datum into the accessed memory cell. A datum generated by a memory cell on the bitline pair B-$\overline{B}$ appears as both a current difference $\Delta i = i_B - i_{\overline{B}}$ and as a voltage difference $\Delta v = v_B - v_{\overline{B}}$. When ϕ_{sel} and ϕ'_{sel} activate both sense amplifiers, and when MT1 and MT2 are on, Δi and Δv are sensed and amplified by the respective current and voltage-mode sense amplifier. When the latch in the voltage-mode amplifier takes a stable state, ϕ_{sel} deactivates the current sense amplifier, and on the bitline pair B-$\overline{B}$ the datum appears as a large voltage swing. Consequently, this datum is either rewritten in the accessed memory cell, or replaced by a new datum which may have appeared on the writelines WR and $\overline{\text{WR}}$ when MT1 and MT2 are turned off.

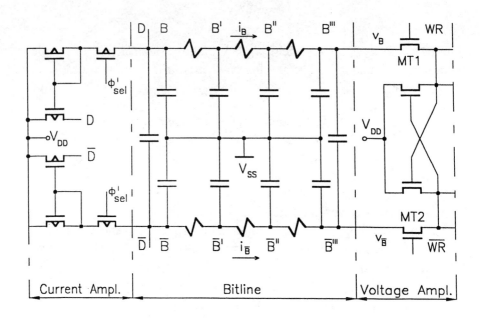

Figure 3.50. A current-voltage sense amplifier distributed on the bitline terminals.

Before sensing of a datum this circuit needs precharge and equalizing, which may be implemented either through MT1 and MT2, or through the current-mode amplifier, or by adding extra precharge devices. The precharge delay is, usually, timed simultaneously with the word access delay, thus, it does not slow the memory operation.

3.4.6 Crosscoupled Positive Feedback Current Sense Amplifier

An elegant implementation of positive feedback and unity gain requires only four equal sized transistors (Figure 3.51). All transistors assumed to be identical, and to operate in their saturation regions, and when clock ϕ_{sel} turns transistors M3 and M4 on, identical input voltages

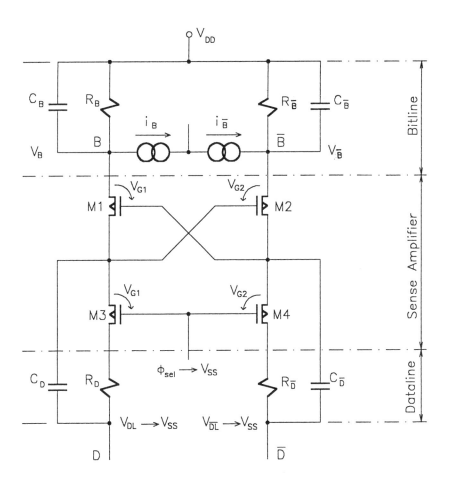

Figure 3.51. Crosscoupled positive feedback current sensing circuit. (After [324].)

$V_B = V_{\bar{B}} = V_{G1} + V_{G2}$ appear on the bitlines B and $\bar{B}$. V_{G1} is the gate-source voltage for both M1 and M3, and V_{G2} is the gate-source voltage for both M2 and M4. The input voltage $V_{G1} + V_{G2}$ changes very little and only for a short transient time, when an accessed memory cell induces a current difference Δi between i_B and $i_{\bar{B}}$. The current difference may change V_{G1} and V_{G2} in opposite directions, but the feedback at unity gain provides a $V_{G1} + V_{G2} \approx$ Constant for small Δi's. Because i_B and $i_{\bar{B}}$ pass through devices M3 and M4, a somewhat reduced Δi appears also on the current transporting data lines D and $\bar{D}$. Thus, a current sensing and current conveyance can be obtained at very little sense voltage variations. The current sensing and conveyance do not require an extra precharge and equalization because the circuit inherently returns to its equilibrium state $V_B = V_{\bar{B}} = V_{G1} + V_{G2}$, where V_B and $V_{\bar{B}}$ are the voltages on the bitlines B and $\bar{B}$.

The analysis of this circuit [324] may be simplified by considering that the identical bitline voltages $V_B = V_{\bar{B}} = V_{G1} + V_{G2}$ cause a virtual short circuit between the inputs. This virtual short makes plausible the appearance of a very low input impedance Z_i (Table 3.3)

$$Z_i = \frac{2(g_{m3,4} - g_{m1,2})}{g_m^2}.$$

Here, g_m is the common transconductance of all devices at the fully balanced ideal state of the circuit, and indices 1, 2, 3 and 4 designate the g_m in transistors M1, M2, M3 and M4. It follows that in the ideal case, when $g_{m1} = g_{m2} = g_{m3} = g_{m4} = g_m$, impedance Z_i approaches zero, and if $g_{m1} = g_{m2} > g_{m3} = g_{m4}$ then Z_i is negative.

A negative Z_i may cause instability. Stability at little current loss can be obtained by keeping Z_i positive and by choosing the bitline load resistance $R_B = R_{\bar{B}} = Z_1/2$. Using the equation of Z_1, the DC stability condition may be expressed as

$$R_B g_m > \left| \frac{g_{m3,4} - g_{m1,2}}{2} \right| ,$$

which indicates the importance of nonzero bitline load resistance.

The presence of bitline capacitance $C_B = C_{\bar{B}}$ and dataline capacitance $C_D = C_{\bar{D}}$ may cause signal ringing in the output current $i_o(t)$ when an accessed memory cell generates a rapid input current charge (Figure 3.52).

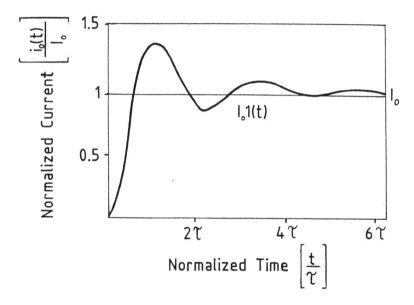

Figure 3.52. Output current signal as a response to an ideal current step on the input.

A simple approximation of the current response $i_o(t)$ to an ideal I_o-amplitude current step-function with $I_o l(t)$ may be obtained by applying the Laplace transformation method, at the assumptions that all parameters are linear, devices M1, M2, M3 and M4 are identical, effects of substrate bias voltages V_{BS}-s are negligible, and the drain-source conductance of the transistor devices $r_d \ll 1/g_m$;

$$i_o(t) = AI_0(1-e^{\frac{t}{\tau}})(\cos \omega t + \frac{1}{\omega \tau}\sin \omega t),$$

where in practice

$$A \doteq 1, \quad \tau = \frac{2C_B}{g_m} \quad \text{and} \quad \omega = g_m \left(\frac{1}{C_B C_D} - \frac{1}{4C_B^2}\right)^{\frac{1}{2}}.$$

The equation of $i_o(t)$ clearly indicates that damped sinusoid current swings can appear on the outputs, the swings have a frequency $f = \omega/2\pi$, and the time constant for the switching τ is smaller than the time constant of the bitline $\tau_B = R_B C_B$. Although a small τ result quick rise and fall times t_r and t_f, but the total sense delay may be long because of the time needed to attenuate the signal swings. The attenuation time T_a may crudely be estimated by $T_a = 5\tau$.

In designs, rather than calculating T_a, the velocity of transient damping v_{td} in the units of N/sec is practical to use. N indicates the time duration necessary to decrease the amplitude of the first overshot I_1 to $I_1/e \approx I_1/2.71$. The velocity of amplitude damping v_{td} can be predetermined by

$$v_{td} = X\frac{N}{\sec} = |\sigma_i|,$$

where $|\sigma_i|$ is the real part of the complex root $\rho_i = \sigma_i + j\omega_i$ of the Laplace transformed transfer function that describes the feedback circuit (Section 3.4.10), and σ_i is a function of g_m, R_B, C_B, R_D and C_D and, in a lesser degree, of other parameters.

3.4.7 Negative Feedback Current Sense Amplifiers

The negative feedback current sense amplifier (Figure 3.53) features small offset, small input resistance and small voltage swings [328]. Because this circuit amplifies an input current difference $\Delta_i = |i_B - i_{\bar{B}}|$ to an output voltage swing $\Delta v_o = |v_{o1} - v_{o2}|$ the circuit, in essence, is a transfer impedance amplifier. For the explanation of the amplifier operation perfect circuit symmetry is assumed prior to the activation of the amplifiers DA_1 and DA_2. Both amplifiers DA_1 and DA_2 are activated simultaneously when the accessed memory cell generates a certain current difference Δi_i. An increasing bitline current i_B through MD1 would increase the drain-source voltage V_{DS1} of MD1, but an increased V_{DS1} creates an increase in the input voltage $v_{i1} = V_{DS1} - V_{ref}$ on the input node B of the differential amplifier DA_1. As DA_1 amplifies this increase in v_{i1}, the gate voltage V_{GS1} of MD1

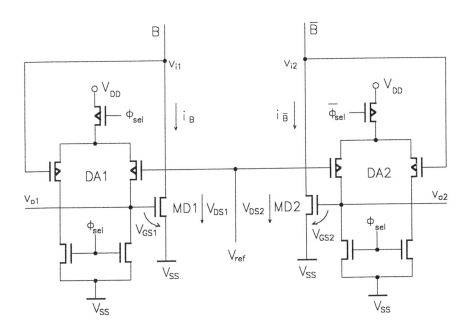

Figure 3.53. A negative feedback current sense amplifier.

increases and, thereby, counteracts the growth in V_{DS1} and v_{i1}. Simultaneously, a decreasing bitline current i_B through MD2 would cause a reduction in input voltage v_{i2} on node B, but this reduction is lessened by the feedback through DA_2. While DA_1 and DA_2 amplifies Δi_i, v_{i1} and v_{i2} changes little, but the output voltage difference Δv_o, which appear on the gates of MD1 and MD2, depends on Δi_i. Assuming that MD1 and MD2 operate in the saturation region and that the saturation currents of MD1 and MD2 are the same, the current balance may be approximated as

$$\frac{\beta}{2}(V_{GS1} - V_{T1})^2 - \Delta i_i = \frac{\beta}{2}(V_{GS2} - V_{T2})^2 ,$$

where V_{T1} and V_{T2} are the threshold voltages of devices MD1 and MD2. From this equation Δv_o is

$$\Delta v_0 = V_{GS2} - V_{GS1} = (2\Delta i_i / \beta)^{\frac{1}{2}} + \Delta V_T \text{ and } \Delta V_T = |V_{T1} - V_{T2}|.$$

The equation demonstrates that the sense amplifier converts Δi_i to Δv_o with a gain of $(2/\beta)^{1/2}$, and because $(2\Delta i_i/\beta)^{1/2} >> \Delta V_T$, the circuit operation suppresses the ΔV_T caused offset.

The effective operation of MD1 and MD2 requires the aid of two amplifiers DA_1 and DA_2. Because the implementation of DA_1 and DA_2 requires compromise in silicon area, this circuit may gain applications in memories where the constraints for sense-amplifier regions are not stringent, e.g., in SRAMs, ROMs and PROMs.

3.4.8 Feedback Transfer Functions

Generally, a feedback sense circuit may be partitioned into a (1) signal generator, (2) measuring element, (3) executor element, (4) error signal former and (5) reference signal generator (Figure 3.54). In this circuit example, the signal generator is an accessed memory cell and the bitline, the measuring element is a voltage divider, the executor element is a one-transistor current amplifier, the error signal former is a differential voltage amplifier, and the reference signal generator is a voltage divider. This separation of constituent elements, and the presumption that in small signal sensing the sense circuit operates as a closed loop linear system, allow for the use of transfer functions [329] in the analysis of feedback current and voltage amplifiers.

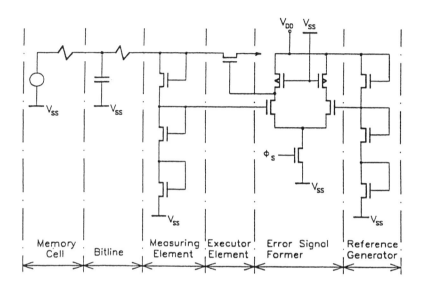

Figure 3.54. Separation of constituent elements in a feedback current sense circuit.

A transfer function Y(p) is the quotient of the Laplace-transformed output signal $S_o(p)$ and input signal $S_i(p)$ at zero initial conditions:

$$Y(p) = \frac{S_o(p)}{S_i(p)}$$

Y(p) may characterize a complete circuit, a subcircuit or a circuit element. If the subcircuits or the circuit elements are separated so that their individual input/output interfaces do not present any load to each other,

Configuration	Transfer Function
Series	$Y_A(p)Y_B(p)$
Parallel	$Y_A(p) + Y_B(p)$
Negative Feedback	$Y_A(p)/[1 + Y_A(p)Y_B(p)]$
Positive Feedback	$Y_A(p)/[1 - Y_A(p)Y_A(p)]$

Table 3.4. Basic configuration and their transfer functions.

the unloaded entities can be represented by blocks. These blocks may be coupled in arbitrary configuration. In linear systems, four basic configurations series, parallel, negative and positive feedbacks, set the basic rules for determining $Y(p)$ of complex networks (Table 3.4).

In the following, $Y(p)$ is applied to demonstrate how the bitline voltage v_B, the output resistance r_o, and the stability of a current sense amplifier are influenced by feedbacks. Nonetheless, in the design of feedback sense circuits $Y(p)$ may also be applied in general mathematical and in specific transient analyses.

3.4.9 Improvements by Feedback

In a parallel-regulated current sense circuit (Figure 3.55a) the effects of positive feedback, on the bitline voltage v_B and on the output resistance r_o, can be shown on the low-frequency equivalent (Figure 3.55b). This low frequency equivalent assumes that all circuit elements are characterizable by linear, time-invariant, concentrated parameters, that the transistors operate in saturation regions, and the capacitances have no influence on the control of v_B levels and of r_o. Although bitline and output capacitances C_B and C_o are determinative in the transient behavior of the bitline signal $v_B(t)$, the effects of feedback on the basic v_B and on its long-term changes Δv_B, and on the basic r_o, can be made plausible on a low-frequency model that disregards all capacitances in the circuit.

Sense Amplifiers 253

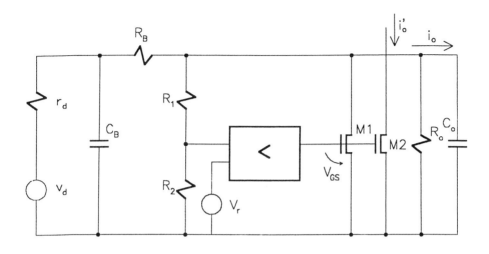

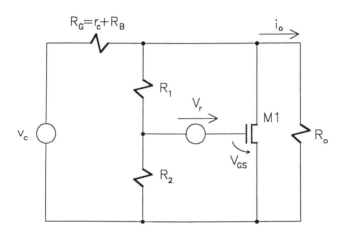

Figure 3.55. A parallel-regulated feedback sense amplifier (a) and its simplified low-frequency equivalent (b).

In the low-frequency equivalent circuit, transfer functions $Y(p)$, $Y_A(p)$ and $Y_B(p)$ can be simplified to time-independent terms Y, Y_A and

Y_B, and by using the block representation of the circuit (Figure 3.56) the transfer function of the complete circuit can be expressed as

$$Y = Y_A Y_B / (1 + K Y_A Y_B) \ .$$

Assuming that voltage divider draws negligible current, the component functions of Y are $Y_A = g_m R_G$, $Y_B = R_o R_G/(R_o + R_G)$, and $K = R_2/(R_1 + R_2)$. Here, g_m is the transconductance of M1, R_G is the equivalent generator resistance coupled to the bitline, R_o is the equivalent load resistance on the sense amplifier output, R_1 and R_2 are the resistances in the voltage divider.

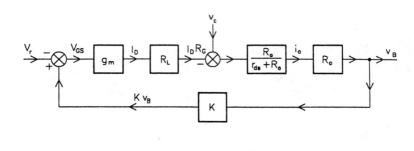

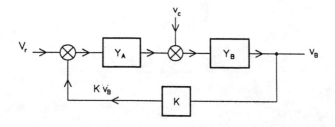

Figure 3.56. Block diagrams of the parallel-regulated feedback sense amplifier.

The bitline voltage v_B as a function of the generator voltage v_C and reference voltage V_R can be obtained from the equations of Y, Y_A, Y_B and K as

$$v_B = \frac{g_m R_L V_r + v_c}{1 + K g_m R_L + \frac{R_C}{R_o}},$$

A partial differentiation of v_B by v_c gives the bitline voltage change Δv_B as a function of the generator voltage change Δv_c;

$$\Delta v_B = \frac{\partial v_B}{\partial v_C} \Delta v_C \approx \frac{1}{1 + K g_m R_L} \Delta v_c .$$

Similarly, the partial differentiation of v_B by the output current i_o provides the output resistance r_o as

$$\Delta r_l = \frac{\partial v_B}{\partial i_o} \Delta i_o \approx \frac{R_L}{1 + K g_m R_L} \Delta i_o .$$

The equation of Δv_B and r_o indicate that both the bitline voltage variation and the output resistance of a sense amplifier can significantly be reduced by the application of a feedback.

By the implementation of an additional or double feedback (Figure 3.57) with a gain of F, parameters Δv_B and r_o may further be improved:

$$\Delta v_B = \frac{1}{1 + D K g_m R_L} \Delta v_L \quad \text{and} \quad \Delta r_i = \frac{R_L}{1 + D K g_m R_L} \Delta i_o .$$

Small bitline voltage variations Δv_B-s increase the sensitivity of the sense amplifier and allow for early signal detections. Small sense amplifier output resistances r_o-s decrease the switching times. Consequently, memory access and cycle times can greatly be improved by the use of feedbacks in sense circuits. Feedbacks may be implemented in a great variety of parallel and series configurations, but their use in sense circuit is limited by size and stability considerations.

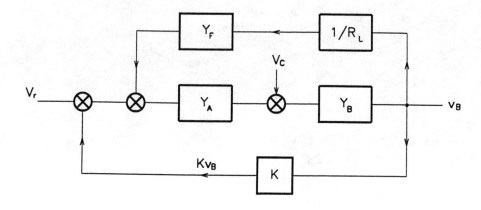

Figure 3.57. Double feedback.

3.4.10 Stability and Transient Damping

The application of feedback in a closed loop sense circuit raises the question of stability. A closed loop electric circuit is stable, if it is able to reestablish its original equilibrium state by itself after a single event signal disturbed its equilibrium. In a sense circuit, the single event may be a signal that is generated by the accessed memory cell or a noise signal that may be coupled into the circuit through capacitances.

When the sense circuit is described by transfer functions, the criterion of stability is that all real parts Δ_i-s of the roots $\rho_i = \sigma_i + j\omega_i$; of the equation $1+Y_A(p) Y_B(p)) = 0$ must be positioned left from the $j\omega$ axis on the complex plane (Figure 3.58), i.e., $\text{Re}\rho_i = \sigma_i < 0$. Here, $Y_A(p)$ and $Y_B(p)$ are the transfer functions of the constituent subcircuits. For certain sense circuits, an extended stability criterion $\sigma_i < \Delta\sigma$, where $\Delta\sigma$ is a safety amount that may be imposed to allow time for the attenuation of eventual signal ringing (Section 3.4.6).

Stability conditions and signal ringings in sense amplifiers can also be investigated, of course, with other well known methods including Ruth-Hurwitz, Mihailov, Nyquist, Bode, Kupfmuller, etc. criteria [330].

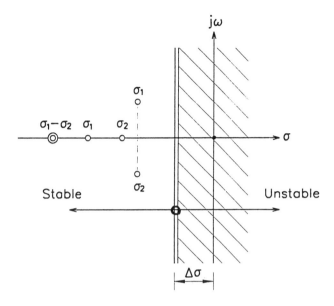

Figure 3.58. Root locations at stable operation on the complex plane. (Source: [329].)

3.5 OFFSET REDUCTION

3.5.1 Offsets in Sense Amplifiers

Offset is particular and inherent to differential sense amplifiers, and it is the voltage or current difference which appears between the two output node potentials or between the two output currents, when an identical input voltage or bias current is applied to the two inputs. The offset voltage or current has to be counteracted by the memory-cell-generated signal for correct sense operations. Theoretically, differential sense amplifiers are electrically balanced symmetrical circuits. In practical implementations both the transistors and the passive elements have slight parameter differences in spite of the utmost design efforts to assure their symmetricity. These parameter differences, and the resulting sense amplifier offsets, are distributed spatially throughout the chips, wafers and lots (Figure 3.59), and the signal generated by a memory cell has to act against and neutralize the appearing maximum offset before the sensing of a data signal could start. Thus, the offset limits the sensitivity, i.e., the minimum data signal amplitude that the circuit can detect, and it delays the effective start of data sensing. To improve both sensitivity and sensing speed the offsets should be kept small by minimizing the imbalances between the halves of a differential sense circuit.

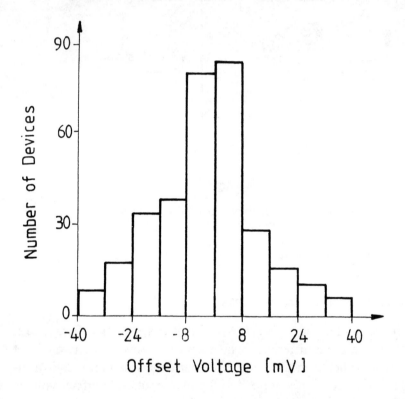

Figure 3.59. A distribution of offsets.

Imbalances may result from the effects of semiconductor fabrication, voltage and current biases, temperature changes, radioactive irradiations and others, and occur as nonuniform variations in threshold voltages ΔV_T, gain factors $\Delta \beta$, leakage currents ΔI_L, load resistances ΔR_L, load capacitances ΔC_L, transistor inherent gate-drain, gate-source and drain-source capacitances ΔC_{GD1}, ΔC_{GS}, and ΔC_{DS}, as well as in a variety of other design parameters. A great deal of reduction in parameter variations can be obtained by improvements (1) in processing (e.g., increasing the accuracy of mask alignments, ion implantation dose, plasma etching, ion millings, diffusion control, annealing, etc.), (2) in transistor device and interconnect designs (e.g., using environmentally insensitive materials, stable oxide-semiconductor and oxide-polysilicon interfaces, etc.) and (3) in starting material (e.g., eliminating nonuniformities in silicon crystals, avoiding localized damages caused by cleaning and polishing, etc.). Despite immense improvements in CMOS processing, integrated active and passive device, and material technologies, the down scaling of feature sizes increases the ratio of the offsets to the data signal amplitudes which can be generated by a memory cell in a sense circuit.

Circuit designs can greatly reduce the offsets by (1) misalignment tolerant layouts, (2) adding offset compensatory circuit elements to the sense amplifier, and by (3) choosing circuits which have inherent offset compensation. Because the previous discussion of individual sense amplifier circuits describes also the circuit's inherent offset reduction capabilities, where such capabilities exist, the following sections present those approaches which use layout and added circuit elements to offset control.

3.5.2 Offset Reducing Layout Designs

Misalignment tolerant layouts may be designed by dividing the component elements of the sense circuit into subelements (Figure 3.60), and place the subelements as diagonal pairs with reversed drain and source electrodes [331] around a common center point. The division to subelements and the common centroid geometry statistically average and partly compensate parameter variations caused by mask misalignments and nonuniformities in the transistor pairs. Wide or long transistors laid out in L shapes provide also some tolerance against mask misalignments and gain applications in buffer amplifiers.

3.5.3 Negative Feedback for Offset Decrease

Sense amplifier layouts may use existing parasitic elements, e.g., wire resistances, R_W and $R_{\overline{W}}$ for offset reduction (Figure 3.61) to implement

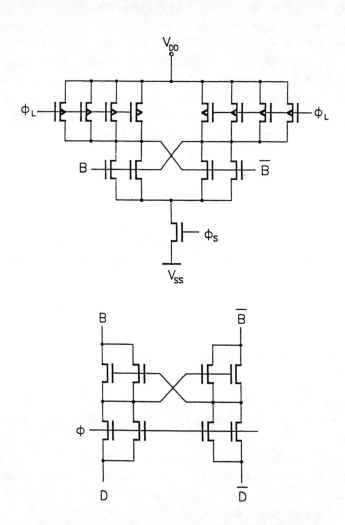

Figure 3.60. Division of driver and load transistors.

Sense Amplifiers

negative feedback. Resistors R_W and $R_{\overline{W}}$ are nearly identical, vary uniformly with environmental changes, and alter independently from transistor parameters such as threshold voltage V_T, mobility μ, oxide thickness t_{ox}, channel width W and length L, in symmetrical layout designs. A layout design that places the bitlines between the input device pair M1-M2 and the source device M3, results in individual negative feedbacks for devices M1 and M2. The feedback modifies the original trans-conductance g_m to a feedback transconductance g'_m as

$$g'_m = \frac{g_m}{1+Rg_m},$$

where $R = R_W = R_{\overline{W}}$ is assumed. If $Rg_m \gg 1$ then $g'_m \approx 1/R$, thereby making g'_m in the sense amplifier nearly independent from the transistor parameter variations.

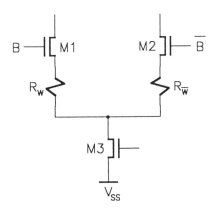

Figure 3.61. Offset compensation by resistances.

Feedback resistance R may also be implemented in forms of polysilicon, implanted, diffusion or junction resistors, or in forms of MOS devices which operate as resistors [319] (Figure 3.62). MOS resistors change less variably than active MOS transistors do, because during circuit operation the bias voltages of MOS resistors change more evenly and, in turn, their threshold voltages and carrier mobilities alter more uniformly than those of active transistors. The percentage of nonuniform variations in g_m may be calculated by

$$g_m(\%) = \frac{\hat{g}_m - \check{g}_m}{\hat{g}_m} 100 = \frac{1}{1 + Rg_m} \cdot \frac{\hat{g}_m - \check{g}_m}{\check{g}_m} 100 \;,$$

where, $\hat{g}_m$ and $\check{g}_m$ are the maximum and minimum transconductances, $R = r_d$ for MOS resistors, and r_d is the dynamic drain-source resistance of the MOS device.

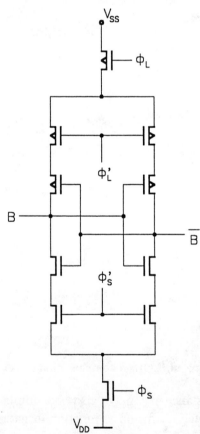

Figure 3.62. Negative feedback implemented by serial active devices. (Source [319].)

Other important reasons to implement feedback resistances are to decrease the parameter dependent gain variations and to stabilize amplification. Assuming that a gain for a simple voltage sense amplifier without feedback is $A \approx g_m r_d$, then the gain with negative feedback may be approached by $A \approx r_d/R$. This expression of A indicates that R decreases the amplification, but the variations in R are much less than in g_m and, thus, R can significantly reduce the changes in A also.

3.5.4 Sample-and-Feedback Offset Limitation

Sample-and-feedback circuit elements are applied to compensate large offsets in sense amplifiers. Tolerance for excessive offsets may be required in memories operating in extreme environments, e.g., radioactive radiation, very high temperature, etc., or for providing high yield when processing parameters vary greatly.

Configurations of sample and feedback circuits may vary, but their operation is based on a common principle [332]. Namely, during initiation a sample is taken from the output nodes (Figure 3.63) by turning feedback

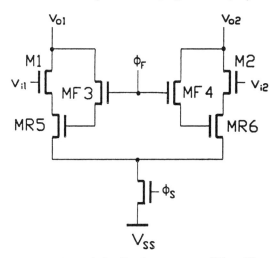

Figure 3.63. A sample-and-feedback sense amplifier. (Source: [332].)

devices MF3 and MF4 on for a very short period of time. When MF3 and MF4 are turned off, the samples are stored on the parasitic capacitances which are present in the gates of regulator devices MR5 and MR6. If the

264 CMOS Memory Circuits

output voltages v_{o1} and v_{o2} are unequal, e.g., $v_{o1} > v_{o2}$, the drain source resistance of MR5 r_{d5} becomes smaller than that of MR6 r_{d6} i.e., $r_{d6} > r_{d5}$, and this tends to equalize v_{o1} and v_{o2} as well as currents i_1 and i_2 in the sample-and-feedback differential sense amplifier circuits.

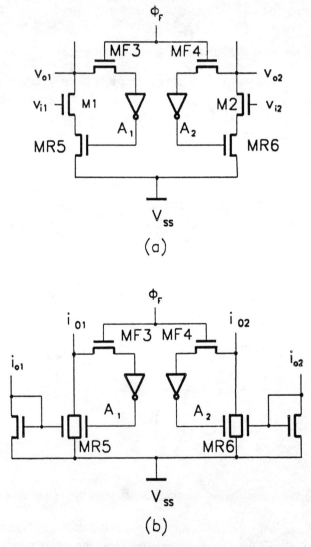

Figure 3.64. Sample-and-feedback with offset-amplification in a voltage (a) and in a current (b) sense amplifier.

In this amplifier the nonuniformities of feedback devices MF3 and MF4 are ineffective, because the amplitude V_1 of clock signal ϕ_F exceeds both v_{01} and v_{02} by more than their bias V_{BG} dependent threshold voltage $V_T(V_{BG})$. Thus, through MF3 and MF4 no voltage drop can occur, and after a transient time v_{01} and v_{02} occurs also on the gates of MR5 and MR6. Although some additional voltage and current differences may be introduced by MF3 and MF4, but the feedback reduces this small additional offset term together with all other offset causing issues.

The efficiency of offset reduction may be increased by inserting linear amplifiers A_1 and A_2 (Figure 3.64) into the sample and feedback loop. Because the sample is taken in a much shorter time interval than the transient time of the feedback operation, the circuit is stable even in designs which may violate the stability criteria of continuous feedback systems (Section 3.4.10). Thus, stability considerations in sample and feedback circuits do not compromise designs for high amplification and speed. In random access memories, the time required for offset compensation is simultaneous with the decoding delay for word access, and does not occur as an additional delay component. Offset compensating may well shorten sense delay times by making possible to detect smaller signals at shorter signal development times. Yet, offset compensation introduces extra transistors in which should be applied only upon careful evaluation of the tradeoffs between reduced offsets and increases in power dissipation and layout area.

3.6 NONDIFFERENTIAL SENSE AMPLIFIERS

3.6.1 Basics

Nondifferential sense amplifiers are those nonsymmetrical circuits which detect and amplify signals which are generated by an accessed memory cell on a single amplifier input node. Topologically, nondifferential amplifiers can not be divided into two mirror-image parts, and their operations and designs are not restricted by offset considerations. Although in a number of applications nondifferential amplifiers have demonstrated access and cycle times which are competitive with those provided by differential amplifiers, yet, the inherent advantage of

266   CMOS Memory Circuits

differential sensing in sensitivity and noise immunity, leaves only a small abating segment for nondifferential data sensing.

Historically, nondifferential sense amplifiers have been used to detect and amplify signals provided by nonvolatile memory cells, and to compensate charge sharing effects by preamplification in random access memories. In future CMOS memories, nondifferential amplification may extensively be applied for impedance transformations, and for signal sensing on long interconnect lines.

Most of the nondifferential sense amplifiers are adopted from analog circuit techniques. Thus, they may be categorized and analyzed as common source, common gate, common drain and combination sense amplifiers [333].

3.6.2 Common-Source Sense Amplifiers

In memories, the basic common-source sense amplifier (Figure 3.65) is usually precharged before its active operation starts on its input, or on both its input and output, by a precharge voltage V_{PR}. Before the start of the operation, V_{PR} provides an input voltage $v_i = V_{PR}$ that places the quiescent

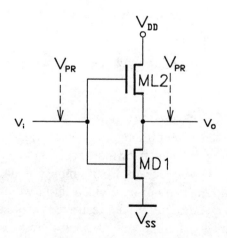

Figure 3.65. Basic common-source sense amplifier.

operation point Q near to the lower knee or to the center of the linear part of the circuit's input-output voltage transfer characteristics, where both devices, the n-channel MD1 and the p-channel ML2, operate in their saturation regions (Section 3.2.1). During a sense operation, the output voltage v_o increases when v_i decreases. At a $v_i = V_{PR} - \Delta v_i + V_N$, where Δv_i is the change in v_i and V_N is the cumulative noise voltage, a log.0 can be detected. A detection of a v_i-decrease, i.e., a discharge of the load capacitance $C_L(t)$, is preferred, because the discharge through an n-channel transistor is faster than the charge through a p-channel transistor, if their sizes are the same and their channel length L>0.12μm.

As long as both the n-channel MD1 and the p-channel ML2 operate in their saturation zones the voltage amplification A_v is high. A_v may be obtained by using the linear small-signal low-frequency model of this amplifier as

$$A_v = g_{m1}(r_{d1}\|r_{d2}),$$

where g_{m1} is the transconductance of MD1, r_{d1} and r_{d2} are the drain-source resistances of MD1 and ML2 respectively. Resistance combination $r_{d1}\|r_{d2}$ and a constant K determine a rather low initial output impedance $Z_o(O) \approx K(r_{d1}\|r_{d2})$. The input-impedance Z_i is very high $Z_i = v_i/I_L$, because the combined amount of the leakage currents I_L appearing on the input node is very little in small sized sense amplifier devices.

The output-impedance $Z_o(t)$ and the load capacitance $C_L(t)$ produces a $\tau = Z_o(t)C_L(t)$ which may be approximated as $\tau_f \approx r_{d1}C_L$ for discharge and $\tau_r \approx r_{d2}C_L$ for charge of C_L. The time independent parameters r_{d1}, r_{d2} and C_L may also be used in the operator impedances:

$$Z(p) = r_{d1} + \frac{1}{pC_L} \quad \text{and} \quad Z(p) = r_{d2} + \frac{1}{pC_L}.$$

Since the operator impedances used here and in the previous transient analysis of the bisected simple differential voltage sense amplifier (Section 3.3.2.3) are similar, the approximate results obtained for switching times and delays can be applied also to this common source sense amplifier.

A common-source sense amplifier (Figure 3.66), that is used mostly in read-only-memories, includes a drive transistor MD1, a current reference device MR2 and current source load devices ML3, ML4. The reference current I_{ref} establishes a gate voltage V_{GS3} on ML3 and V_{GS4} on ML4. Due to $V_{GS4} = V_{GS3}$, device ML3 operates as a nearly constant current source that provides a very high virtual load resistance for the driver MD1. High load resistance results in great voltage amplification A_v when MD1 operates in the saturation region. Assuming that MD1, ML3 and ML4 function in their saturation regions, and that the drain-source resistance r_{d2}

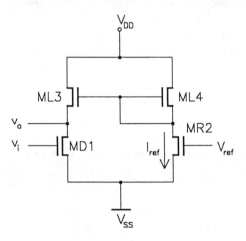

Figure 3.66. Common source sense amplifier used in read-only memories.

of MR2 is linear, and exploiting that the drain currents of MD1 and ML3 are the same, the output voltage v_o as a function of the input voltage v_i may be estimated by the equation

$$v_o = \frac{\beta_{3,4}(V_{GS3} - |V_{TP}|)^2 (1 + \lambda_3 V_{DD}) - \beta_1 (v_i - V_{TN})^2}{\beta_{3,4}(V_{GS3} - |V_{TP}|)^2 \lambda_3 + \beta_1 (v_i - V_{TN})^2 \lambda_1} ;$$

where $\beta_1 = \beta_2$ and $\beta_3 = \beta_4$ for most of the designs, and parameters β and λ are the gain and the channel length modulation factors, V_{GS}, V_{TP} and V_{TN} are the gate-source, p-channel and n-channel threshold voltages, and subscripts 1, 2, 3 and 4 indicate devices MD1, MR2, ML3 and ML4.

Another four-device amplifier (Figure 3.67) uses positive feedback and self-bias to obtain shorter sensing delays. Prior to a sense operation, control signal ϕ turns ML4 on, and the feedback through MD2 equalizes the input voltage v_i and output voltage v_o, so that $v_i = v_o$. When a decreasing v_i is generated by a memory cell, ϕ turns ML4 off, and v_o falls rapidly due to the positive feedback effect provided by the operation of MD2. The sizes of MD2 and ML4 depend on the speed and load requirements, and greatly effect the sizes of MD1 and MD2. Nevertheless, the effective total size of this amplifier is usually small, because it does not need extra precharge circuit and because its wiring is simple.

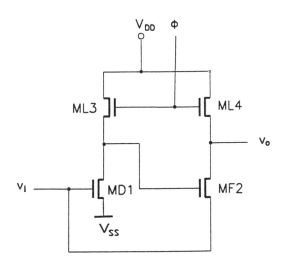

Figure 3.67. Positive feedback and biasing in a nondifferential sense amplifier.

3.6.3 Common-Gate Sense Amplifiers

Common gate amplifiers are applied in random access memories as preamplifiers or, in more complex implementations, as current sense amplifiers (Section 6.3). The equivalent of a presense circuit that comprises a one-transistor common-gate amplifier circuit (Figure 3.68)

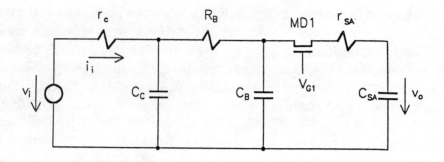

Figure 3.68. Presense circuit equivalent with a one-transistor common-gate amplifier.

includes an input signal generator providing voltage v_i, the output impedance of the accessed memory cell $Z_c = r_c + 1/j\omega C_C$, the bitline impedance $Z_B = R_B + 1/j\omega C_B$, an amplifier device MD1, and the input impedance of the sense amplifier $Z_{SAi} = r_{SA} + 1/j\omega C_{SA}$. Here, r_C, R_B, r_{SA} and C_C, C_B, C_{SA} are the equivalent resistances and capacitances of the output of the accessed memory cell, bitline, and the input of the active main sense amplifier. The common-gate amplifier has a small input impedance Z_i, that can be calculated by using the low-frequency small-signal model for transistor MD1 in the equivalent presense circuit, and the analysis of this circuit yields

$$Z_i = \frac{r_{SA} + R_s + r_{d1} g_{m1} R_s + r_{d1}}{1 + g_m r_{d1}},$$

where $R_s = R_B + r_C$, and r_{d1} and g_{m1} are the drain-source resistance and transconductance of transistor MD1. Active device MD1 provides also a considerable voltage gain A_V as indicated by

$$A_v = \frac{r_{SA}(1 + g_{m1} r_{d1})}{r_{SA} + r_{d1} + (1 + g_m r_{d1}) R_s}.$$

In the equations of Z_i and A_v the assumption is applied that MD1 is biased to operate in the saturation region. The initial bias is imposed by the precharge voltage V_{PR}. A mid-level V_{PR} may be generated by charge redistribution on a dummy cell and terminated bitline (Section 4.2.4), and a mid-level V_{PR} provides good amplification for both log.0 and log.1 data. If the detection of log. 0 is sufficient, then high-level V_{PR} can be used in this circuit.

The analysis of the common-gate preamplifier circuit may aid the design of a charge-transfer preamplifier (Figure 3.69). Although the charge-transfer and the one-transistor common-gate preamplifiers are similar in configurations, they differ in operation concepts. In charge-transfer operation [334], initially, the input capacitance of the sense amp-

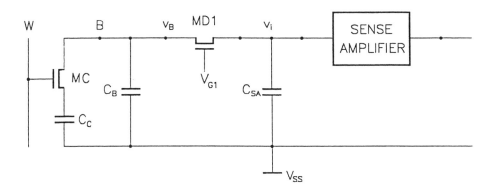

Figure 3.69. Charge transfer preamplifier.

lifier C_{SA} and the bitline capacitance C_B are precharged. C_B is much larger than the input sense amplifier's input capacitance C_{SA} and the cell capacitance C_C. At t_o time, capacitance C_B is brought through the device MD1 to the bitline voltage $v_B(t_o) = V_{G1} - V_T(V_{BG})$, where V_{G1}, V_T and V_{BG} are the gate, threshold and backgate-bias voltages of MD1. When MD1 turns off, the accessed dynamic memory cell generates a voltage charge, and the potentials of the cell capacitor C_C and the bitline capacitance C_B

evens quickly, because $C_C \ll C_B$. At this t_1 moment, the bitline voltage $v_B(t_1)$ is

$$v_B(t_1) = \frac{C_B}{C_B + C_C}[V_{G1} - V_T(V_{BG})] < V_{G1} - V_T(V_{BG}) ,$$

After t_1, device MD1 turns on, and a current begins to flow from C_{SA} to C_B until time t_2. At t_2, C_B is charged back to the bitline voltage

$$v_B(t_2) = V_{G1} - V_T(V_{BG}) ,$$

and the input voltage of the charge transfer preamplifiers $v_i(t)$ decreases by $\Delta v_i = v_i(t_o) - v_i(t_{o1})$, because at perfect charge distribution the total amount of the charge remain the same

$$C_{SA} v_i(t_0) + (C_C + C_B) v_B(t_0) = C_{SA} v_i(t) + (C_C + C_B)(t_2) .$$

Substituting $v_B(t_o)$ and $v_B(t_2)$ into this charge-equivalence expression, Δv_i appears to be independent of C_B

$$\Delta v_i = v_i(t_2) - v_i(t_0) = \frac{C_C}{C_{SA}}[V_{G1} - V_T(V_{BG})] .$$

Here, an amplification of the bitline voltage charge occurs during the time $t_2 - t_o$, because $v_B(t_1)$ is charged back to $v_B(t_o)$ and net charge-change in C_B is zero.

The voltage change may, however, be slow because the time constant of the bitline circuit τ_B is increased by the drain-source resistance of MD1 to an estimated

$$\tau_B \approx \frac{2(C_C + C_B)^2}{C_C \beta [V_{G1} - V_T(V_{BG})]} .$$

Although τ_B can be decreased by making β larger, but a large β requires large area and that increases the parasitic capacitances in the circuit.

Moreover, the presence of charge amplifiers in the sense circuit can also increase imbalances, magnify offsets and, thereby, reduce sensitivity and operational speed.

To improve both sensitivity and speed in charge transfer amplifiers, positive feedbacks may be applied (Section 3.3.6.2).

3.6.4 Common-Drain Sense Amplifiers

A common-drain or source follower sense amplifiers (Figure 3.70) is usually applied as impedance transformers to minimize signal reflections and as level shifters to adjust signal levels among the operation margins in sense circuits.

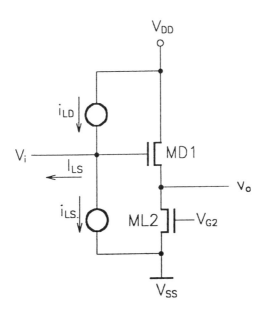

Figure 3.70. Common drain amplifier.

The circuit's low-frequency input-impedance Z_i is very high, and it is determined by the leakage currents between the gate and the ground nodes i_{LS} and between the gate and the supply nodes i_{LD}, and by the input voltage v_i;

$$Z_i = \frac{v_i}{I_L} \quad \text{and} \quad I_L = i_{LD} - i_{LS} \ .$$

The output-impedance Z_o may be obtained from the low-frequency small-signal Thevenin equivalent of the common-drain amplifier as

$$Z_o = r_{d1} \parallel r_{d2} \parallel \frac{1}{g_{m1}} \ ,$$

where r_d and g_m are the drain-source resistance and transconductance of transistors, and subscripts 1 and 2 indicate devices MD1 and ML2. During a sense operation device MD1 operates in the saturation region while ML2 acts as a resistor r_{d2}, and the voltage gain A_v is less than unit

$$A_v = g_{m1} Z_o < 1 \ .$$

To estimate the circuit's transient switching times, the operator impedance $Z(p)$ and the Laplace-transform method may be used so that

$$Z(p) = \frac{p r_{d1} r_{d2} C_L + r_{d1} + r_{d2}}{p r_{d1} C_L + 1} \ ,$$

and the reversed Laplace transform gives the output current $i_o(t)$ and voltage $v_o(t)$

$$i(t) = \frac{V_1}{r_{d2} e^{-\frac{t}{\tau}}} \quad \text{and} \quad v_o(t) = V_1 (1 - e^{-\frac{t}{\tau}}) \ ,$$

where V_1 is the amplitude of an ideal voltage step $V_1 1(t)$ that is applied as the input signal $v_I = v_i(t)$ to the source follower. From $i_o(t)$ and $v_o(t)$ the signal fall, rise and propagation delay times, t_f, t_r and t_p, may crudely be approximated. By applying common-drain amplifiers, short delay times on the bitlines and wordlines can be obtained, because the choice of an r_{d2} that matches the characteristic impedance of the driven transmission line Z_o, i.e., $Z_o \approx r_{d2}$, can prevent signal-reflections (Section 4.1).

4

Memory Constituent Subcircuits

Subcircuits of memories, apart from the memory cells and sense amplifiers, are similar to those component circuits which are used in traditional digital and analog circuits. State-of-the-art requirements in combining very high circuit performances and packing densities, nevertheless, place the constituent subcircuits of CMOS memories in the forefront of the progress. For CMOS memory designs and analyses, this chapter provides a unique insight to the quasi-stationary transmission-line-like behavior of the array wires, to the memory specific aspects of the peripheral circuits, and to the reductions of the harmful signal reflections, and distortions, reference- and timing-inaccuracies and power-line bounces.

4.1 Array Wiring

4.2 Reference Circuits

4.3 Decoders

4.4 Output Buffers

4.5 Input Receivers

4.6 Clock Circuits

4.7 Power Lines

4.1 ARRAY WIRING

4.1.1 Bitlines

4.1.1.1 Simple Models

A memory bitline or dataline unites the write-read nodes or the read-only nodes of an arbitrary number of memory cells, and ties them to a sense amplifier and to a write amplifier, or to a combined sense-write amplifier and to precharge devices (Section 3.1.1). At a data-sense or read operation, an accessed memory cell generates a signal on a bitline (Figure 4.1a). The accessed cell may be represented by a voltage or a current

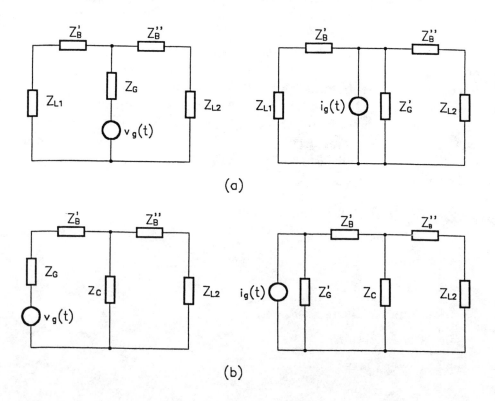

Figure 4.1 Impedance equivalents at read operation (a) and at write operation (b).

signal generator $v_g(t)$ or $i_g(t)$ and by a generator impedance Z_G or Z'_G. Impedance Z_{L1} that combines the impedances of the decoupler device, sense and write amplifiers and precharge devices terminates one end of the bitline, and its other end is closed by a design- and operation-dependent impedance Z_{L2}. At a write operation, the write amplifier generates a signal on a bitline (Figure 4.1b). The write amplifier may be modeled by generators $v_g(t)$ or $i_g(t)$ and by an impedance Z_G or Z'_G. Z_G or Z'_G is combined of the impedances of the decoupler device, write and sense amplifiers and precharge devices, Z_C is imposed by the accessed memory cell, and Z_{L2} is the same as that for read operation. In both read and write models, the bitline impedance Z_B represents the impedances of the unaccessed memory cells and the bitline, and Z_B is divided into both parts Z'_B and Z''_B by the physical location of the accessed memory cell.

The impedance of a bitline Z_B that connects n write-read nodes of access devices to memory cells, comprises (Figure 4.10) the distributed

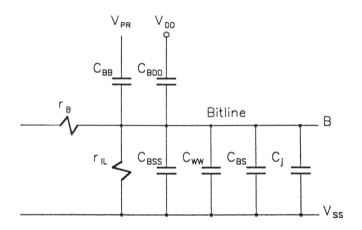

Figure 4.2. Components of a bitline impedance.

resistances of the bitline $R_B \approx n \times r_B$ and those of the leakage current paths $R_{IL} \approx n \times r_{IL}$; the distributed capacitances of the bitline to other bitlines $C_{BB} \approx n \times c_{BB}$, to the ground $C_{BSS} \approx n \times c_{BSS}$ and to the powerlines $C_{BDD} \approx n \times c_{BDD}$; the capacitances of the bitline to crossing wordlines

$C_{WW} = n \times c_{WW}$, to the gates $C_{BG} = (n-1) \times c_{BG}$ and to the sources of the turned-off access transistors $C_{BS} = (n-1) \times c_{BS}$; and the p-n junction capacitances coupled to the bitline $Cj \approx (n+k) \times c_j$. Here, r and c represent specific resistances and capacitances referred to a unit bitline length, and $k = 5...12$.

In practical implementations the bitline with the unselected access devices of the memory cells and with the parasitic elements may be modelled by lumped circuit elements or by transmission lines [41] (Section 4.1.3). Whether a lumped-element or a transmission-line model should be used for analysis, depends on the ratio of the rise-time t_r and fall-time t_f to the propagation delay t_p of an impulse which is generated on the bitline-input and which propagates through the bitline. A rule of thumb for the choice of bitline model is that a transmission-line model should be used if either one of both ratios $t_r/t_p < 2.5$ or $t_f/t_p < 2.5$, and lumped models approach the accuracy of transmission-line models if both $t_r/t_p > 5$ and $t_f/t_p > 5$.

For first order approximations $t_r = t_f = 2.2\ R_B\ C_{BG}$ and $t_p = l/v$ may be used. Here, l is the length of the bitline, $v = c/\sqrt{\varepsilon_r}$ is the propagation velocity of the signal in the transmission line, c is the speed of the light, and ε_r is the permittivity. $\varepsilon_r \approx 3.9$ for the silicon-dioxide dielectric material that is commonly used to isolate the bitlines from the other semiconductor, metal and polysilicon materials.

Bitline models which use Π or T type of passive RC networks in ladder configurations (Figure 4.3), result increased accuracy approximations in computing the switching times t_r and t_f. In these approximations, bitline resistance R_B and bitline capacitance C_B include all bitline resistances and capacitances, associated with the bitline (Figure 4.2), and N is the number of Π or T circuit elements used in the model. By application of a ladder made of Π or T circuit elements in a circuit analysis programs, at N=3 the relative error for the computation of transient responses can be kept less than 1.1%. Similarly, small error may appear with the use of the pocket-calculator estimate [42] for t_r and t_f if open circuit termination $Z_L \to \infty$ can be assumed;

$$t_r = t_f \approx t_{0.9} \approx 1.02\ R_B\ C_B + 2.2\ (r_g\ c_g + c_g\ R_B + r_g C_B),$$

where $t_{0.9}$ is the time to reach the 90% of the switched impulse's amplitude from the time of zero amplitude, r_g and c_g are the equivalent generator resistance and capacitance of the accessed memory cell or, alternatively, of the write-signal generator. Because $R_B \ll r_g$ and $c_g \ll C_B$, the reduction of the bitline capacitance C_B and of the generator resistance r_g are the effective methods to improve switching times t_r and t_f.

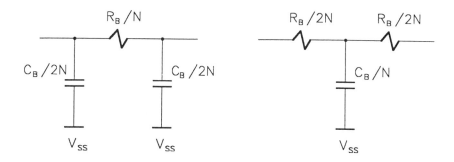

Figure 4.3. Π and T type of RC networks in bitline models.

Transmission-line models (Section 4.1.3) allow to predict switching characteristics of bitline signals as function of both the location on the bitline (x) and the time of observation (t), and to investigate the effects of signal-reflections caused by the impedances Z_G and Z_L which terminate the bitline. Signals generated or received on the bitline by different memory cells, travel different lengths x_1 and x_2 in different times t_1 and t_2, and the signals get reflected differently on $Z_{L1}(R_{L1},C_{L1})$, $Z_{L2}(R_{L2},C_{L2})$ and $Z_G(R_G,Z_G)$ (Figure 4.4). Although approximations as $R_{L1} \to \infty$ and $R_{L2} \to \infty$ may be justified; the computations of wave forms, which appear at various bitline locations and at diverse time points, are arduous. In a strict sense, the modeling of bitlines by transmission lines is also a simplification, because actually electro-magnetic waves are generated in the dielectric material placed under the bitline. Nevertheless, the thickness of the dielectric material t_{ox} is very small in comparison to the length of the bitline l, i.e., $t_{ox} \ll l/100$, which well warrants the application of the transmission-line theory.

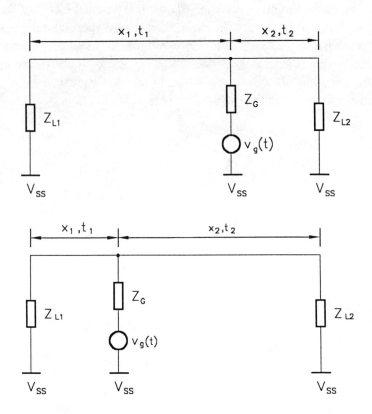

Figure 4.4. Different lengths of signal travels in a bitline.

A sense circuit design that has to take the transmission-line characteristics of the bitline into consideration, should minimize signal reflections. To minimize effects of the reflected waves on the data signal amplitudes, either a wave-impedance termination of the bitline, or an amplitude clamping, or a timed coupling and decoupling of the sense amplifiers to and from the bitline, may be applied.

A termination of the bitline by a load impedance Z_L that is equal with the bitline's characteristic or wave impedance Z_0 results in a zero voltage reflection coefficient $\rho_r = 0$, because $\rho_v = (Z_L - Z_0)/(Z_L + Z_0)$. $Z_L = Z_0$ can be designed by connecting passive elements parallel Z_P and series Z_S with the input impedance of the sense amplifier Z_{SAi} (Figure 4.5). Of course, wave-impedance termination can be designed directly as the sense amplifier's inherent input-impedance. Moreover, the generator impedance of each individual memory cell and the terminating-impedance of the other end of the bitline may also be designed to be equal with Z_0 to implement a reflection-free bitline.

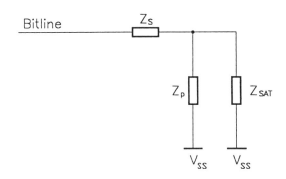

Figure 4.5. Added elements to provide wave-impedance termination.

Reflection caused distortions in data signals may also be mitigated by the applications of the next described signal limiters (Section 4.1.1.2).

4.1.1.2 Signal Limiters

To limit reflection-caused over- and undershots in the bitline signal $v(t)$ clamp circuits may be used. Voltage clamping may simply be implemented by (1) clamp diodes and by (2) inverted pull circuits.

Clamp diodes (Figure 4.6a) apply transistor devices MN1 and MN2 in diode configuration and reference voltages V_1 and V_2. V_1, and V_2 and the threshold voltages V_T (V_{BG1}) and $V_2(V_{BG2})$, and through the backgate biases V_{BG1} and V_{BG2} of MN1 and MN2, determine the clamp voltages V_{C1} and V_{C2}, so that $V_{C1} > V_1 - V_T$ (V_{BG1}) and $V_{C2} > V_2 + V_T$ (V_{BG2}). Since the effective gate-source voltages $V_{GS1} - V_T$ (V_{BG1}) and $V_{GS2} - V_T$ (V_{BG2}) of MN1 and MN2 are small, MN1 and MN2 may require large channel widths to provide high currents. Extended channel widths, however, increase bitline capacitances.

With the applications of inverted pull circuits (Figure 4.6b) the bitline capacitances can be much less than those with the use of clamp-diodes. Namely, clamp transistors MP5 and MN8 can have much larger effective gate-source voltages V_{GS5}-V_T (V_{BG5}) and V_{GS8}-V_T (V_{BG8}) and can have, for the same current, smaller channel widths than diode-devices MN1 and MN2 do. Devices MN3 and MP6 may also be small, because these devices are used exclusively for under- and overshot detection at the voltages $V_{D1} \approx V_1$-V_T (V_{BG3}) and $V_{D2} \approx V_2$+V_T (V_{BG6}). The detected and amplified under- or overshot signal turns either MP5 or MN8 on, and MP5 pulls the bitline voltage toward V_{DD} at a signal undershot, or MN8 pulls the bitline voltage toward V_{SS} at an overshot. Since at a signal reflection, the current through MP5 or through MN8 may not have sufficient time to significantly change the bitline voltage, this circuit may not be able to clamp fast changing signals acceptably.

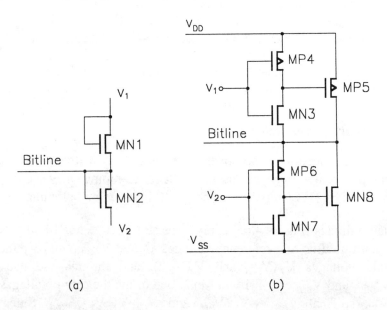

Figure 4.6. Clamping by diodes (a) and by inverted pull circuit (b).

A signal clamp circuit (Figure 4.7), implemented by devices R_L, MN1, MN2, MP3 and MP4 and by the use of the sense amplifier, may also be designed to limit current over- and undershots. Here, current $i_L(t)$ generates a voltage drop $v_L(t)$ across the load resistance R_L. A change in $i_L(t)$ and, thereby, in $v_L(t)$ may alter the gate voltages of the turned-off transistor devices MN1 and MP3. At an overshot MN1, and at an undershot MN3, turns on, and the highly conductive device shunts the input current of the sense amplifier. Devices MN2 and MP4 operate in their triode region to allow for effective regulation of the threshold voltages V_{T1} (V_{BG4}) and V_{T3} (V_{BG3}) of the devices MN1 and MP3. V_{T1} (V_{BG1}) and V_{T3} (V_{BG3}) can be regulated by changing their backgate bias voltages V_{BG1} and V_{BG3} through the gate voltages V_1 and V_2. Voltage and current clamp circuits are seldom used in memories, because reflections can economically be avoided by wave impedance terminations, and the effects of reflections can be minimized by careful timing more effectively at less power dissipation than that could be done by clamping.

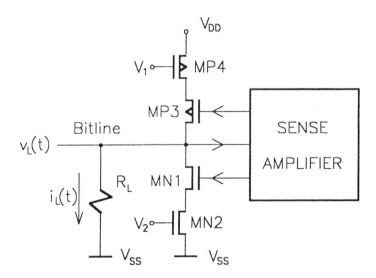

Figure 4.7. Clamp circuit implementation with the use of the sense amplifier.

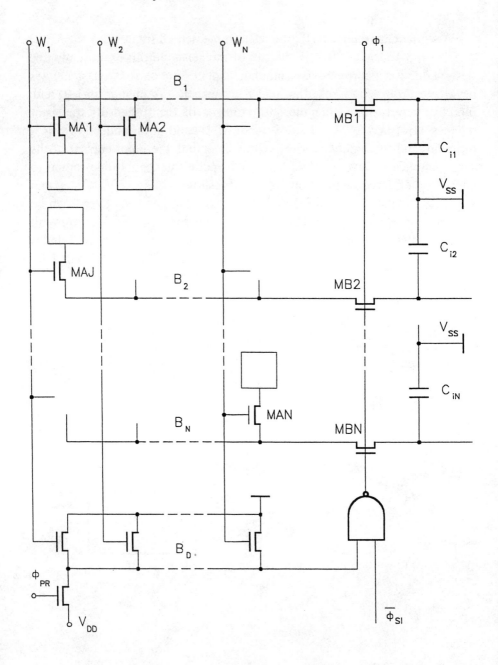

Figure 4.8. Dummy bitline aids timing in an array.

Careful timing couples the sense amplifier to the bitline when the impulse generated by a memory cell arrives to the input of the sense amplifier, and decouples the sense amplifier from the bitline before a reflected signal can appear on its input. Generally, the signals travel in different times from the accessed memory cells to the input of the sense amplifier, because of the different locations of the memory cells on the bitline. To mimic the different signal transmission times and to provide correct timing for each memory cell location a dummy bitline B_D can be used (Figure 4.8). The signal propagation time on B_D copies the signal delays on the bitlines $B_1, B_2, ... B_N$, and tracks the parameter variation caused signal delay variations as well. When a word decoder signal activates a word of memory cells, the signal generated by the accessed dummy memory on B_D travels to the dummy sense amplifier SA_D which turns the couple-decouple devices MB1, MB2,... MBN on for the time T_D. T_D is shorter than the signal propagation time between the last access device MAN and the sense amplifier input. On the sense-input capacitors $C_{i1}, C_{i2}, ... C_{iN}$ trap the signal for the time required for amplification. At the end of time period T_D the word decoder deactivates all memory cells which are accessible by this decoder circuit. The correct timing for sense-amplifier activation tracks nearly all uniform parameter variations which occur in the array.

Signal amplitude variations caused by transmission line effects, i.e., signal propagation time and reflections in the bitline circuit, may result in unduly long times for data sensing or in read or write errors. Thus, high speed data sensing may require the use of wave-impedances, or clamping, or activation timing in bitline circuits. In most designs, activation timing is provided by clock signals which are derived from a master clock, and which activate sensing and other circuit functions in worst-case timing computed by circuit simulation programs.

4.1.2 Wordlines

4.1.2.1 Modelling

A wordline is the low-resistance wire that interconnects the gates of the access transistors of memory cells. The memory cells are arranged in a row, and the data set stored in a row of memory cell is called, somewhat misleadingly although, a word. A wordline circuit includes (1) a buffer amplifier placed between a decoder output and the wordline, and (2) the wordline with its parasitic resistances and capacitances, and, eventually, (3) wordline enable devices and (4) a signal accelerator circuit. In simple

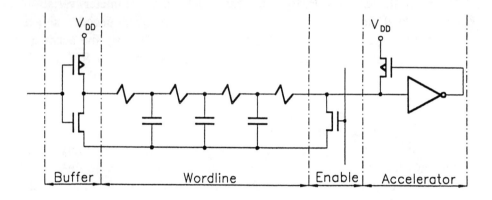

Figure 4.9. Simple equivalent of a wordline circuit.

wordline equivalent circuits (Figure 4.9), the buffers are represented by a signal generator $v_g(t)$ or $i_g(t)$ and by a generator impedance Z_G or Z'_G, the wordline is symbolized by Z_W and the eventual enable devices, accelerator circuit, and any wordline impedance is combined in impedance Z_L.

The impedance of a wordline Z_W, that ties the gates of n access devices to memory cells, comprises (Figure 4.10) the distributed resistance of the wordline $R_W \approx n \times r_w$; the distributed capacitances of the wordline to other wordlines $C_{WW} \approx n \times c_{WW}$, to the ground $C_{WSS} \approx n \times c_{WSS}$ and to the powerlines $C_{WDD} \approx n \times c_{WDD}$; the capacitances of the wordline to the crossing bitlines $C_{WB} = n \times c_{WB}$, to the channel-areas $C_{WC} = n \times c_{WC}$ and to the drains and sources of n access transistors $C_{WD} = n \times c_{WD}$ and $C_{WS} = n \times c_{WS}$.

Transistor leakage currents from the transistor gates and junction leakage currents are negligibly small in most of the wordline circuits. Here, r and c indicates respective specific resistances and capacitances referred to a unit wordline length.

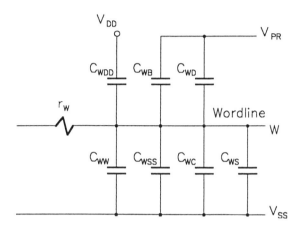

Figure 4.10. Components of a wordline impedance.

A wordline may be modelled, similarly to the bitline models, by passive RC networks (Section 4.1.1.1) and by transmission lines (Section 4.1.3). To both the passive RC and transmission line models the equivalent parameters can be obtained by the use of the here-introduced wordline-impedance components (Figure 4.10).

The signals generated by the buffer on the wordline must be fast to keep access times as short as possible. Signal switching and propagation times on the wordline, however, may be long due to the significant capacitive load and nonzero resistance of the wordline and due to its transmission-line like behavior. To provide fast signal transients on the wordline, the buffer should switch high currents. Nevertheless, the current-drive capability of the buffer depends on the buffer's size A, and A is limited by the row- or wordline-pitch. Furthermore, an enlargement in A by a large factor, e.g., 10, over the minimum buffer size A, results in very little reduction in signal propagation time t_s [43] in comparison to an assymptotic minimum propagation time t_s (Figure 4.11). This is mainly because the driver's output capacitance C_D gets larger with increasing buffer sizes, and C_D adds to the total wordline capacitance C_W. At little increase in A and C_D, the use of bootstrap buffers or boosting of the wordline can provide fast wordline drive signals, and additionally, avoid threshold voltage drops (Section 3.1.3.2)

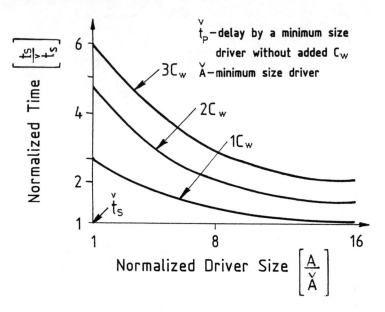

Figure 4.11. Signal propagation time versus aspect ratio and load capacitance.

The wordline capacitance usually kept on minimum by designing the access devices of the memory cells to be of minimum gate size. Because in a wordline the gates are connected to each other without contacts to minimize array area, the rather high resistance polysilicon gates are often replaced by or combined with polysilicide and polysalicide materials to reduce wordline resistance.

4.1.2.2 Signal Control

Wordline resistance and capacitance, which appears as a load for the wordline buffer, may be decreased by dividing the wordline into sectors and shorting the sectors by stripes made of metal or other low-resistance material. From the various striping schemas those are preferred which divide both the wordline resistance and capacitance as well (Figure 4.12). The efficiency of the division is limited, of course, by the area of the con-

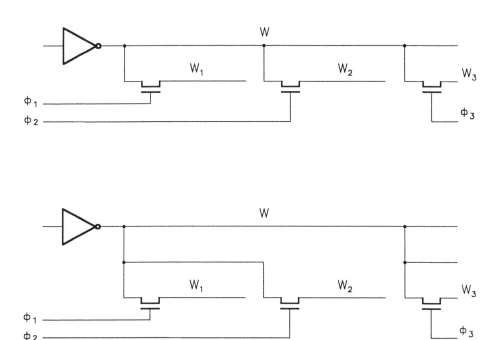

Figure 4.12. Schemas for wordline division.

tacts, and transistors added to the array, by the number of the available interconnect layers, and by the effects of the extra capacitances and resistances of the shunting stripes and the transistors and wordline signals. By similar division bitline performance may also be improved.

Further performance increase may be achieved by application of an accelerator circuit which is placed to the undriven end of the wordline (Figure 4.13). This accelerator circuit amplifies the rising edge of the wordline signal $v_w(t)$, and beyond a threshold-amplitude, device MP1 provides a rapidly increasing drain current $i_{D1}(t)$ and, thereby, a shorter pull-up time. The influence of signal pull-down times on the wordlines are noncritical in most of the designs and, therefore, a design with wordline pull-down acceleration can rarely be justified in CMOS memories.

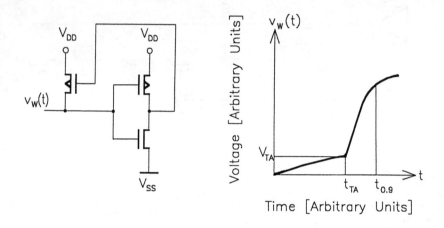

Figure 4.13. Signal accelerator circuit on the wordline.

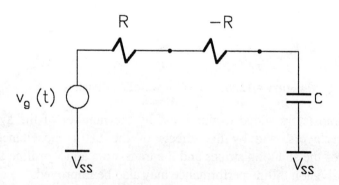

Figure 4.14. Application of negative resistance.

Unconventional memory designs may apply negative resistance -R to compensate the wordline driver's internal resistance r_{dr} and the wordline resistance R_W in the word-select circuits (Figure 4.14). Combined with wordline capacitance C_W resistances r_{dr} and R_W are determinative in the switching times through time constant $\tau = (r_{dr} + R_W) C_W$. τ can be reduced to $\tau = (r_{dr} + R_W - R) C_W$ by applying -R. Negative resistance -R can be obtained by numerous methods, e.g., lambda and tunnel diodes (Section 2.9.2), but the most amenable ones to CMOS applications are the crosscoupled inverters.

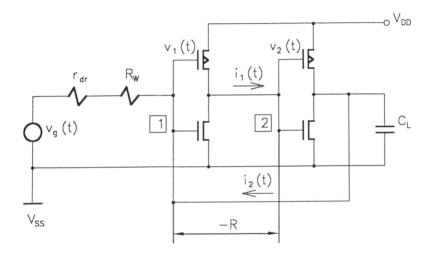

Figure 4.15. A crosscoupled inverter pair used as negative resistance.

A pair of crosscoupled inverters (Figure 4.15) can provide the negative resistance -R [44] between nodes [1] and [2]. It is assumed, at t_o time, that node voltages $v_1(t)$ and $v_2(t)$ are the same, i.e., $v_1(t_o) = v_2(t_o) = V_F$, where V_F is the flipping voltage or switching threshold voltage. After t_o, a generator $v_g(t)$ provides a voltage difference $\Delta v(t) = |v_1(t) - v_2(t)|$ which is amplified by the inverters. This is equivalent with a reduction in the generator resistance R_g of the voltage generator $v_g(t)$, which means a negative resistance -R(t) between nodes [1] and [2]. R(t) is nonlinear, and depends on the node voltages $v_1(t)$ and $v_2(t)$. With varying voltages $v_1(t)$ and $v_2(t)$ the charge and discharge currents $i_1(t)$ and $i_2(t)$, and the inverter output resistances $r_{o1}(t)$ and $r_{o2}(t)$, change as well. For an initial stable state $R(t) = r_{o1}(t_o) \| r_{o2}(t_o)$ is an acceptable approach. From t_o, resistance $R(t)$ can be approximated by piece-wise linear models until $t=t_s$ when the circuit takes a stable state. Between $t=t_o$ and $t=t_s$, the resistance $R(t)$ varies considerably, which variation limits the effectiveness of negative resistances in the reduction of wordline-signal delays.

Switching times of wordline signals may be reduced, moreover, by the application of a negative capacitance -C (Figure 4.16) generated by the Miller effect of a noninverting amplifier A_M [45]. Because CMOS operational amplifiers are slow and require large layout area and power, the use of negative capacitances is very unlikely in wordline circuits.

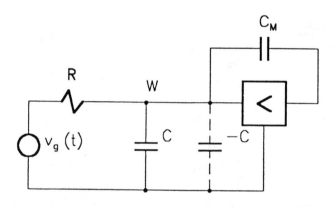

Figure 4.16. Application of negative capacitance.

Wordline signals must completely turn off the access transistors of the memory cells to minimize the chance of leakage-current caused pattern sensitivity. To minimize subthreshold leakage currents, either the threshold voltages of the access transistors are increased substantially, or the wordline is kept close to the ground potential by a high-current wordline-enable transistor during the unselected period of the wordlines. Fast changing wordline signals, furthermore, may induce significant crosstalking signals (Section 5.2.2) in the long parallel running wordlines.

Crosstalking may add to the effects of the reflected signals and their combination increases the probability of multiple selections of memory cells, and aggrandizes spurious currents in the bitline and degradations in operating margins.

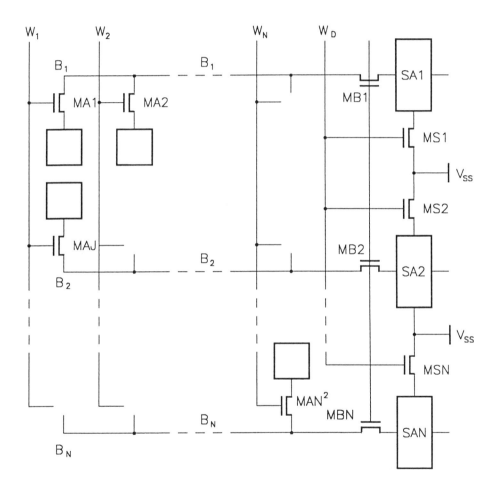

Figure 4.17. Dummy wordline assists timing in an array.

The bitlines are separated from the sense amplifiers, and the sense amplifiers are inactive when a wordline turns the access devices of the memory cells on. After the selected row of memory cells generate signals on the bitlines, either a selected bitline is coupled to a sense amplifier common to all bitlines, e.g., in SRAMs, or each bitline is coupled to a sense amplifier, e.g., in DRAMs, and the sense amplifier or amplifiers are activated. An early activation of sense amplifiers, however, may result in excessive power dissipation and in incorrect data reading. To avoid false reading of data the start of sensing must be timed properly, e.g., by adding a dummy wordline to the array (Figure 4.17). After turning a dummy cell on, the voltage on the precharged dummy wordline changes. This voltage change propagates with a delay to an amplifier that turns on the sense amplifiers. The delay tracks the uniform device-parameter variations in the wordlines.

Reverse-phased signal reflections in a selected wordline may increase the drain-source resistances of the access devices or may turn them off before the data signals exceed the noise level. Furthermore, over- and undershoots appearing in word-select signals may cause excessive hot-carrier emissions, device breakdowns and high level crosstalking. Unselected wordlines may inadvertently be turned on and off by reflected word-disable signals. Thus, reflections may decrease performance and reliability, or may impair memory operations. Memory designs minimize the signal reflection or their effects in wordline circuits by applying wave-impedance termination, amplitude clamping and activation timing the same was as in bitline circuits (Section 4.1).

4.1.3 Transmission Line Models

4.1.3.1 Signal Propagation and Reflections

If the memory design has to take in account transmission line effects, the analysis should be based on the general model of a dx long portion of a transmission line (Figure 4.18) which describes the characteristics of the propagating voltage v and current i signals by the classic telegraph equations [46]:

$$-\frac{\partial v}{\partial x} = Ri + L\frac{\partial i}{\partial t} \quad \text{and} \quad -\frac{\partial i}{\partial x} = Gv + i\frac{\partial v}{\partial t} .$$

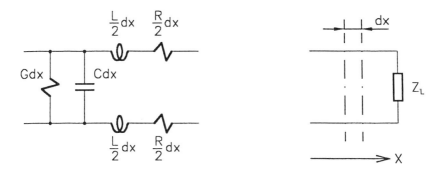

Figure 4.18. Model circuit for a dx long portion of a transmission line.

Here, R, L, G and C are the resistance, inductance, conductance and capacitance of a dx long increment of the transmission line. For a transmission line, that is endless and open in one direction and that is closed by a load impedance Z_L on its single end, a solution shows that both the distance from the line-end x and the time of the observation t are determinative in both the voltage v(x,t) and current i(x,t) along the line:

$$v(x,t) = \frac{V_0}{1+\rho}(e^{j\omega t - \gamma x} + re^{j\omega t + \gamma x}) ,$$

$$i(x,t) = \frac{V_0}{(1+\rho)Z_0}(e^{j\omega t - \gamma x} - re^{j\omega t + \gamma x}) .$$

In these equations, V_0 is the voltage signal amplitude across Z_L; $Z_o = [(R+j\omega L)/(G+j\omega C)]^{1/2}$ is the complex characteristic or wave impedance; $\rho = (Z_L - Z_0)/(Z_L + Z_0)$ is the complex reflection coefficient; $\gamma = \pm[(R+j\omega l)(G+j\omega C)] = \alpha + j\beta$ is the complex propagation coefficients; $\omega = 2\pi f$ is the circle-frequency and f is the frequency of a sinus signal; α is the amplitude attenuation and β is the phase factor. R, L, G, C, Z_L, V_o, f and l are usually given, thus, by applying the equations of v(x,t) and i(x,t) and Fourier-integration, the voltage and current at any point of time and at any distance from the end as well as parameters ρ and Z_o can be computed.

If the wave impedance Z_o matches the load impedance Z_L, i.e., $Z_o = Z_L$, then no signal reflection occurs, i.e., $\rho = 0$, the signal energy is absorbed by the load Z_L that terminates the transmission line.

If a transmission line would not have any loss, i.e., $R = O$, $G = 0$, a digital pulse would propagate along the line without any distortion since all of the signal's component frequencies would travel with the same v velocity. In this idealized case neither the wave velocity v nor the attenuation α depends on the frequency, and for this reason a signal in the lossless transmission-line may conveniently be represented by a Fourier-integral. The Fourier-integral representation of a step function $V_o 1(t)$ is

$$F\ V_o 1(t) \rightarrow V_o \left(\frac{1}{2} + \frac{1}{\pi} \int_0^\infty \frac{\sin \omega t}{\omega} d\omega\right) \ .$$

Since the value of a $V_o \sin\omega t$ signal at an arbitrary x location is $V\sin\omega(t-x/v)$ then at arbitrary x after the inception of $V_o 1(t)$ the voltage $v(x,t)$ is

$$v(x;t) = V_o \left(\frac{1}{2} + \pi \int_0^\infty \frac{\sin \omega (t - \frac{x}{v})}{\omega} d\omega\right) \ .$$

The equation of $v(x,t)$ demonstrates that a step function suffers no distortion after any time of propagation in a transmission line that has no resistive and conductive loss.

In case, the line's terminating impedance Z_L is frequency independent, i.e., purely ohmic $Z_L = R$, any digital signal is reflected by Z_L without distortion and with an amplitude determined by the reflection coefficient ρ. Thus a signal at any ρ for any x location at any point of time can be approximated by a geometric approach which simply sums up the reflections at any point determined by any x and t coordinates (Figure 4.19).

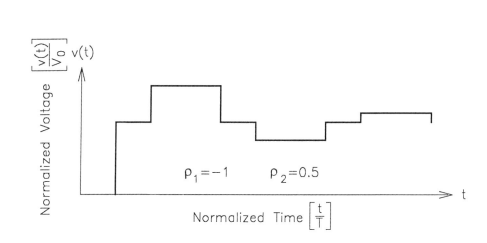

Figure 4.19. Reflection diagram for a lossless transmission line.

For lossless transmission lines which are terminated by purely ohmic impedances, the voltage reflection coefficient is $\rho_v=(R-Z_0)/(R+Z_0)$, and the current reflection coefficient is $\rho_i=\rho_v \approx (Z_0-R)/(Z_0+R)$. Therefore, $\rho_v=1$ and $\rho_i=-1$ when the line is terminated by an open circuit, and the open line reflects a voltage impulse in the same phase and same amplitude, while it reflects a current input in the opposite phase and same amplitude. When the line is closed by a short circuit the reflection coefficients are $\rho_v=-1$ and $\rho_i=1$ and the reflection phases are the opposite as for an open circuit termination. With ρ_v and ρ_i the amplitude V_0 and I_0 can be calculated for any time $t=l/v$ and any location $x=vt$ along the line.

If a transmission line is loaded by $R_L \to \infty$ and the signal generator resistance $R_G \to 0$, which may be a crude model for a bitline that is connected to a voltage sense amplifier, then the full voltage and current amplitudes V_0 and I_0 appear in every period of $t = 4T$ (Figure 4.20). Here, T is the propagation time of the wave from one end to the other end of the transmission line.

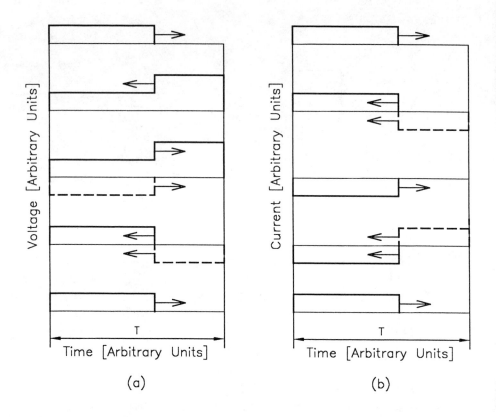

Figure 4.20. Voltage (a) and current (b) impulse reflections when $R_G \to 0$ and $R_L \to \infty$. (Source [46].)

If a transmission line is coupled to $R_L \to 0$ and $R_G \to 0$, which situation may be applied as an approximative model for a sense circuit applying a current sense amplifier, then V_o occurs in every $t = 2T$, but I_o increases beyond limits as it should be in an ideal transmission line (Figure 4.21).

Of course, in nonideal transmission lines, where $R > 0$ and $G > 0$, the ever present losses limit the increase of reflected signal amplitudes and the idealization of transmission line provides not much more than an insight to the phenomena occurring in the circuit. Nevertheless, these simplifications in circuit operation allow for uncomplicated analysis of the phenomena appearing in high-speed sense circuits.

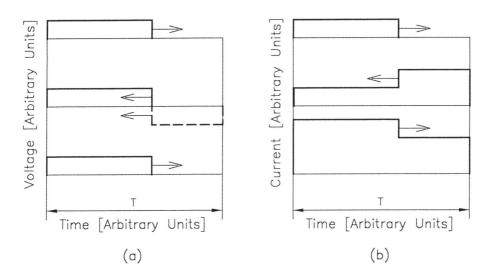

Figure 4.21. Voltage (a) and current (b) impulse reflections when $R_G \to 0$ and $R_L \to 0$. (Source [46].)

4.1.3.2 Signal Transients

Transient analyses of signals appearing in transmission lines, which have l length and which are terminated by a generator impedance Z_G and by a load impedance Z_L, may conveniently be performed by applications of Laplace transforms. By Laplace transforming the telegraph equations (Section 4.1.3.1) at the conditions that initially at t=0 both v(x,0)=0 and

$i(x,0)=0$, the voltage $v(x,t)$ and the current $i(x,t)$ can be transposed to the complex p plain as

$$V(x,p) = V_0 \frac{Z_L \operatorname{ch}\gamma\,(l-x) + Z_0 \operatorname{sh}\gamma\,(l-x)}{(Z_G + Z_L)\operatorname{ch}\gamma l + Z_Z \operatorname{sh}\gamma l},$$

$$I(x,p) = \frac{V_0}{Z_0} \frac{Z_0 \operatorname{ch}\gamma\,(l-x) + Z_L \operatorname{sh}\gamma\,(l-x)}{(Z_G + Z_L)\operatorname{ch}\gamma l + Z_Z \operatorname{sh}\gamma},$$

where

$$Z_Z = Z_0 + \frac{Z_G Z_L}{Z_0}.$$

Since parameters l, V_0, Z_0, Z_G, Z_L and γ as well as the Laplace transformed of the practically used generator functions are readily obtainable, the Laplace transformation of $v(x,t)$ and $i(x,t)$ to $V(x,p)$ and $I(x,p)$ is convenient. The reverse Laplace transformation of $V(X,p)$ and $I(x,p)$ to $v(x,t)$ and $i(x,t)$, however, may be difficult; and to alleviate the complications in the analysis the use of restrictions in the parameters of the transmission line and of the terminating impedances are necessary.

A purely capacitive load impedance $Z_L = C_L$ can often be used to model the input of a CMOS voltage amplifier, and the response signal for a generated voltage step is very important for the design of timing and eventual circuit elements which compensate undesirable signal irregularities. For an ideal voltage step $v_G(t) = V_0 1(t)$ the Laplace transformed is $V_G(p) = V_0/p$, and the operator impedance of a capacitive load is $Z_L(p) = 1/pC_L$. Combining the expressions of $V_G(p)$, and $Z_L(p)$ with the equation of $V(x,p)$ and after some mathematical manipulations, the reverse Laplace transform that gives the voltage $v(x,t)$ in four time intervals can be obtained as a function of the location of the observation x, time of the observation t, input signal amplitude V_0, attenuation factor α, length of the transmission line l, and velocity of the signal propagation v;

$$v(x,t) = 0 \text{ if } 0 < t \frac{x}{v}, \quad v(x,t) = v_0 \text{ if } \frac{x}{v} < t < \frac{2l-x}{v},$$

$$v(x,t) = 2V_0 \left[1 - e^{-\alpha(t-\frac{2l-x}{v})}\right] \text{ if } \frac{2l-x}{v} < t < \frac{2l+x}{v}, \text{ and}$$

$$v(x,t) = V_0 \left[1 - 2e^{-\alpha(t-\frac{2l-x}{v})} + 2e^{-\alpha(t-\frac{2l+x}{v})}\right] \text{ if } \frac{2l+x}{v} < t < \frac{4l-x}{v}.$$

The shape of the voltage signal across the load capacitor C_L over a time of $t = 9T$ (Figure 4.22) shows over- and undershots. Here, T is the time-period during the signal travel l distance.

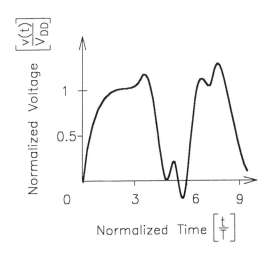

Figure 4.22. Voltage waves across a load capacitor terminating a lossless transmission line.

Signal undershots may also appear when a capacitance C_G is discharged from a V_G initial potential through a transmission line that is terminated by a short circuit $Z_L = 0$ (Figure 4.23) on its other end. With the assumptions of $Z_G = C_G$ and $Z_L = 0$, the behavior of the bitline circuit of a dynamic memory, that consists of a storage capacitor, a bitline microstrip and a current sense amplifier, may be approximated.

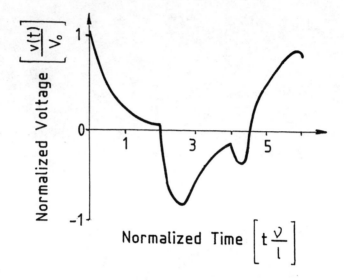

Figure 4.23. Signal shapes when a short-circuit terminated lossless transmission line discharges a capacitor.

In sense circuit analyses, a resistive generator impedance $R_G > 0$ may be added to the ideal step-function $V_o l(t)$ generator for improved approximation of the delay t_d, rise t_r, fall t_f and eventual sap t_s times. Sap times for attenuations of the output-signal swings have to be considered when $0 < R_G < Z_o$ (Figure 4.24a), but when $R_G > Z_o$ the output signal has no over- or undershots (Figure 4.24b) in a lossless transmission line.

For transmission lines with little losses, i.e., $R \ll j\omega L$ and $G \ll 1/j\omega C$ a polynomial approximation to the propagation coefficient γ reveals that the wave velocity v increases with increasing frequencies ω

$$v \approx \frac{1}{\sqrt{LC}}\left[1 - \frac{1}{8\omega^2}\left(\frac{R}{L} - \frac{G}{C}\right)^2\right],$$

but the attenuation α is independent of ω

$$\alpha \approx \frac{R}{2}\sqrt{\frac{C}{L}} + \frac{G}{2}\sqrt{\frac{C}{L}} \ .$$

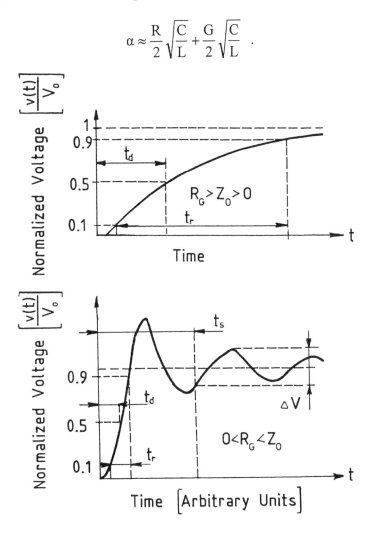

Figure 4.24. Output waveforms when $0 < R_G < Z_0$ (a) and when $R_G > Z_0 > 0$ (b).

The higher frequency components' higher velocity manifests itself in gradual "flattening" of an ideal impulse as the impulse travels away from the signal source along the transmission line that has finite but small losses (Figure 4.25).

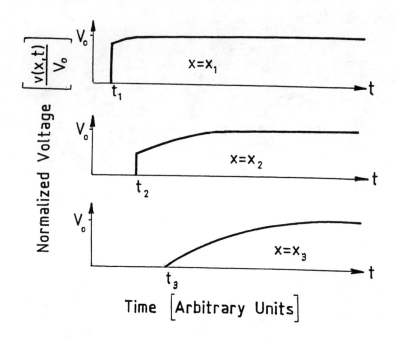

Figure 4.25. Impulse flattening in a transmission line with little losses.

Transmission lines with little losses and with capacitive-resistive generator and load impedances, model most of the interconnects, including bitlines, wordlines and other wires in a CMOS memory chip, acceptably. In interconnect lines, which are plagued with significant losses, as the impulse propagates away from the generator an exponential decrease in impulse amplitude adds to the eventual distortions. Significant losses in interconnect lines, however, are avoided by technological and design measures in both on- and off-chip wirings to obtain high operational speed and low power dissipation and little noise sensitivity.

Off-chip interconnects and chip-pins may combine considerable load inductance L with relatively little capacitive and resistive components.

Depending on the factor $K=Z_0/Lv$ an inductive load on a transmission line may substantially distort an ideal generator step-impulse (Figure 4.26).

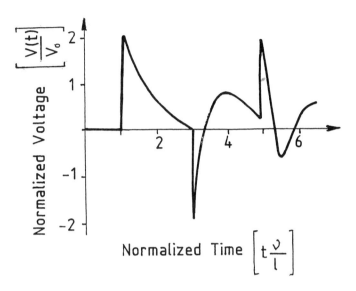

Figure 4.26. Impulse distortion by an inductive load.

The computation of the distortions in an ideal impulse in the general case, when all elements R, G, C and L appear in the transmission line and in the terminating impedances, are complex and requires the use of computers. In CMOS memory designs, the use of a general model is seldom necessary. Yet, the modelling of bitlines, wordlines, decoder lines, input and output wirings as lossless and little-lossy transmission lines, i.e., as quasi-stationary electric circuits, is very important to the approximation of delay and switching times, signal over- and undershots and general signal forms in CMOS memories which have large storage capacities and fast operations.

4.1.4 Validity Regions of Transmission Line Models

The transmission line models which are widely applied to analyze bit-, word-, decoder-, and signal-lines in a memory chip, include two imiportant assumptions; (1) the effects of the wire-inductances are negligible and (2) the propagation time of the signal along a wire of length l is constant. While assumption (1) is well justified by the operation and performance of nearly all implemented CMOS memories, but the validity of assumption (2) to large CMOS memory chips may be questioned [47] because the propagation velocity of the signals depend on the transmission line's length l, specific resistance r and specific capacitance c, as well as on the magnitudes of the terminating generator resistance R_G and load capacitance C_L (Figure 4.27). Depending on the parameters l, r, c, R_G and

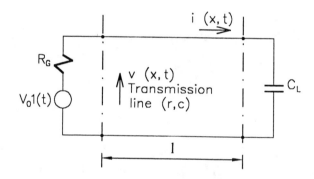

Figure 4.27. Model for determining validity regions.

the propagation of a bit-signal along a wire of length l may be computed by the application of

(a) a synchronous model with constant time irrespectively of l [48],

(b) a capacitive model with a delay increase of log l with increasing l [49],

(c) a diffusion model with a delay increase of l^2 with increasing l [410].

The model to apply for a particular technology and design may be determined in a logarithmic diagram (Figure 4.28) where the relative deviation of the propagation delay $\varepsilon(\gamma_r,\gamma_c)$ from the propagation delay that can be obtained from the idealized synchronous capacitive model $t_p \approx R_o C_o (1+\gamma_r+\gamma_c)$ is plotted as a function of $\gamma_r = rl/R_D$ and $\gamma_c = cl/C_L$.

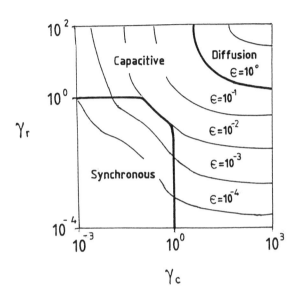

Figure 4.28. Validity regions of delay models. (After [47].)

Computations of γ_r and γ_c with parameters of 0.15-2μm CMOS technologies, results $\gamma_r = 10^{-4}...10^0$ and $\gamma_c = 10^{-2}...10^3$. These results indicate that in practical memory designs the long interconnects can be approached either by synchronous or by capacitive models.

To operate in the synchronous operating region the design has to satisfy both conditions for maximum wire length l_s;

$$\hat{l}_s \leq 10 \frac{\hat{R}_G}{\check{r}} \quad \text{and} \quad \hat{l}_s \leq 10 \frac{\hat{C}_L}{\check{c}}$$

where R_G and C_L are the maximum generator resistance and the load capacitance, and r and c are the specific resistance and capacitance of the transmission line. Expressions for l_s indicate that an operation in the capacitive delay region can be avoided by designing the wire length $l < l_s$, by subdividing the main memory cell array into smaller subarrays, or by the insertion of repeater amplifiers at fixed R_G and C_L. In sense circuits, R_G, and C_L, can be varied only in limited ranges, because the design of the memory cell, bitline and the sense amplifier and the processing technology predetermine these parameters. Nevertheless, adjustments in the amplifiers which are coupled to the word-, decoder- and other long lines can well be used for delay time decrease.

For delay reduction in the unlikely case when the transmission line would operate in the diffusion region the methods described for avoiding operation in the capacitive region can also be used in addition to the eventual development of integrated nondispersive transmission lines.

The theoretical foundation for the three regions of delay-length computation (Section 4.13) may be obtained from the definition of capacitance and resistance in transmission lines

$$\frac{\partial v}{\partial x} = -ri \quad, \quad \frac{\partial i}{\partial x} = -c\frac{\partial v}{\partial t} \quad,$$

whence

$$\frac{\partial^2 v}{\partial x^2} = rc\frac{\partial v}{\partial t} \quad, \quad \frac{\partial^2 i}{\partial x^2} = rc\frac{\partial i}{\partial t} \quad.$$

These are instances for the classical Poisson, or also called diffusion, or heat equations which may be solved by the method of variable separation of homogenous boundary conditions [411]. The final solutions describe the location and time dependency of the voltage v(x,t) and current i(x,t), and an expansion of the propagation delay times $t_p = f(r, c, l, R_G, C_L)$ into Taylor-series assists to obtain the plots of $\varepsilon(\gamma_r, \gamma_c)$, at the assumption that both the driver and driven transistors have the same minimum size. By increasing the transistor width, and if width is proportional to the capacitive load for all transistors, a nearly constant delay can be approached not only in the synchronous, but also in the capacitive region.

4.2 REFERENCE CIRCUITS

4.2.1 Basic Functions

A reference circuit provides a voltage, current or charge reference level for determination of log.0 and log.1 information when a signal level generated by a memory cell is compared to a reference level by a sense amplifier. In practice, the reference level is a range of quantities rather than a single quantity, because of the effects of parameter variations resulting from the effects of semiconductor processing, supply voltages, temperature and radioactive radiations. These parameter variations influence the operating and noise margins in a sense circuit significantly (Section 3.1.3). To keep the margins wide, either the parameter and reference-level variations should either be minimized, or designed so that the reference tracks the changes of the operating and noise margins, or both. From the plethora of reference circuits, the next sections describe those which have been applied mostly and have future potentials for applications in CMOS memories.

4.2.2 Voltage References

Voltage reference circuits, which are applied in most CMOS memories, include voltage dividers, threshold voltage droppers and complex stabilized voltage regulators.

Voltage dividers may be resistive or capacitive. In resistive dividers at precharge operation a series of resistors (Figure 4.29a) or MOS transistors (Figure 4.29b) charge storage capacitor C_{L1} to an intermediate reference precharge voltage V_{PR} when switch device MS1 is turned on and precharge transistor MT2 is off. During precharge, when MS1 is off and MT2 is on, reference voltage V_R may significantly be subdivided by charge distribution between C_{L1} and sense circuit capacitance C_{L2}; and may further be reduced by the threshold voltage V_T of precharge device MT2 if the gate-source voltage V_{GS} is designed as $V_{GS} \leq V_{DS} - V_T (V_{BG})$, where V_{DS} and V_{BG} are the drain-source and backgate-bias voltages of MT2.

312   CMOS Memory Circuits

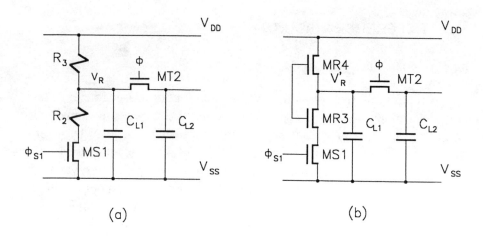

Figure 4.29. Voltage dividers implemented in resistors (a) and transistors (b) provide precharge levels.

Capacitance C_{L1} are formed preferably of polysilicon lines, which can be placed underneath the metal lines distributing the supply voltages V_{DD} and V_{SS}. Transistors MR3 and MR4 can also be positioned under V_{DD} and V_{SS} lines, because they are long and narrow devices. Long divider resistors R_1, R_2, and long transistors MR3 and MR4 allow for increased reference level accuracy if they are used with a wide high-current MS1 in the voltage division, because with long R_2, R_3, M3 and M4 and with wide MS1 in the expressions of the reference voltages, i.e.,

$$V_R = (V_{DD} - V_{SS})\frac{R_2 + r_{d1}}{R_2 + R_3 + r_{d1}} \text{ and } V'_R = (V_{DD} - V_{SS})\frac{r_{d1} + r_{d3}}{r_{d1} + r_{d2} + r_{d3}},$$

the resitances R_2, R_3, r_{d3} and r_{d4} have small percentage length-fluxuations, and R_2, R_3, r_{d3} and r_{d4} are much large thatn r_{d1}. Here, subscripts 1, 3, and 4 indicate transistors MS1, MR3 and MR4. High-resistance dividers consume small power, while low-resistance dividers can provide smaller variations in reference level and in precharge impulse amplitudes than high-resistance dividers.

In a capacitive divider (Figure 4.30), the impulse amplitude obtained by switching device MS1 is reduced in accordance with the ratio of divider capacitances C_{C1} and C_{C2}, and the resulting voltage V_R is approximately $V_R = (V_{DD} - V_{SS}) C_{C2}/(C_{C1}+C_{C2})$. At precharge, V_R is further reduced by the eventual charge distribution when capacitor C_{L3} is coupled to the divider and, in some designs, by a threshold-voltage-drop through the device MT2.

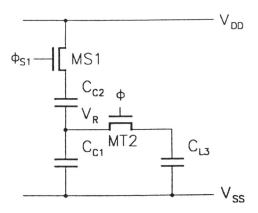

Figure 4.30. Capacitive divider in a reference circuit.

Threshold-drop references use one of the basic characteristics of the MOSFET transmission gate, i.e., the reference voltage $V_R = V_{GS} - V_T(V_{BG})$ if $V_{GS} \leq V_{DS}$. V_{GS}, V_T, V_{BG}, and V_{DS} are the gate-source, threshold, backgate-bias, and drain-source voltages, respectively. The reference voltage V_R may be obtained directly as a V_T drop (Figure 4.31a) or as a difference of two unequal V_T-s in devices MN1 and MN2 (Figure 4.31b). In both cases, the voltages V_T and V_{PR} follow the on-chip uniform threshold voltage variations which may be induced by the effects of semiconductor processing, supply voltage and temperature variations, and by certain radioactive radiations. In these simple threshold-drop reference circuits, the effective gate voltages $[V_{GS}-V_T(V_{BG})]$-s of transistors MN1 and MN2, may be small, and the charge time of capacitor C_L, therefore, may be unacceptably slow.

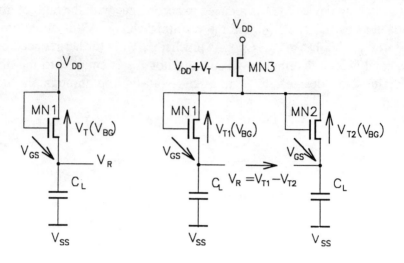

Figure 4.31. Threshold-drop references.

Faster charge times and tracking of V_T can be provided by increasing V_{GS} to $V_{GS} \approx 2V_T$ [412] (Figure 4.32). For identical n-channel devices MP1 and MP2, at $V_{GS} \approx 2V_T$ the current balance may be approximated as

$$\beta(2V_T - V_T)V_{DS} - \frac{1}{2}V_{DS}^2 \approx \frac{\beta}{2}(2V_T - V_T)^2 .$$

From the current balance the reference voltage is

$$V_R = V_{DS} \approx V_T.$$

The gate voltage $V_{GS} \approx 2V_T$ is provided by series connected p-channel devices MP3, MP4, and MP5, when clock ϕ_1 is high and ϕ_2 is low, because during this period the drains of MP3, MP4 and MP5 are individually coupled to their own gates by conductive transfer devices. In standby mode, when ϕ_1 is low and ϕ_2 is high, devices MP1-MP5 are biased for radiation worst-case (Section 6.1.2) so that their drains and sources are at V_{SS} and their gates are tied to V_{DD}. The reference voltage is $V_R = V_{DS} - V_T(V_{DD}, T, RD_T, V_{BG}) - \Delta V$, where V_T is a function of the supply voltage V_{DD}, temperature T, and radiation total dose RD_T, and backgate bias V_{BG}, and ΔV is the error in tracking the threshold voltage V_T.

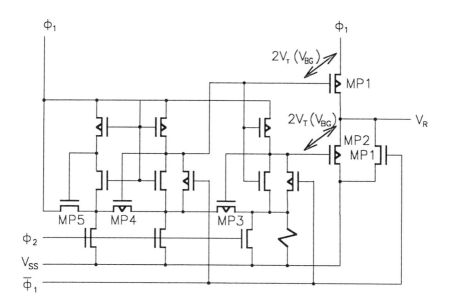

Figure 4.32. Threshold voltage change tracking reference circuit. (Source [412].)

Temperature stabilized reference voltage can be obtained from the combination of a $V_{GS} = 2V_T$ biased MOS transistor and negative feedback circuit (Figure 4.33). In this circuit [413] the reference voltage V_R may be expressed as

$$V_R = V_T + \frac{1}{R g_m},$$

where threshold voltage V_T and transconductance g_m are parameters of the output device MT1, and R is the feedback resistance. V_T has negative, while $1/Rg_m$ has positive temperature coefficient. The compounded effects of both the positive and negative temperature coefficients can reduce the temperature dependency of V_R to less than 200 ppm/°C.

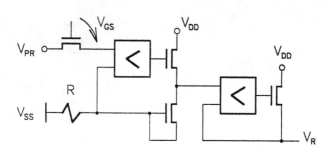

Figure 4.33. Temperature stabilized voltage reference.

Generally, increased V_R stability for temperature and other environmental effects can be obtained by applications of the voltage regulator principles. In a series reference regulator circuit (Figure 4.34), the regulator device MR1 is placed between the input and the output, the output voltage is divided by resistors R_1 and R_2 and compared to the basic reference signal V_B. After a linear amplification the error signal E' is coupled to the gate of the executor device MR1, and MR1 counteracts any changes in V_R. The sensitivity S and the output resistance of the circuit R_o may be approximated from the block model of the circuit (Figure 4.35) as

$$S = \left.\frac{\partial V_{PR}}{\partial V_R}\right|_{I_o = \text{const}} \approx \frac{1}{1+g_m r_d A d} \quad , \quad R_o = \left.\frac{\partial V_{PR}}{\partial I_o}\right|_{V_R = \text{Constant}} \approx \frac{r_d}{1+g_m r_d A d} \quad .$$

Here, V_{PR} is the stabilized reference output or precharge voltage, I_o is the output current; r_d is the drain-source resistance and g_m is the transconductance of device MR1; A is the voltage gain of the amplifier; and d is the division ratio provided by resistances R_1 and R_2. By increasing g_m, A and d, in accordance with the equations, both S and R_o can significantly be reduced, but they can not be made zero.

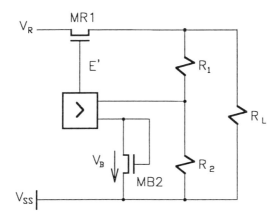

Figure 4.34. Series-regulated reference circuit.

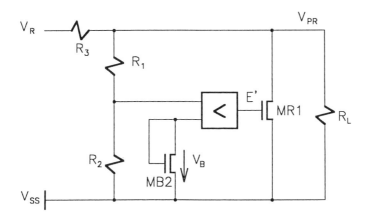

Figure 4.35. Parallel-regulated reference circuit.

In a parallel-regulated reference circuit (Figure 4.35) both the regulator device MR1 and the voltage divider are placed between the output and the ground. The difference between the divider's output voltage V_D and a basic reference voltage V_B provides the error signal E, and E is amplified to E' by a gain of A before it reaches the gate of shunt-device MS1. MS1 draws increased current when V_R increases, the increased current tend to decrease the output voltage V_{PR}, and vice versa.

In the parallel-regulated voltage source, as in the previously outlined series voltage regulator, with increasing g_m, d and A both the S and R_o may greatly be decreased and the stability of V_{PR} may significantly be improved. The stability of both the series and parallel regulated reference circuit may be mathematically examined by application of the Ruth-Hurwitz, Nyquist, Mihailov, Bode, or Kupfmuller methods [330].

Some early designs adopted band-gap reference circuits from the bipolar technology, but bipolar references have seldom been used in CMOS memories, because of the additional process steps required to fabricate bipolar transistors in a CMOS chip and because of their poor voltage regularity.

4.2.3 Current References

From the wide range of current sources, mostly the current mirror and feedback type of circuits are applied for current references in sensing schemes.

In current mirror references (Figure 4.36) a simple unloaded chain of transistors MR1 and MR2 provides approximately constant gate-source voltages V_{GS1}, V_{GS2} and V_{GS3}. When all devices MR1, MR2 and MS3 operate in the saturation region, the output reference current I_{RO} varies mainly with the current I_{R2} of MR2, and I_{RO} changes very little with the alterations of the load current $\pm \Delta I_L$ of source device MS3. The size of MS3 determines the output current I_{RO} and output resistance $R_o \approx r_{d3}$ (Section 3.4.3). For fast sensing I_{RO} should be high, which requires a high gain-factor ratio of $\beta_q = \beta_2 / \beta_3$ and, in turn, a large silicon layout area.

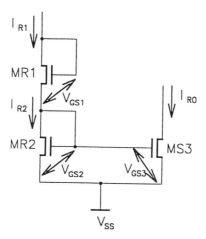

Figure 4.36. Simple current-mirror reference circuit.

High current ratio I_{R2}/I_{R0} can be obtained without high gain factor ratio β_q by using a modified Widlar current source (Figure 4.37). In this circuit, the gate-source voltage of V_{GS2} of MR2 is substantially smaller than the

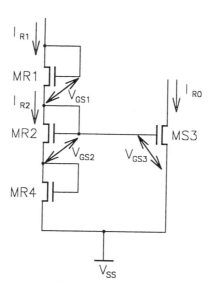

Figure 4.37. Modified Widlar current source.

320 CMOS Memory Circuits

gate-source voltage V_{GS3} of MS3. The high and constant V_{GS3} provides a high and constant output current I_{R0}, as long as all transistors operate in the saturation region.

Both series and parallel feedback current source circuits (Figure 4.38) can be derived from feedback voltage sources (Section 4.2.2) by regulating output currents rather than output voltages. Series and parallel current regulators may be combined for more efficient stabilization (Figure 4.39).

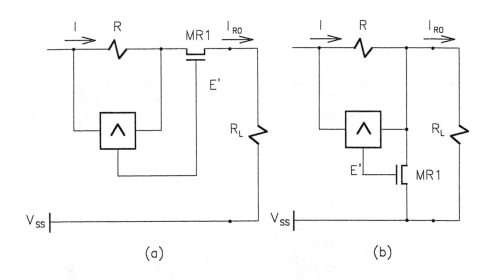

Figure 4.38. Current regulation in series (a) and parallel (b) configurations.

For the feedback current sources, the analysis and design considerations are the same as for the earlier discussed current sense amplifiers (Sections 3.4.3 and 3.4.4) and voltage sources (Section 4.2.2).

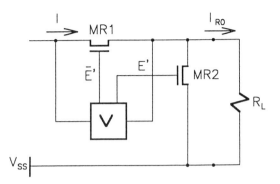

Figure 4.39. Combination of series and parallel current regulation.

4.2.4 Charge References

Charge reference circuits use the principle of charge distribution and redistribution on switched capacitors. In a circuit that contains a finite number of capacitors and switches, in absence of generators and power sources, the total amount of charge stored on the capacitors of the circuit before the activation of switches $\Sigma Q(t_0)$ equals with the total amount of charges after the activation of switches $\Sigma Q(t_1)$ plus the amount of charges lost during the considered amount of time $\Delta Q(t_1 - t_0)$ i.e.,

$$\sum Q(t_0) = \sum_{i=1}^{n} C_i v_i(t_0) = \sum_{i=1}^{n} C_i v_i(t_1) + \Delta Q(t_1 - t_0).$$

Here, $v_i(t)$ is the voltage on a capacitor C_i at a t time, and $Q_i(t)$ is the charge stored in capacitor C_i at a t time.

Charge distribution and redistribution are widely applied in dynamic differential sense circuits (Figure 4.40) to compare data to a reference level on a pair of bitlines. In dynamic memories, a binary datum is stored on a cell capacitor C_C, all capacitance coupled to the bitlines C_{B1} and C_{B2} are precharged to V_{PR}, and the capacitor of a dummy cell C_D is applied as a

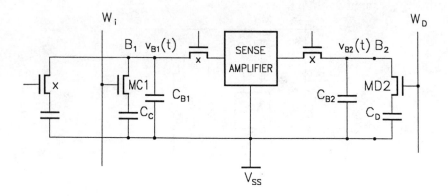

Figure 4.40. Charge reference in a dynamic differential sense circuit.

charge reference to generate a reference voltage $V_R = v_D(t_1)$. Here, $v_D(t)$ is the time function of the voltage on capacitor C_D. If the time-dependent voltage on the capacitor C_C is $v_C(t)$, the time after complete precharge and before turning access devices MC1 and MD2 or is t_0, and the time after turning MC1 and MD2 on and after the complete charge redistribution is t_1; then the voltage difference $\Delta v(t_1)$ between the bitline nodes at the time of t_1 is

$$\Delta v(t_1) = |\Delta v_{B1}(t_1) - \Delta v_{B2}(t_1)|.$$

Voltage difference $\Delta v(t_1)$ may be approached by applying the charge equivalence principle. From the total amounts of charges at the time t_1 bitline voltages $v_{B1}(t_1)$ and $v_{B2}(t_2)$, if zero charge-loss is assumed, i.e., $\Delta Q(t_1-t_0)=0$, then $\Delta v(t_1)$ may be expressed as

$$v_{B1}(t_1) = \frac{C_C}{C_C + C_{B1}} v_C(t_0) + \frac{C_{B1}}{C_C + C_{B1}} v_{B1}(t_0) \,,$$

$$v_{B2}(t_1) = \frac{C_D}{C_D + C_{B2}} v_D(t_0) + \frac{C_{B2}}{C_D + C_{B2}} v_{B2}(t_0) \,,$$

and initially the bitlines are precharged to $V_{PR} = v_{B1}(t_o) = v_{B2}(t_o)$. By subtracting $v_{B2}(t_1)$ from $v_{B1}(t_1)$ the voltage difference $\Delta v(t_1)$, i.e., the differential input voltage for the sense amplifier, can be obtained

$$\Delta v(t_1) = \frac{C_C}{C_C + C_B} v_C(t_0) - \frac{C_D}{C_D + C_B} v_D(t_0),$$

where $C_B = C_{B1} = C_{B2}$.

Clearly, a $\Delta v(t_1) \neq 0$ appears if $C_C \neq C_D$ or if $v_C(t_0) \neq v_D(t_0)$. Thus, the magnitude of either one or both C_D and $v_D(t_0)$ can be used to control the reference voltage V_R. In most of the CMOS memories V_R is chosen so that $V_R = Q_R/C_D$ provides operating margins which are approximately the same for sensing log.0 and log.1, i.e., $V_R-V_O = V_1-V_R$. Here, Q_R is the reference charge, V_O is the maximum of the logic low level, and V_1 is the minimum of the logic high level. Knowing V_O, V_1, C_C, C_B and the desired $V_R=v(t_1)$, and by setting $v_C(t_0) = V_D(t_0) = V_R = V_{PR}$; the dummy capacitance C_D can be approximated. Alternatively, by setting $C_D = C_C$ the initial dummy cell voltage $v_D(t_0) = V_{PR}$ can be approached. In designs where $C_D \neq C_C$, the dummy capacitance is about $C_D \approx C_C/2$. Sense circuits using $C_D = C_C$ closely track parameter variations in both C_C and MC1, and they feature, therefore, higher sensitivity, faster operation and greater environmental tolerance than circuits applying $C_D \approx C_C/2$ do.

One of the primary aims in sense circuit designs is to generate an acceptably large signal $\Delta v(t_1)$ for the sense amplifier. Although the amplitude of $\Delta v(t_1)$ depends mainly on the ratio between C_C and C_B, an optimization of $\Delta v(t_1)$ by appropriate charge reference design can significantly improve the performance of a dynamic memory.

4.3 DECODERS

The address information to locate memory cells in an array are transmitted in codes to reduce the number of chip-to-chip and chip-internal interconnects. The codes applied in CMOS memories are almost exclusively of binary types, because of their area efficiency and their inherent amenability to memory-array implementations. Nevertheless, some military and high reliability memories may apply other addressing codes also.

The addressing of a memory cell in a two-dimensional XY array of $n \times n = n^2$ number of memory cells by the simple binary code needs $2 \log^2 n$ addressing bits and two one-out-of-n decoders. Three one-out-of-n decoders may be used in very large memory chips in three-dimensional XYZ addressing schemas.

Addressing decoders, most commonly, are implemented in rectangular NOR and NAND forms (Figure 4.41). In both NOR and NAND decoders the output lines are precharged by devices MP1-MP4, while high-resistance leak-transistors MP5-MP8 compensate the leakage-current caused changes in logic levels when the decoders are inactive.

The application of a full-complementary decoder (Figure 4.42) is beneficial, where the row or column pitch allows for accommodation of double amount of transistors, and where low power dissipation and large noise and operating margins are required. Particularly, memories operating in radiation hardened or in other severe environments, apply full-complementary decoder circuits.

Theoretically, rectangular decoder implementations provide neither the smallest area nor the fastest operation for one-out-of-n decoders. Nonetheless, the structural similarity between a rectangular decoder and an array of memory cells, and the adjustability of the decoder output lines to the row and column pitches of a memory cell array, makes rectangular schemas the smallest and fastest operating decoder implementations to memory cell arrays.

The implementation of a tree-decoder (Figure 4.43) can provide speedy operation at reduced layout area in some special memories. In tree-configurations, high decoding speed can be obtained because only one threshold voltage V_T drop appears between an input and an output, and because buffers may conveniently be inserted in the layout. The layout can be designed in a small area, because only a single address line, rather than two, the true and complement lines, dare needed for decoding, and because no drain, source or gate contacts are required to implement the tree circuit.

Memory Constituent Subcircuits 325

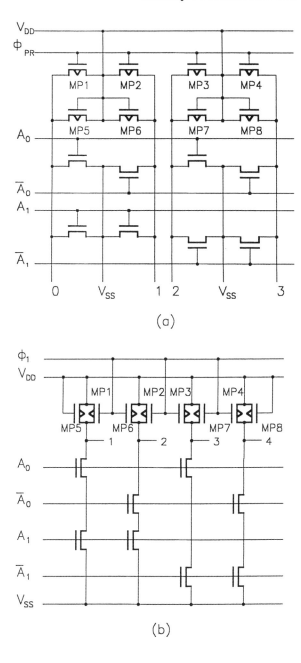

Figure 4.41. Rectangular NOR (a) and NAND (b) decoders.

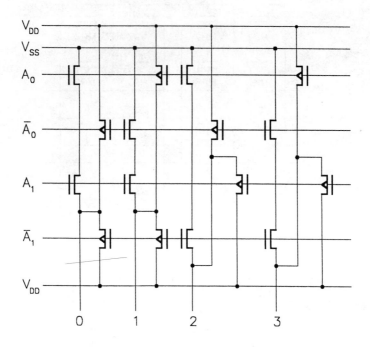

Figure 4.42. Full-complementary decoder.

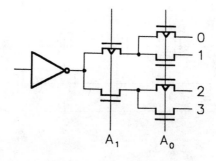

Figure 4.43. Tree-decoder configuration.

In implementation of large memories the application of a precoder (Figure 4.44) can substantially reduce both the layout area and access time. Namely, the use of a predecoder cuts the number of transistors that load an address buffer in normal rectangular decoders. Moreover, predecoding allows for an efficient layout design when decoder segmentation for subarrays is required (Figure 4.45).

The analysis and design of one-out-of-n decoder circuits are similar to those of wordlines and memory cell arrays and transmission line models may have to be applied where the number of outputs n is large (Section 4.1).

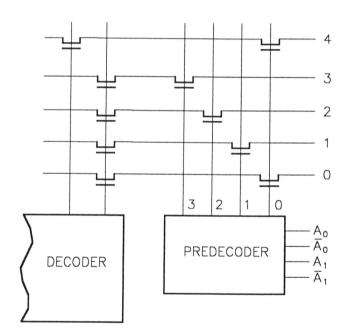

Figure 4.44. Two-to-four rectangular predecoder applied to a decoder.

328   CMOS Memory Circuits

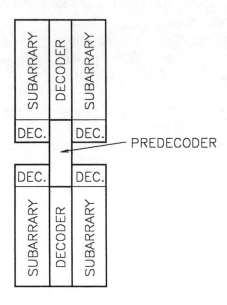

Figure 4.45. Predecoder enables subarrays.

4.4 OUTPUT BUFFERS

The output buffers of a memory convert the chip internal logic levels and noise margins to those required for driving the inputs of chip-external circuits in digital systems. Memory output circuit operations in certain temperature and supply voltage ranges, have to satisfy requirements in both DC and AC conditions, which are specified at the outset of the design.

The DC operating conditions of the memory outputs (Figure 4.46) define a minimum output voltage V_{OH} (current I_{OH}) for the logic high level at a given current (voltage), and a maximum output voltage V_{OL} (cur rent I_{OL}) for the logic low level at a given current (voltage). By these output levels, the inputs of another integrated circuit have to be driven, and for the inputs the minimum logic, high voltage V_{IH} (current I_{IH}) at a given current (voltage), and the maximum logic low voltage V_{IL} (current I_{IL}) at a given current (voltage), are usually provided. The differences V_{OH} - V_{IH} and V_{IL} - V_{OL} result the margins in which the occurrence of noise signals can be tolerated.

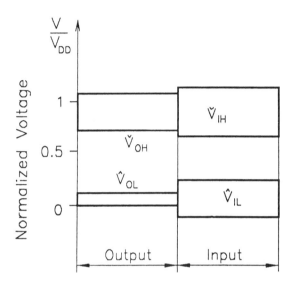

Figure 4.46. DC output and input logic levels at room-temperature.

The AC operating conditions for the outputs determine the properties of the signal-transients which are to be performed by the output buffers at given DC signal levels, and comprise the required rise time t_r and fall time t_f of the output signal when the output pin is connected to a specific load impedance. For an output buffer the load equivalent impedance is modeled usually by a capacitive-resistive circuit (Figure 4.47), but modeling for high speed operations requires the inclusion of inductive circuit elements also.

The primary objective of the output-circuit design is to provide the required output signal levels for log.0 and log.1 at a switching speed which approaches the memory's chip-internal performance, or which maintains a predetermined performance. High speed performance, in driving a large capacitance, can be achieved by applying scaled buffer stages between the wide output devices and the minimum-size chip-interim transistors. Scaling factors may be optimized for speed, power or area by theoretical approaches [414], but in practice the factorless backward-scaling proved to be the most useful approach.

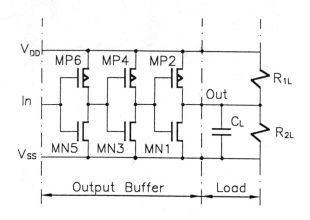

Figure 4.47. Simple output buffer with a load equivalent circuit.

In the application of the factorless backward-scaling technique to the design of the simple output buffer, first the size of the two output transistors MN1 and MP2 are determined so that MN1 and MP2 are capable to drive the load impedance with the required fall and rise times t_f and t_r. Then, the input impedance of MN1 and MP2 is considered as the load impedance for devices MN3 and MP4, and the sizes of MN3 and MP4 are designed to provide the same t_f and t_r as the required ones to drive the chip-external load. The sizes of MN3 and MP4 are smaller than the sizes of MN1 and MP2, because for MN3 and MP4 the capacitive load is smaller than that for MN1 and MP2. Similarly, the sizes of MN5 and MP6 are also smaller than those of MN3 and MP4 when they are designed to provide the same t_f and t_r on the input impedance of MN3 and MP4. To approach the chip-internal t_f and t_r of minimum sized logic circuits, a backward scaling of sizes through three or two stages is sufficient in most cases.

Output circuit operations often comprise requirements for a high-impedance output-state in addition to provide standardized log.0 and log.1 levels at certain switching speed. In such tri-state output circuits (Figure 4.48) the backward scaling involves the sizing of both logic gates and inverter circuits.

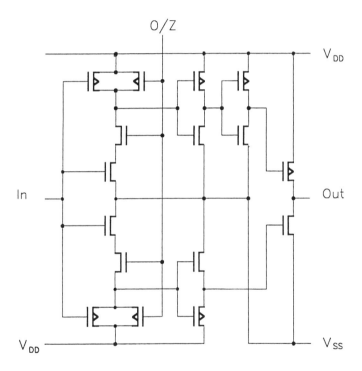

Figure 4.48. Tri-state output buffer.

The operation of complementary inverter circuits include phases when both the n- and p-channel devices are turned on. During these phases, the output buffer generate noise currents, ground- and supply-bounces, bulk-potential variations and increased substrate currents. To decrease these undesired currents and voltage changes the output transistor may be turned on during two distinct impulses (Figure 4.49) rather than by a single complementary signal.

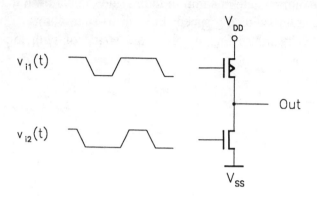

Figure 4.49. Avoiding direct current between the power supply poles.

The output signal, in numerous applications, is required to stay unchanged on the output-pin until a different datum appears. For that, the datum may either be stored in a minimum-sized latch placed between the sense amplifier and the output logic circuit, in a positive-feedback sense amplifier in some designs.

Output buffer designs for fast operating systems may have to cope with signal reflections. To minimize signal reflections the output impedance Z_{Go} should be the same as the wave impedance of the driven transmission wire Z_o (Section 4.1.3). Wave-impedances for complementary common-source and source-follower outputs may economically be approximated by a series combination of a resistor R and a transistor, in which the transistor has a very small drain-source resistance $r_d < R/k << Z_D$, where k = 12-25, so that $R + r_d \approx Z_o \approx Z_{or}$. For small wave-impedances, e.g., $Z_o = 50\Omega$, this approach may be impractical, because a small drain-source resistance, e.g., $r_d = 2.5 \pm 0.5\Omega$, requires the implementation of large p- and n-channel transistors. Transistor-only drivers providing nearly wave-impedance outputs $Z_{Go} \approx r_d \approx Z_o$ can be designed with the application of analog or digital control circuits. In CMOS memories the digital control of the output impedances is the preferred approach because the economical implementation of its constituent circuit elements is economical.

In an exemplary digital controlled output circuit (Figure 4.50), parallel-connected devices MN1-MN5 and MP7-MP11 determine the output impedance Z_{Go} [415]. Devices MN6 and MP12 are replicas of transistors MN1 and MP7, respectively, and reference impedances Z13 and Z14 are designed to approximate the $Z13=Z14=Z_o$ condition. The voltage drops on

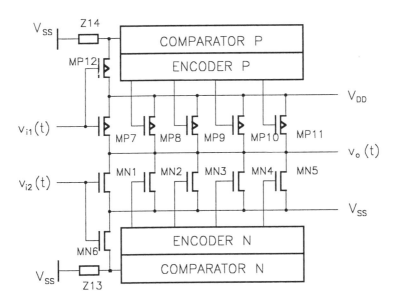

Figure 4.50. Digital controlled output buffer providing a near wave-impedance interface.

Z13 and Z14 are compared to voltage references in Comparator N and Comparator P, and their digital outputs are coded by Encoder N and Encoder P. Depending on the code used, the sizes of devices MN2-MN5 and MP8-MP11 may be the same or weighted. Devices MN2-MN5 and MP8-MP11 are activated in agreement with the codes representing the instantaneous voltage drops on Z13 and Z14, so that the combined drain-source resistances of MN1-MN5 and MP7-MP12 approximate the wave impedance, i.e., $r_{d1-5} \approx Z_o$ and $r_{d7-12} \approx Z_o$, during and after the change of the output signal level. Throughout a signal switch either the N-channel or the P-channel output devices are activated. Other variations of impedance controlled output circuits may unify the comparators with the encoder circuits, may combine a linear amplifier, a linear integrator with a digital time-window quantizer, or may adopt various other design approaches from the abundance of digital and analog circuits.

CMOS memory designs may be required to accommodate a simultaneously operating multiplicity of output buffers to increase the communication bandwidth between the memory and the computing circuits. Simultaneous multiple output circuit operations greatly enlarge power dissipation and noise generation. A reduction in both power consumption and noises may be achieved by the applications of low weight codes, most economically by the implementations of Berger codes (Section 5.7.4.4) to the consecutive sets of the output data. An encoder-decoder circuit for an N-bit output set (Figure 4.51) may comprise a digital comparator DCOMP, a majority vote logic MVL (Section 5.6.5), an inverting/noninverting circuit I/NI, an encoder-decoder circuit for a Berger code ENC/DEC, and a flip-flop FF. The DCOMP circuit compares the upcoming N-bit data $v_{o1}(t_1)...v_{oN}(t_1)$ with the present output data $v_{o1}(t_o)...v_{oN}(t_o)$, e.g., by 2-input XOR gates. Each XOR gate feeds the result of the comparison into the MVL. The MVL circuit indicates the number of output bits which differ at time t_1 from those at time t_o, i.e. ΔN. If $\Delta N > N/2$, then FF generates a flag signal and the I/NI circuit inverts each output datum, otherwise the output data remains noninverted. Thus, the possible number of the output-signal transitions can be reduced to N/2 or to less than N/2. Further reduction in the number of simultaneous output-signal transitions are provided by the ENC/DEC circuit that encodes the output data-set into a low-weight Berger code (Section 5.7.4.3). If the data terminals are bidirectional, i.e. they serve as both outputs and inputs, then the ENC/DEC circuit also decodes the incoming data-set from the Berger code to a weighted binary code.

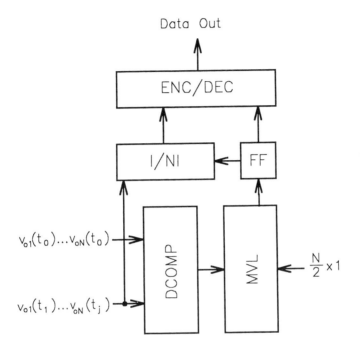

Figure 4.51. Coding schema for N simultaneously operating outputs and inputs.

All three types of output buffer operations, the phase-shifted, impedance-controlled and the coded ones, introduce additional delays into the signal transfer, and require extra silicon area for implementations. Nevertheless, the signal delay extensions and the area increases are small in many of the CMOS memory designs, and the obtainable combination of low-power, high-speed and reliable operation greatly outbalances the drawbacks of the use of complex output buffer circuits. The designs of output buffer circuits for CMOS memories are essentially the same as those for other digital CMOS integrated circuits.

4.5 INPUT RECEIVERS

Input receivers convert the chip-external logic levels and noise margins to those required to the memory operation chip-internally, and provide the data signal characteristics which are necessary for the safe operation of the chip-internal circuits. Prerequisites for circuit designs may be obtained from the DC and AC operating conditions. The DC operating conditions for an input receiver require signal-level and noise-margin conversions in the opposite direction to those which are performed by an output buffer (Section 4.5). Since the chip-external capacitances coupled to an input receiver are much larger than the receiver's chip-internal load capacitance on its output, the AC conditions and the AC design goals of an input receiver are also the reverses of those of an output buffer. For the inputs of an input receiver, the worst-case characteristics of the chip-external incoming signal are given, and the receiver has to generate an output signal with certain minimum and maximum logic levels, and with given rise and fall times, for the chip-internal digital circuits.

Traditional input circuits apply cascaded inverter-chains, in which the input logic levels and noise margins are adjusted by using substrate bias for modification of the threshold voltage in the first inverter. Threshold voltages may also be adjusted by ion implantation of the channel region without the use of voltage bias between source and substrate nodes.

A differential voltage amplifier (Section 3.3) coupled to a reference voltage source (Figure 4.52) may also apply nonstandard- often zero-threshold devices for input signal detection. Here, devices MN1, MN2 and MP3 form a low pass filter to avoid detection of spurious signals, and the output signal of the differential amplifier is further amplified to provide the required signal levels and transient times.

To those inputs, on which the input signal transients are expected to be particularly slow, Schmidt triggers [416] may be used to reshape the input signal. A Schmidt trigger (Figure 4.53a) is a threshold switch that applies positive feedbacks selectively to each the rising and the falling signal

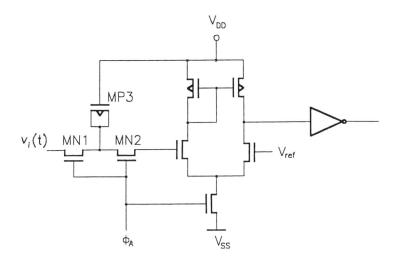

Figure 4.52. Differential amplifier in an input receiver.

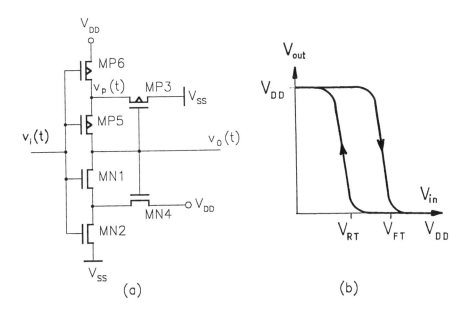

Figure 4.53. A Schmidt trigger circuit (a) and the hysteresis curve (b)

transient. The selective feedback allows to design the DC output-input characteristics $v_{out}=f(v_{in})$ of the circuit as well as the forward and reverse trigger voltages V_{FT} and V_{RT} and the hysteresis voltage $V_H=V_{FT}-V_{RT}$ (Figure 4.53b). V_{FT} can be set by the relative sizes of the n-channel devices MN1, MN2 and MN3, and V_{RT} can be controlled by the sizes MP4, MP5 and MP6. When the input voltage $v_i(t)=0$, devices MN3, MP4 and MP5 are turned on and the other devices are turned off. Device MN1 becomes conductive as $v_i(t)$ rises beyond the n-channel threshold voltage V_{TN}, and device MN2 is in cutoff until $v_i(t)=V_{FT}$. When $v_i(t)>V_{FT}$, MN1 and MN2 pull the output voltage $v_o(t)$ from V_{DD} toward V_{SS}, MP6 lowers $v_p(t)$, and MN3 gets less conductive. In turn, the lower $v_o(t)$ makes MP6 more and MN3 less conductive. When MN3 is turned off and MP5 is in cutoff, $v_p(t)$ is decreased to the p-channel threshold voltage V_{PN}, and the current through MN1 and MN2 results $v_o(t)=v_n(t)=V_{SS}$. Equating the saturation currents of MN1 and MN3, i.e., $I_1=I_3$, gives an approximation to the forward trigger voltage

$$V_{FT} \approx (V_{DD}+V_{TN})\beta_{qn}^{\frac{1}{2}}/(1+\beta_{qn}), \beta_{qn} = \beta_1/\beta_3,$$

and, similarly, to the reverse trigger voltage

$$V_{RT} \approx (V_{DD}-V_{TN})\beta_{qp}^{\frac{1}{2}}/(1+\beta_{qp}^{\frac{1}{2}}), \beta_{qp} = \beta_4/\beta_6,$$

where β is the gain factor, and indices 1, 3, 4 and 6 designate devices MN1, MN3, MP4 and MP6. The expressions of V_{FT} and V_{RT} are approximation, because they disregard the varying effects of back gate bias voltages, channel length modulations, carrier mobilities, and other parameters, on the signal development.

For input signals which are to be stored and reshaped a level-sensitive latch (Figure 4.54) can be used. The reference level V_{ref} may be set at TTL logic threshold $V_{TTL} = V_{ref}$ or at CMOS logic threshold $V_{CMOS} = V_{ref}$. Clock signal ϕ_A, input voltage $v_i(t)$ and V_{ref} control the currents $i_1(t)$ and $i_2(t)$, and $i_1(t) \neq i_2(t)$. When ϕ_A goes high and $v_i(t)$ gets lower, both currents $i_1(t)$ and

$i_2(t)$ decrease, but the current difference $|i_1(t)-i_2(t)|$ increases rapidly because of the regenerative action of the circuit. When the currents stop the circuit latches the datum consistent with the input level.

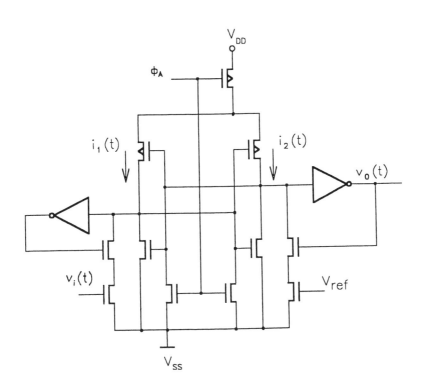

Figure 4.54. Level-sensitive latch.

To avoid reflection by the high input impedance of the input receiver Z_i, a chip-external or a chip-internal wave-impedance Z_o, that shunts Z_i, may be applied. $Z_o = Z_i$ provides a reflection coefficient $\rho=0$ on the input of the receiver circuit (Section 4.1.3).

Input receivers are applied to detect and amplify all type of signals including data, address, control and clock signals. Address input circuits, however, may be required to generate a chip-enable CE signal when a transition or change in the address information occurs and to perform as an address transition detector (ATD). A wide variety of ATD circuits can be combined from digital logic gates; yet the economical designs combine the input receiver with ATD functions and use the static memory cells (Figure 4.55) or simple flip-flops which are designed actually for the memory core

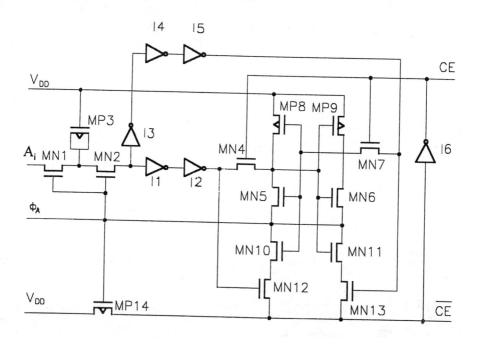

Figure 4.55. Circuit combining input receiver and address transition detector.

or overhead circuits. The shown ATD circuit includes an input receiver circuit MN1, MN2, MP3, I1-I5, a memory cell MN4-MN7, MP8, MP9 and a one-bit digital comparator circuit MN10-MN13, MP14, I6. The input receiver provides the binary address information A_i to both the memory cell and to the comparator in the form of a digital signal that is converted into chip-internal standards. If the A_i signal represents the same datum as the one that is stored in the memory cell, the potential on the precharged node $\overline{CE}$ does not change. If the A_i signal does not match the datum stored in the memory cell, node $\overline{CE}$ is going to be discharged, the new A_i information is written in the memory cell, and a chip-enable signal is generated. Nodes CE and $\overline{CE}$ may be common to all address inputs $A_1...A_n$.

In general, input receiver designs for CMOS memories are similar to those for CMOS digital circuits, e.g., [417].

4.6 CLOCK CIRCUITS

4.6.1 Operation Timing

In both system and chip levels, memory circuits may be designed to operate in synchronous or in self-timed (asynchronous) mode. Synchronous design associates sequence and time through the use of a system- or chip-wide clock signal as a reference. Self-timing does not rely on a reference clock signal, but starts circuit operation when an output signal event, which is generated by a circuit, appears in the input of the circuit under consideration, and concludes circuit oepration when the circuit under consideration creates its own output signal event that indicate the accomplishment of the operation. In self-timed designs the output signal event initiates the operation of other circuits, while in synchronous designs a circuit external clock signal leads the next operation off.

In systems, a memory operates synchronously when a chip enable input is driven by a central clock signal directly, or indirectly by a derivative of the central clock signal. Self-timing is applied, when a signal change in the address or data activates the memory without the use of any chip-external clock signal.

Chip internally, most of the memory designs use synchronous discipline, because it makes possible to combine high operational speed and clear control on the timing design for the constituent circuits. Self-timed designs may reduce power dissipation, but they are less controllable by available design tools and provide longer memory access times than clocked synchronous designs do.

Clock impulses which are distributed within a memory chip, may significantly be delayed (skewed) and distorted due to the effects of the parasitic resistances and capacitances distributed along the clock lines and due to properties of the electromagnetic wave propagation in the clock line (Sections 4.1.3 and 4.1.4). Delays and distortions of clocks may be analyzed by lumped resistor-capacitance, transmission-line and diffusion models (Section 4.1) to obtain the deviations from intended timing.

Perfect simultaneity in timing within a memory chip can only be designed theoretically. In praxis, the design considers the clocking simultaneous as long as the clock skew does not interfere with the planned operation of the circuit. In memory circuits, approximately equal clock skews and, thereby, regional simultaneousness, are rather easy to provide, because of the symmetricity of the memory architectures. A double mirror-symmetric architecture is inherently amenable to lay out equal delay paths for the clocks when the generator G is placed into the center of the chip (Figure 4.56). If the generated clocks propagate with equal speed in all the

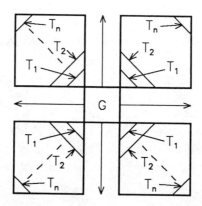

Figure 4.56. Equitime lines in a double mirror-symmetric architecture.

four x, -x, y and -y directions, then events of equal delays appear along diagonal equitime lines $T_1, T_2,...T_n$. Equitime lines occur parallel with one of both edges of the array, when a synchronized multiplicity of local generators $G_1, G_2...G_k$ provide clocks to the symmetrically arranged subarrays of a large memory chip (Figure 4.57).

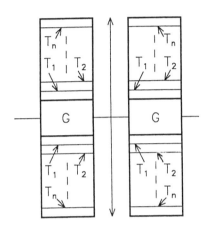

Figure 4.57. Equitime regions in symmetrically arranged subarrays.

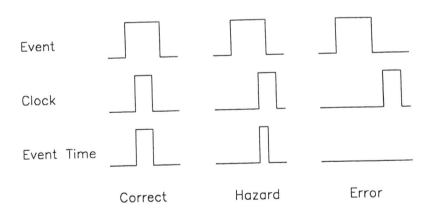

Figure 4.58. Correct, hazardous and erroneous timing.

344 CMOS Memory Circuits

When a subarray, that is placed far from the clock generator, requires clocking at a certain equitime line T_i the chances are significant that the clock impulse arrives at $T_I \pm \Delta t$ rather than at T_i. Δt represents the clock phase shift, that can be so large that the desired event can not be timed simultaneously with the logic event planned to a certain T_i. Moreover, logic error that result from clock skew may impair the operation of the effected circuit (Figure 4.58). Therefore, very large memory circuits may require the adjustment of clock phase and establishment of additional "zero" time reference.

4.6.2 Clock Generators

Memory designs use very high number of clock impulses, e.g., 150, for ensuring precise sequence in subcircuit operation, and for careful timing that facilitates memory operations at worst-case variations of processing and environmental parameters in the chip. Clock generation in the chip is implemented nearly in all of the designs, because it greatly reduces the number of chip-to-chip interconnects, makes the complexity of

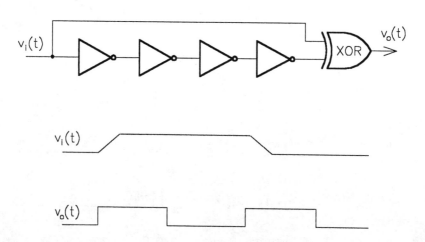

Figure 4.59. Cascade of inverters in an address transition detector.

system design acceptable, and can be designed to track the variations of some parameters. From the wide variety of clock generating circuits CMOS memories apply those which base their operation on inverter chains, simple flip-flops or memory cells and a few logic gates. A cascade of inverters is often used for delay and shaping of signals, e.g., in address transition detector (Figure 4.59), and for generation of nearly symmetrical timing signals, e.g., in ring-oscillators (Figure 4.60).

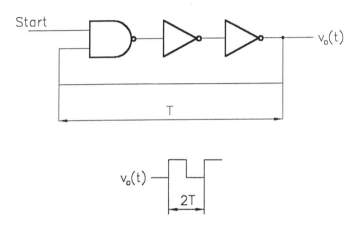

Figure 4.60. Clock generation by a ring-oscillator.

Arbitrary signal lengths and delays can be obtained by combining set-reset SR flip-FLOPS or memory cells with inverter chains (Figure 4.61). The time delays introduced by the inverter chains and flip-flops, change as threshold voltage, gain factor, supply voltage and other CMOS device parameters vary due to processing and environmental influences. As long as the parameter changes are approximately uniform on the chip, the timing provided by these circuits adjusts to the changes occurring in the addressing, data write and read operations. In designs for very fast operations, some of the inverter chains may be replaced by transmission lines formed of interconnect lines (Section 4.1).

346 CMOS Memory Circuits

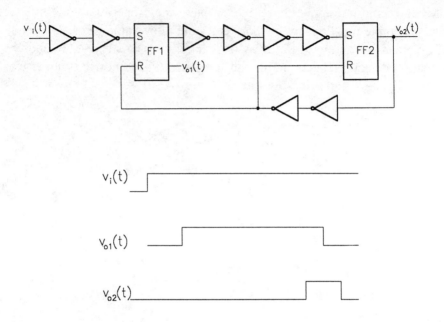

Figure 4.61. Generation of various clock signals.

Clock frequency dividers may preferably be formed of the well known binary-counter, shift-register-based nonlinear counter (Figure 4.62), and Johnson counter (Figure 4.63) for divisions by 2^n, 2^n-1, and n, respectively, there n is the number of stages in the divider. Common features of these three frequency dividers are the nearly hazard free operation, the applicability of memory cells designed for the data storage array, and small layout area.

All clock generator circuits introduced here can apply the memory cells and a variety of repetitive circuit elements which are designed actually for other memory circuits.

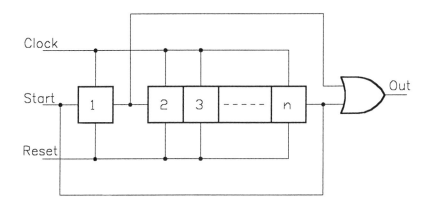

Figure 4.62. Nonlinear counter.

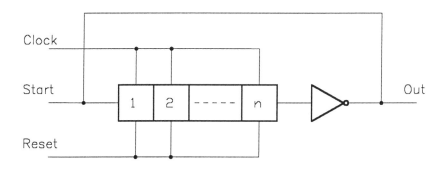

Figure 4.63. Johnson counter.

4.6.3 Clock Recovery

To recover clock-phase shifts and to reestablish timing reference, many memory designs employ simple logic gate combinations, changes in operation modes from synchronous to asynchronous and back to synchronous, and phase corrections by phase locked loop PLL circuits. Apart from the use of traditional logic gate circuits, applications of Muller C, delay mimicking and digital PLL are the most docile approaches to clock recoveries in CMOS memories.

A Muller C circuit (Figure 4.64) [418], called also as join, last-of, or rendezvous circuit, is a bistable device which provides a log.1 on output MC only after all the inputs A and B and the output MC are log.1, and MC gives log.0 output only after all variables A, B and MC are log.0. The latched output responds to the last one of a set of signals changing in the same direction and, thus, it can indicate the accomplishment of a set of logic operations. The indicator signal may be used to start a new sequence of clocks or another set of operations.

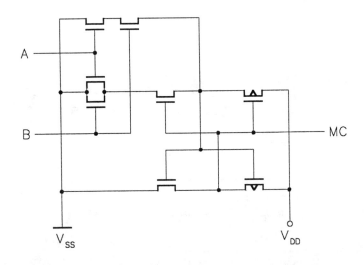

Figure 4.64. Muller C circuit. (Source [418].)

A new sequence of clocks may also be initiated by a delay line that mimics the worst-case delay of the circuit (Figure 4.65). The delay line is preferably a replica, or a dummy slice, of a column or a row of the memory cell array or decoder (Section 4.1.1). The dummy elements copy the signal delays occurring in the controlled circuit, and provide parameter tracking which assure circuit operation in wide ranges of parameter variations.

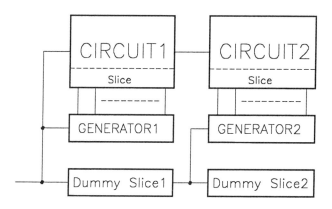

Figure 4.65. Delay mimicking.

Most of the very large memory circuit designs employ phase-locked loops PLLs to control clock-phase shifts and to reestablish a timing reference. PLL theory and operation are extensively studied [419] and evolved to be an important branch in the communication technology. Albeit PLLs aroused many unrealized expectations, simple PLLs can very well be used to reduce clock-skews, i.e., to resynchronize clocks. In memories PLLs correct clock-skews by exploiting a fundamental property of PLLs, that is the capability to adjust the phase of a signal to a reference signal and to lock the adjusted phase by a feedback loop.

The loop includes a (1) phase detector PD, (2) low-pass filter LPF, and (3) voltage controlled oscillator VCO (Figure 4.66). In simple implementations, the PD compares the digital output signal of the VCO y(t) to a digital reference signal r(t) and generates a digital phase error signal e(t) on its output. Phase error signals are separated from their high-frequency components by the LPF. On its output, the LPF provide an analog voltage signal $v_a(t)$ that varies its amplitude in correspondence with the magnitude of the phase errors. The amplitude variations of $v_a(t)$ tune the VCO, and the loop locks when both the reference and output signal have the same frequency and phase.

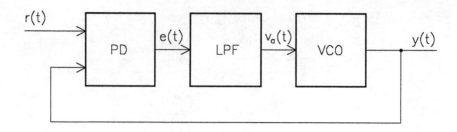

Figure 4.66. Basic phase-locked loop.

In correction of a phase error the loop needs a so-called latency time to stabilize itself. At a signal acquisition, the shape and the timing of the output signal may fluctuate without a change in frequency, and this type of fluctuation is referred to as jitters. Latency and jitters limit the applicability of PLLs in CMOS memories.

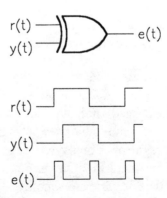

Figure 4.67. Primitive phase detection.

Memory Constituent Subcircuits 351

In CMOS implementation, a phase detector may be as primitive as one logic gate (Figure 4.67), and a voltage controlled oscillator may be a simple ring-oscillator with transistor-capacitor tuning elements (Figure 4.68). For designs of low-pass filters resistive-capacitive RC elements in Π and T types of ladder configurations can be applied most conveniently, and most of the filters developed in the linear circuit technology [420] can be adopted to CMOS memory designs.

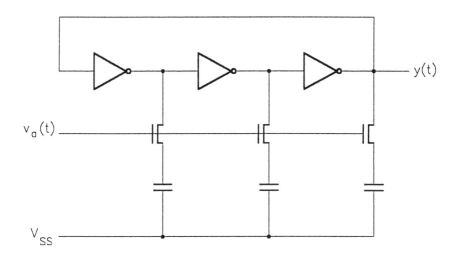

Figure 4.68. Simple voltage controlled oscillator.

The nearly linear operation characteristics of the CMOS PLL circuits, i.e., $e(t) \approx K_1|\Phi_y - \Phi_r|$ for the PD, $v_a(t) = K_2 e(t)$ for the LPF, $\Phi_y = K_3 \int_0^t v_d(t)\, dt$, and $y(t) = A \cos(\omega_F(t) + \Phi_y)$ for the VCD, allow the use of analog analysis and design methods. Here, K_1, K_2, and K_3 are circuit dependent constants, Φ_y and Φ_r are the phase angles of the output and reference signals, A is the amplitude of the ring oscillator signal and $\omega_F = 2\pi f_F$ is the frequency of the VCO at $v_a(t) = 0$. The input and output signal frequencies are the same $\omega_{in} = \omega_{out}$ when the PLL is in locked state, because then $|\Phi_y - \Phi_r|$ is constant and $d\Phi_y/dt = d\Phi_r/dt$. Simple CMOS PLLs are so-called second-order analog systems [421] because their transfer-function H(p) has two poles

$$H(p) = \frac{\Phi_{out}(p)}{\Phi_{in}(p)} = \frac{\Phi_y(p)}{\Phi_r(p)} = \frac{\omega_N^2}{p_n^2 + 2\xi\omega_N p + \omega_N^2},$$

where

$$\omega_N = (\omega_{LPF} K_1 K_3)^{\frac{1}{2}}, \quad \xi = \frac{1}{2}\left(\frac{\omega_{LPF}}{K_1 K_3}\right)^{\frac{1}{2}},$$

and $\omega_{LPF} = 2\Pi/RC$ represents the frequency in the low-pass filter that is determined by the constituent resistive R and capacitive C elements. For near optimum operation parameter $\xi = \sqrt{2}/2$ should be aimed, which provides a nonperiodical and fast system transient. The optimization of ξ, however, is limited by the interdependency of parameters ξ, ω_N, ω_{LPF}, K_1, K_2 and K_3. Other system parameters, such as loop bandwidth, attenuation of high frequencies, frequency and phase capture ranges, also impose limitations and, together with inherent frequency- and phase-noises, coerce tradeoffs in PLL designs. Further tradeoffs are introduced by the mandatory satisfaction of stability criteria (Section 3.4.10) in the designs of the PLL feedback loop. To alleviate the effects of design tradeoffs a variety of system and circuit technical approaches have been developed and published, e.g., [422], for communication devices.

4.6.4 Clock Delay and Transient Control

Clock delays may be reduced most simply by inserting buffers periodically into the long clock line (Figure 4.69) [423]. The clock delay T_{cl} on a line that is buffered by single inverters may be approximated as

$$T_{cl} \approx (2.2 r_t C_l)^{\frac{1}{2}} + (r_l c_t)^{\frac{1}{2}},$$

while T_{cl} for a line that uses tapered inverters may be calculated by

$$T_{cl} \approx K e^{r_0 c_0} \ln\left(\frac{C_l}{KC_0}\right) + \frac{R_l C_l}{K}.$$

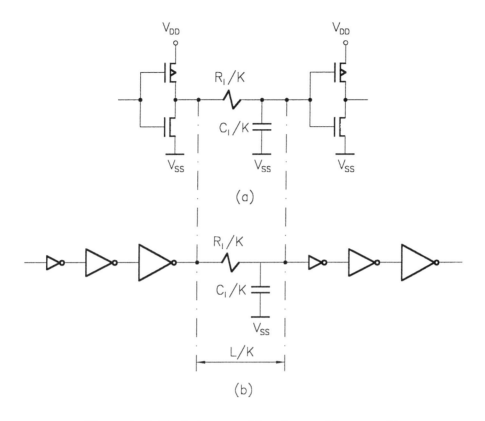

Figure 4.69. Single inverters (a) and tapered inverters (b) reduce clock delays (Source [423].)

Here, r_t and c_t is the total output resistance and load capacitance of a line buffer output capacitance, r_o and c_o are the output resistance and the load capacitance for a minimum size inverter, R_l and C_l are the line resistance and capacitance, and K is the division factor for a line of L lengths, i.e., L/K is the distance between two line buffers. As a function of K the clock delay T_{cl} curves minimum (Figure 4.70) because the inverters' inherent delay and their parasitic capacitances counteract the faster switching times and shorter signal propagation delays gained by the division of lines and by the amplification through the buffers. The decrease in delay T_{cl} depends strongly on the line parameters L, R_l and C_l also, and improvements in clock skews may be little.

354   CMOS Memory Circuits

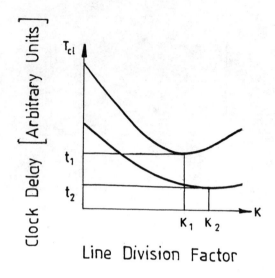

Figure 4.70. Clock delay as a function of the line division factor.

In addition to clock skews, signal reflections (Section 4.1.3), crosstalkings (Section 5.2.2), ground and supply bounces (Section 5.2.4), may create signal distortions which enlarge the voltage- and time-windows for failures in signal detection. To reduce failure possibility, the clock lines may have to be terminated by its wave-impedance, carefully shielded by ground and supply lines, and reshaped by signal detector and amplifier circuits. A simple reshaping circuit applies a resistor-capacitor as low-pass filter to smooth out the signal transitions, and a differential amplifier that switches logic levels at a reference voltage (Figure 4.71). The voltage amplitude of the clock signal CL may be optimized for minimum delay.

To control the shape and delay of a clock signal a variety of circuits are developed in CMOS technology, nevertheless, the described circuits seem to be the most amenable approaches to memory designs. Approaches like driving the clock network from a multiplicity of pads or driving each subarrays by individualized external clocks, can significantly reduce clock skew and latency, but the extension of pin count and system clock network may only be acceptable in specific high performance systems.

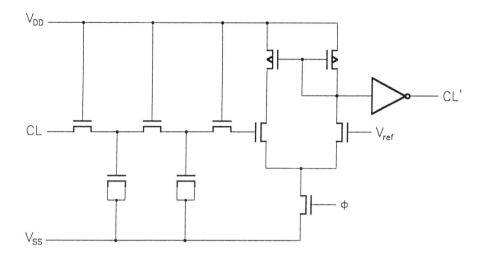

Figure 4.71. Clock signal reshaper circuit.

4.7 POWER-LINES

4.7.1 Power Distribution

Power distribution within a memory chip is provided by a circuit of low-impedance interconnect wires. The nonzero impedance of the interconnects causes undesired voltage drops (current-resistance IR drops) on the supply lines, temporary changes of ground and supply voltages (ground and supply bounces), appearance of spurious signals in the supply network (power bus noises) (Section 5.2.4), and electromechanical degradations in supply-line materials (electromigration) [424]. All these power supply problems and the reduction of their harmful effects are extensively analyzed in general electronics and integrated circuit techniques [425]. Nonetheless, some techniques are particularly amenable to memory applications.

In memory arrays IR drops may significantly decrease the sense circuits' operating margins, and the currents generated by potential differences in the power network may oppose the current induced by the datum stored in the accessed memory cell (Section 3.1.3.3). Thus, the datum may be misread. To reduce operating margin degradations and misreading probability a Winston-bridge like configuration for power distribution (Figure 4.72) is suggested. In this configuration the voltage drop from nodes V_{DD} to nodes 1,2,...n are the same as from V_{DD} to nodes 1', 2',...n' if the wires have uniform resistance per length unit. Consequently, a voltage drop $V=V_{DD}-V'_{DD}$ generates no DC current between node pairs 1-1', 2-2'.

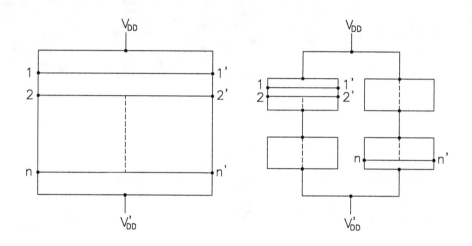

Figure 4.72. Power lines in bridge configurations.

Ground and supply node bounces may impose significant reductions in operating and noise margins, induce substrate currents, disturb both read and write operations, and may cause data loss in memory cells. Each time a memory circuit element changes logic state, a current impulse, mostly a small one, propagates through the power lines. Large impulses, which are capable to influence memory functions, can be generated by the output buffers, especially when they switch date simultaneously. The output buffers with the power lines and package pins, and with all the chip-internal circuits, constitute a rather complex network that can reasonably be modelled by a combination of transmission lines (Section 4.1.3) and lumped circuit elements. For qualitative analysis, the power supply circuit may be represented by a concentrated parameter network (Figure 4.73). In this model network, it is assumed that the pin and the chip-external wire inductances L_{11} and L_{21}, the chip-external load capacitances C_{11}, C_{21},...C_{2k},

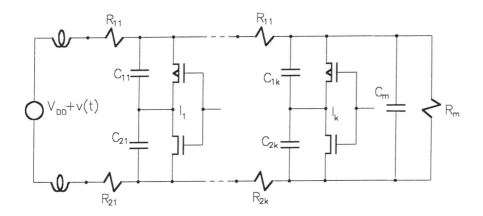

Figure 4.73. Model of the power circuit.

the wire resistances R_{11}, R_{21}...R_{2k} and the output buffers I_1, I_2,...I_k domineer the transient behavior. Capacitance C_m and resistance R_m represent the impedance of all chip-internal memory circuits with the exception of the impedances of the output buffers. As a further simplification, the impedances can be transformed in a serial inductor-register-capacitor LRC circuit and the buffer operation may be idealized as a switch between two resistors R_1 and R_2 (Figure 4.74). The switching transient may be obtained by applying the operator impedance of the circuit and by using Laplace-transform. The reverse Laplace-transform results the time dependent voltage $v_c(t)$ across the capacitor C:

$$v_c(t) = V_{DD} e^{-\frac{t}{\tau}} \left[(K_{R1} - K_{R2}) ch\beta t + \frac{1}{\tau\beta}(K_\tau - K_{R1} - K_{R2}) sh\beta t \right] + V_{DD} K_{R2},$$

where

$$\tau = \frac{2R_2 CL}{RR_2 C + L}, \quad K_{R1} = \frac{R_1}{R_1 + R}, \quad K_{R2} = \frac{R_2}{R_2 + R}, \quad \beta = \left(\frac{1}{\tau^2} - \frac{1}{K_2 LC} \right)^{\frac{1}{2}}.$$

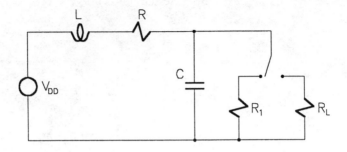

Figure 4.74. Simplified power circuit equivalent.

The equation of $v_c(t)$ describes the approximate signal form that appear in the chip internal power lines, and indicates the effects of the elements L, R, C, R_1 and R_2. Low chip-external inductivity L, small power-line resistance R, and reduced difference between the on and off resistances R_1 and R_2 in the output device, decrease the amplitude of the bounce signal. Switching from a small R_2 to a large R_1 results in a large disturbance signal. Moreover, this disturbance signal has sinus-hyperboloid components, and the signal transient rings with β frequency, and the signal amplitude decreases exponentially with the time constant τ.

In crude approaches, the exponential characteristics can be replaced by linear approaches, and the signal swingings can be disregarded [426]. Presuming that the current signal appearing on the output of the buffer i(t) can be approximated by an equilateral triangle (Figure 4.75), the voltage signal's switching or flight time t_s is measured from $0.05V_{DD}$ to $0.95V_{DD}$, N buffers switch simultaneously and share M ground or supply connections, and each of the M ground or supply connections has L inductance and C load capacitance; then the induced noise voltage amplitude V_L can be approximated as

$$V_L \approx \frac{N}{M}\left(\frac{4LC}{t_s^2}\right)V_{DD} \ .$$

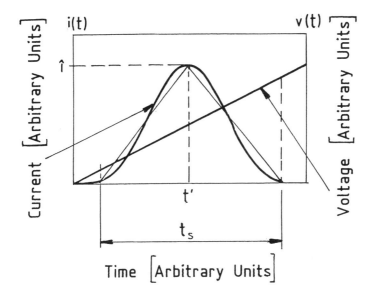

Figure 4.75. Approximation for the shape of the switching current.

The expression of V_L indicates that V_L, at given inductance L and capacitance C, can be reduced by minimizing the number of simultaneous data switches N, increasing the number of ground and supply pins M, extending flight time t_s and decreasing supply voltage V_{DD}.

Parameters, L, C, t_s, N, M and V_{DD} can limitedly be varied to decrease the amplitude of the power-line noise V_L, because these parameters are rather tightly controlled by the fabrication and packaging technologies, and by the operation, performance and pin-out requirements of the memory. At no effect on technology and at little influence to memory characteristics, power line noises can substantially be decreased by careful layout and circuit designs.

4.7.2 Power-Line Bounce Reduction

Layout designs can effectively reduce ground- and power-line bounces by keeping apart the output buffers' power and ground lines and their p and n wells from all the other memory circuits, and by applying multiple pins for the ground and supply connections to the separated output buffers and to all the other memory circuits (Figure 4.76). Furthermore, ground and supply wiring configurations may be designed to establish local current loops for those high-current buffers which are loaded with large

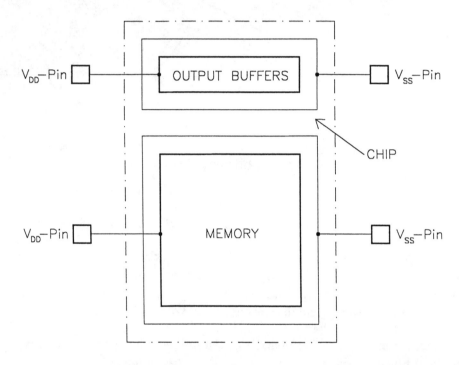

Figure 4.76. Architectures to reduce power-line bounces.

capacitances and inductances (Figure 4.77a) rather than to combine the output buffers and loads with the memory-global power route into common current loops (Figure 4.77b). Local current loops for the output buffers and loads decrease not only the ground- and power-line bounces but the switching times of the output-drivers as well. Although, longer output signal switching-times may decrease the amplitudes of ground- and power-line bounces, output buffer operations with decreased switching-speeds are unacceptable in most memory designs.

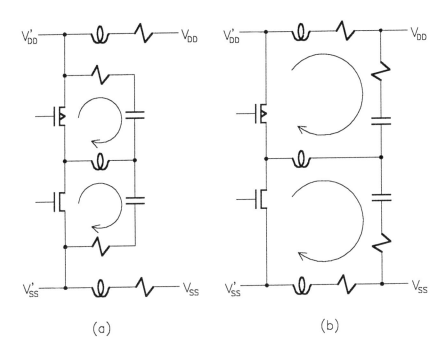

Figure 4.77. Designs of output-buffer to power-line connections for local current loops (a) and global current loops (b).

The use of differential sense amplifiers in memories conveys the idea of the application of differential data processing and differential data output circuits. Albeit the implementations of a symmetrical differential output buffers (Figure 4.78) are costly, because each differential output needs two output pads D_i and D_j, yet in a differential structure the bounce-signals can nearly neutralize each other. Furthermore, the differential signal transmission at reduced and optimized amplitudes (Section 4.1.1) can greatly improve both speed and power performances.

362 CMOS Memory Circuits

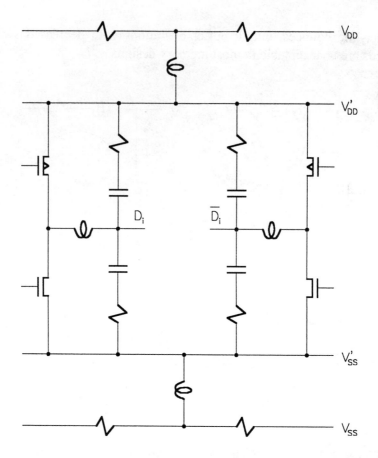

Figure 4.78. Differential output-bounce suppression.

The effects of fast data switching in the output buffers may also be mitigated by reducing the inductances and resistances of the power lines. Namely, both the wire inductance- and resistance-per-unit-length decrease with increasing wire width, and reduced power-line bounces can be obtained by applying wider power lines. Additionally, wide power lines lessen power-line bounces also by capacitance increase, and reduce electromigration by lessening current densities.

The continuous DC current density J, allowed by electromigration may be computed as a function of the maximum time-to-failure $\hat{t}_F$ [424]

$$J = \left(\frac{K_A e^{\frac{Q}{K_B T}}}{\hat{t}_F} \right)^{\frac{1}{N}},$$

where factor K_A depends on the wire-material's grain size, thermal-gradient and microstructure, $N \approx 1.5$ for memory designs, $Q \approx 0.6$ eV for pure Al and Al-Si alloys and $Q \approx 0.8$ eV for Al-Cu alloys, K_B is the Boltzmann constant, and T[°K] is the chip-internal temperature of the memory circuit.

5

Reliability and Yield Improvement

Reliability greatly effects the application area, environments, and costs, while yield strongly influences the manufacturing costs of CMOS memories. Both the reliability and yield of CMOS memories can highly be enhanced, in addition to various circuit and process technological approaches, by implementations of circuit redundancies. For CMOS memory circuits the issues; how redundancy effects reliability and yield, what the principal noises, failures, faults and errors are, what methods can be applied to control noises, reduce failures, repair faults and correct errors, how much redundancy is needed, and how to obtain fault-tolerance by fault-repair and error control code implementations, are described in the present chapter.

5.1 Reliability and Redundancy

5.2 Noises in Memory Circuits

5.3 Charged Atomic Particle Impacts

5.4 Yield and Redundancy

5.5 Fault-Tolerance in Memory Designs

5.6 Fault-Repair

5.7 Error Control Code Applications in Memories

5.8 Combination of Error Control Coding and Fault-Repair

5.1 RELIABILITY AND REDUNDANCY

5.1.1 Memory Reliability

Memory reliability R(t) is expressed by the probability that the memory performs its designed functions with the designed performance characteristics under the specified power-supply, timing, input, output and environmental conditions until a stated time t. Commonly, the reliability as a function of time R(t), at given conditions, is expressed by the cumulative distribution function, i.e., the probability of failure prior to some t time F(t); by the density function of the random variable time-to-failure f(t); by the hazard rate, i.e., the limit of failure rate when the time interval-length Δt approximates zero, $h(t)|_{\Delta t \to 0}$; and by the mean-time-to-failure, i.e. the average time to failure, MTTF [51]:

$$R(t) = 1 - F(t) = \int_0^\infty f(t)dt = \frac{f(t)}{h(t)} = e^{-\int_0^t h(t)dt} = \frac{-dMTTF}{dt}.$$

Failure, here, means an inability to perform a designed function or a parametric characteristic under the specific conditions the device is planned to operate.

The MTTF for a device may be computed through the failure rate $\lambda(t)$. For $\lambda(t)$ the definition is

$$\lambda(t) = \frac{R(t_1) - R(t_2)}{\Delta t R(t_1)}$$

where t_1 and t_2 gives the start and the end of the time interval $\Delta t = t_2 - t_1$, and $R(t_1)$ and $R(t_2)$ are the reliabilities at t_1 and t_2 times. If the failure rate $\lambda(t)$ represents the number of expected failures over a time interval Δt, then

$$MTTF = \frac{1}{\lambda \Delta t}, \text{ and if } t_1 = 0 \text{ then } MTTF(\Delta t) = \frac{1}{\lambda(t)}.$$

Reliability and Yield Improvement 367

Failure rates λ(t)-s vary with the time of device usage t as the traditional "bathtube" curves indicate (Figure 5.1). The shapes of the bathtube-curves change with the level of stress. Nevertheless, for each stress level, all of these curves the failure rate have (1) an initial rapidly decreasing part that represents infant mortality, (2) a central constant segment that corresponds to the useful device life, and (3) a final increasing portion that implies the wear-out in advanced age.

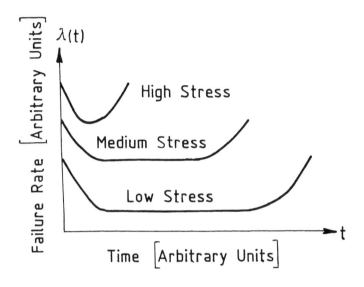

Figure 5.1. Failure rates versus time at various stress levels. (After [51].)

For any of the three portions of a bathtube curve, parameters R(t), λ(t), h(t), MTTF, etc. can be modeled by discrete and continuous standard statistical distributions. The generally applied statistical distributions are the binomial, Poisson, Gaussion, gamma, Weinbull Erlang, long-normal and exponential distributions.

In memory technology the exponential distribution is the most simple one to use, because the infant mortalities can be eliminated by burn-in and screening procedures, and both the hazard and failure rates, h and λ, can be approximated by constants. With the exponential assumption and with constant hazard and failure rates the reliability R(t) may be approximated as

$$R(t) \approx e^{-\lambda(t)} \approx e^{-\lambda t} = e^{-\frac{t}{MTTF}} = e^{-\frac{t}{h}}.$$

Although these approximations are questioned by some, it has been shown that they are adequate for reliability estimates of digital memories.

Generally, memory reliability at a set of operating and environmental is expressed as a triplet of number-ranges representing (1) the confidence level, (2) probability and (3) reliability parameters, e.g., with a confidence of 0.82±0.08, the probability is 0.9±0.1 that the lifetime of a 64-Mbit static memory is 2±0.2 years in a satellite on a geostationary orbit. CMOS memories and their constituent circuits, however, are often represented by doublets: (1) the fraction of the total number of structures, memories, memory arrays, subcircuits, bits, memory cells, etc.; and (2) the (range of) values of the reliability parameter (mean-time-between-failures (MTBF), mean-time-between-errors (MTBE), soft error rate (SER), time-to-failure (TTF), failure-in-time (FIT), average failure-rate, mean up-time, percentage up-time, etc.), e.g., the MTBF of a 256-Mbit dynamic memory CMOS array with built-in error correction is greater than 500,000 hours. Determination of reliability differences between two memories or constituent circuits can only be made if one of both doublet components is the same, and the assumed operating environments are also the same for the compared memories or constituent circuits.

Modeling of memory reliability is somewhat less complex than that of digital logic circuits, because in memories logic-state transition-rates vary moderately, rather than extremely as in logic circuits do. Nevertheless, reliability estimates for memory circuits, with better than first order accuracy, require the extensive use of computer programs, which are based on intricate, mostly on Markovian discrete-state continuous-time, models [52]. At the choice of reliability programs and models the designer should weigh the accuracy against the cost-effectiveness of model tractability. This, in turn, depends on the model's complexity for fault and recovery simulations, technique of numerical approximation to solution, requirements for computer operating system, output data, etc.

Reliability and Yield Improvement

Explicit reliability data are needed to consider the effects of fabrication, transportation, storage, hot-carrier emission, oxide wearout, electrostatic discharge, electromigration, latch-up, mechanical stress, corrosion, radioactive radiations and others in the memory circuit design. For circuit reliability simulation a number of simulation programs [53] are available in which a variety of failure-causing phenomena can be analyzed separately or combined by a multiplicity of models.

5.1.2 Redundancy Effects on Reliability

Apart from the conventional measures in design, fabrication and application, memory reliability can be improved by the application of some form of redundancy. The term "redundancy" probably was first used in information theory in 1920 by Nyquist, who referred to a "useless" sinusoid component signal that "conveyed no intelligence" as redundant. In CMOS memory technology, however, "redundancy" designates circuits which are added to the memory to repeat data storage, access, write, read and other or all memory functions, or to implement error detecting and correcting codes for the improvement of reliability and, what is more, fabrication yield.

A reliability improvement (insurance of operation) can be gained by adding only a limited amount of redundant elements. Beyond a limit, where the reliability increase is balanced by the reliability loss due to the inflated number of elements in the memory, a reliability decrease (nuisance) appears. In monolithic CMOS memories reliability improvements are restricted by compromises in silicon surface area speed, and power to much less amounts of redundant elements than reliability could be limited by the number of elements for the insurance-nuisance equilibrium.

Redundant elements can be applied to the memory in active or standby modes.[54] Active redundancy that duplicates a circuit (Figure 5.2a) improves the reliability of the nonredundant circuit R(t) to the reliability of the duplicated active circuit $R_{DA}(t)$ as

$$R_{DA}(t) \approx e^{-\frac{t}{MTTF}}(2-e^{-\frac{t}{MTTF}}) ,$$

while the reliability of a circuit duplication R_{DS} that uses the redundant circuit in standby mode (Figure 5.2b) may be calculated as

$$R_{DS}(t) \approx e^{-\frac{t}{MTTF}}(1-\frac{t}{MTTF}) \ .$$

The probability of success for active triplicate redundancy in a majority decision configuration (Figure 5.3) $R_T(t)$ when all three circuits are active is

$$R_T(t) = 3(e^{-\frac{t}{MTTF}})^2 - 2(e^{-\frac{t}{MTTF}})^3 \ .$$

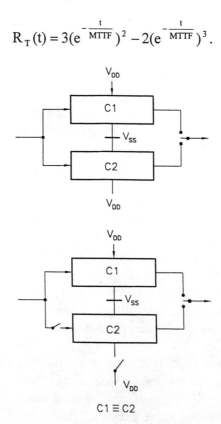

Figure 5.2. Active and standby mode redundancy by circuit duplications.

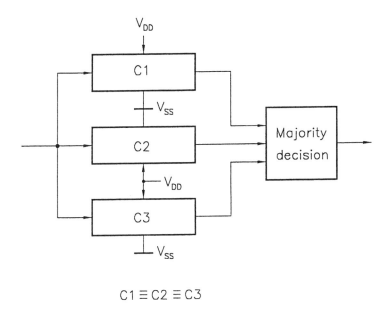

Figure 5.3. Redundancy by active circuit triplication.

The reliability of a memory that is composed of n identical and independent circuits, and m or more of the n circuits must be functioning to provide memory operations, may be expressed by the elementary n-modular redundant model $R_{NMR}(t)$ as [55]

$$R_{NMR}(t) = \sum_{i=m}^{n} \binom{n}{i} R(t)[1-R(t)]^{n-1} \quad .$$

If n-1 of n identical components maintained failure-free in standby position, and one of the n-1 components can be switched into operating with no cost associated to the switching, and if all n components have a constant failure rate λ, then the simple n-modular standby redundant model $R_{NSR}(t)$ with n-stage Erlang distribution may be applied;

$$R_{NSR}(t) = \sum_{i=0}^{n-1} \frac{(\lambda t)^i}{i!} e^{-\lambda t} \,.$$

In applications of redundant memories, the probability that the memory operates as designed at time t, called the instantaneous availability A(t), in presence of repairs [56]

$$A(t) = R(t) + \int_0^t R(t-x) dM(x) \,,$$

is a more descriptive reliability parameter than R(t) alone. Here, M(x) is the expected number of repairs in the time interval [0,x]. In the absence of failure reparation, A(t) becomes A(t) = R(t).

Estimation of reliability parameters R(t), A(t), etc. for systems has evolved as an important branch of mathematics, and many of the computer programs developed for systems can well be applied for analysis of memory reliabilities.

The rather simple analysis of R(t) and A(t) indicate clearly that the memory reliability can be improved without the use of redundancy by reliability increase of the constituent memory circuits and by the use of certain amount of redundant elements to repeat memory functions.

For reliability improvement on-chip redundancy is rarely employed on the chip of commercially applied memories, but commercial memories often incorporate redundant elements for yield increase. Generally, the extent of on-chip redundant elements in commercial memories is constrained by cost-effectiveness, yield and performance considerations, rather than by error rates acceptable in commercial computing and communication systems. Space, military as well as avionics, automobile, and industrial systems, however, demand strict reliability parameters, e.g., life time of 10 years, MTBF of 500,000 hours, etc., which are unlikely to satisfy without the use of redundant elements in addition to the use of worst-case statistical design approaches and to the tight control of CMOS parameters.

In addition to design and processing measures high reliability applications require redundancy in CMOS memory circuits. Redundancy in a memory chip is used mostly to implement (1) error detecting and correcting codes through added memory cells and control circuits, and (2) fault repair by using spare circuit elements and control circuits. The use of error control coding and fault-repair circuits greatly depend on the required and on the primary memory-inherent reliability and yield parameters. Reliability parameters of CMOS memories are influenced by the susceptibility of their component circuits to noises, hot charge-carrier emissions, impacts of charged atomic particles, thermal conditions, mechanical shocks, chemical erosions, radioactive radiations, and by the sensitivity to the other parameters which generally affect the functions of CMOS digital and analog circuits.

From the large variety of events affecting CMOS circuit reliability the following discussion includes three items; (1) noises, (2) impacts of charged atomic particles and (3) radioactive radiations, because their effects and the suppression of the effects may be specific to CMOS memories. Furthermore, for those memory designs in which reliability or yield (Section 5.4) requirements may not be satisfied by direct circuit design, fabrication and handling approaches, the reliability and yield improvement by employing memory-global fault-tolerance features is described (Sections 5.5-5.9).

5.2 NOISES IN MEMORY CIRCUITS

5.2.1 Noises and Noise Sources

Noises are unintentionally generated spurious signals which may cause errors in memory functions by temporary operating margin reductions, incorrect reads, writes and addressings, and by upsets of data stored in memory cells and of data processed in sense and peripheral logic circuits. In encapsulated chips, memory circuits may pick up noises from chip-internal and chip-external sources (Table 5.1).

Noises	
Chip-Internal	**Chip-External**
Crosstalking Power Supply Thermal	Electrical Mechanical Electromagnetic Radioactive

Table 5.1. Chip-internal and chip-external noises.

Since the noise signals, which are generated in CMOS memories by chip-external sources, can be perceived and analyzed the same way as in all other types of integrated circuits, this section focuses exclusively on the memory specific chip-internally generated noises.

Chip-internal noises occur due to capacitive couplings (crosstalk or induced noises), due to the nonzero resistance and inductance of power supply wires (power line noises), and due to the thermal fluctuation of discrete electronic charge elements (thermal noises). Noise couplings by inductive elements inside a memory chip is very small.

In a memory chip noises behave stochastically, yet the characteristics of crosstalk and power line noise signals are predictable.

5.2.2 Crosstalk Noises in Arrays

Crosstalk or induced noises appear when a signal of memory operation affects also circuit nodes other than the intended ones through a variety of coupling mechanisms, mainly through capacitive couplings. The largest capacitive couplings are among the bitlines, wordlines and decoder lines. In memory cell arrays parallel bitlines are placed perpendicularly to parallel wordlines, while in decoders the parallel input lines and the parallel output lines are laid out rectangularly.

The capacitive network of these parallel-rectangular structures (Figure 5.4) may be simplified for approximate computations by considering that the capacitances between direct neighbor lines dominate, i.e., $C_{ij} \gg C_{ik}$ and $C_{ik} \gg C_{il}$, by applying that capacitances $C_{ij}=C_{jk}$, $C_{iG}=C_{jG}$, $C_{iX}=C_{jX}$, etc., and by exploiting that the crosstalking from useful signals in wire X to lines h, i, j and k may be separated from the crosstalking among the parallel wires h, i, j and k. The capacitance between a signal-wire and the ground or the power supply C_{iG} combines parallel with the crossover capacitance C_{iX} when the wire X is on ground or on supply voltage potential.

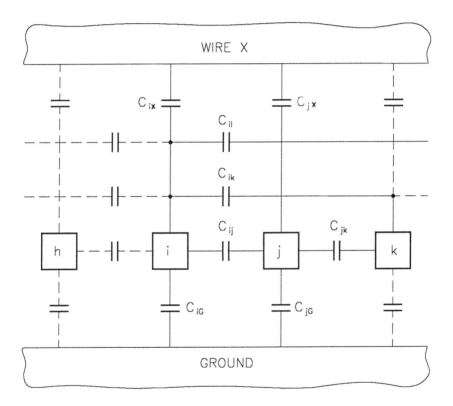

Figure 5.4. Crosstalk capacitances in a memory array.

Although the presence of capacitances among the wires is the cause of crosstalkings, neither the amplitudes nor the time-spans of the induced noise signals depend solely on the capacitances, but also on the impedances of the circuits coupled to the ends of the lines Z_i and Z_o, on the resistance of the line R and, of course, on the waveform of the generator signal $v_g(t)$ driving a signal line. Long lines in a memory array may also operate as transmission lines or microstrips. For line-to-line crosstalking analysis is sufficient, in most of the memory designs, to use a model (Figure 5.5) that includes three Π or T linear passive resistor-capacitance networks to model transmission-line effects. In the Π and T networks, the appropriately divided values of the line capacitances and resistances, e.g., C_{ij} (C_{ij1}, C_{ij2}, C_{ij3}, C_{ij4}), C_{iG} (C_{iG1}, C_{iG2}, C_{iG3}, C_{iG4}), C_{iX} (C_{iX1}, C_{iX2}, C_{iX3}, C_{iX4}) and R (R_1, R_2, R_3) for line i, have to be applied, while impedances Z_i, Z'_i, Z_o and Z'_o terminate the lines and generator $v_g(t)$ drives the selected line in the array circuit.

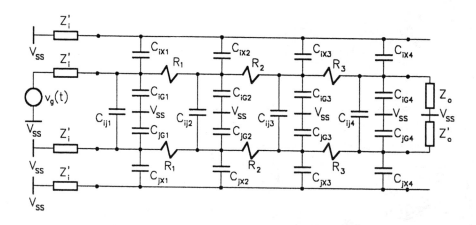

Figure 5.5. Model circuit for crosstalk analysis in an array.

To make plausible the effects of the dominating circuit elements on the crosstalk-signal a rudimentary model (Figure 5.6), which integrates the effects of Z_i, Z_o, R, C_{iG} and C_{iX} and of the generator signal $v_g(t)$ into the generator signal $v'_g(t)$; treats C_{ij} as the coupling capacitance C'; combines

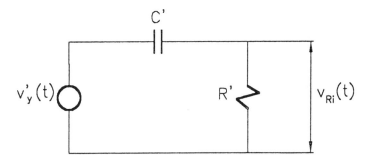

Figure 5.6. Rudimentary model for crosstalk-signal demonstration.

Z'_i, Z'_o and R in R', and disregards C_{jG} and C_{jX}; may be applied. Assuming that an exponential $v_g(t)$ appears in line i

$$v_g(t) = V_g(1-e^{-\frac{t}{\tau_g}}),$$

Here, V_g is the amplitude of the signal generated on line i, and τ_g is the time constant of the generator circuit determined by the impedances of the driver of line i and of the elements connected to line i. In this rudimentary model circuit, the Laplace-transformed of the voltage signal across R', i.e., of the crosstalk signal, may be gained as

$$V_R(p) = \frac{\frac{1}{\tau_g}}{\left(p+\frac{1}{\tau'}\right)\left(p+\frac{1}{\tau_g}\right)} V_g,$$

where $\tau'=C'R'$ and $R'=f(Z'_i, Z'_o, R)$. From $V_R(p)$ an inverse Laplace-transformation gives the time function of the crosstalk signal on line j as

$$V_R(t) = \frac{\tau V_g}{\tau_g - \tau'}\left(e^{-\frac{t}{\tau_g}} - e^{-\frac{t}{\tau'}}\right).$$

The expression of $v_R(t)$ clearly shows that the crosstalk-signal amplitude and waveform (Figure 5.7) depend on both τ_g and τ' and, through those, of all capacitances and resistances of the wires, drivers and terminating circuits. If in this circuit $\tau' \ll \tau_g$, $V_g \ll V_{DD}$ and V_{DD} is the supply voltage, then the amplitude of the crosstalk signal is small. A small τ' can be obtained by creating small line-to-line capacitance C_{ij}, large line-to-ground (line-to-supply) capacitance C_{jX}, small line-to-ground (line-to-supply) terminating impedances Z_i, Z'_i and Z_o, and small line resistance R. A large τ_g is undesired, because it would slow the operation by the effects of impedances Z_i and Z_o, resistance R, and capacitances C_{iG} and C_{iX}. Line-to-line capacitances are controlled, most often, by limiting the lengths of parallel running-signal lines, and by decreasing the fringing capacitances

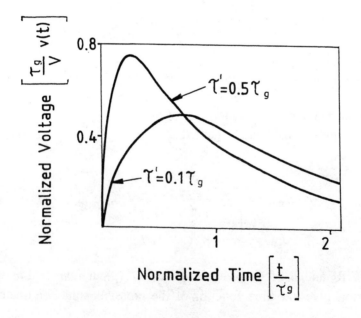

Figure 5.7. Crosstalk signals.

of the wires. Inasmuch as fringing capacitances determine line-to-line capacitances, the process technology has to aim to reduce the thickness of the conductors and to optimize the layout rules for acceptable crosstalking rather than for minimum area or for maximum speed conditions.

Improvements in both noise signal amplitudes and speed can be obtained by coupling drivers with low output resistances and receivers with small input resistances to the effected wires. Designs of low resistance drivers and receivers require wide transistors, yet the widths may be bounded by circuit area restrictions, e.g., row or column pitch. The general avoidance of "floating" circuit nodes, i.e., drain, gate, or source nodes which are disconnected from ground or supply lines, and the use of low-impedance sense amplifiers and line-buffers, are vitally important to keep chip-internally generated noises on small levels in high-density memory circuits.

Noise reduction can also be provided by passive shielding that protects a wire from the other ones' electric fields. Nonetheless, the passive shielding of electrical fields may require extra space, often increases process complexity, and the increased wire-to-ground capacitances may cause longer signal delays and elevated power dissipations.

5.2.3 Crosstalk Reduction in Bitlines

Particularly noise-prone is the read signal on the bitline in high-density memories. Although the folded bitline design exploits the high common-mode-rejection-ratios of the differential sense amplifiers, the capacitive coupling between the bitlines of neighbored sense amplifiers can pick up significant amount of noise signals through wire-to-wire capacitances, e.g., C_{ij} (Figure 5.8).

Significant reduction in noise coupling can be achieved by inter-digitizing the bitline structure (Figure 5.9). In the interdigitized structure, when a sense amplifier SA_i is activated, SA_I's mirror symmetric counterpart SA_j is passive, and the bitline pair connected to SA_j is tied to a supply pole through small resistances. The bitline pair of SA_j shields and greatly reduces the effective capacitive coupling between SA_i and SA_j, but increases the active bitlines cumulative wire-to-ground (supply) capacitances.

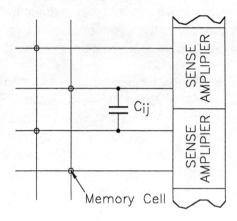

Figure 5.8. Wire-to-wire capacitances in a bitline circuit.

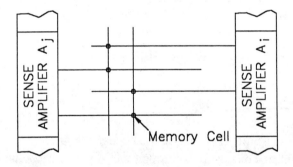

Figure 5.9. Interdigitized bitlines.

The twisted bitline design (Figure 5.10) does not increase wire-to-ground (supply) capacitances, but cancels the induced noise by coupling both the high and low components of the differential signal generated by the activated sense amplifiers through pairs of identical interbitline capacitors, e.g., through $C_{ij1}=C_{ij2}$.

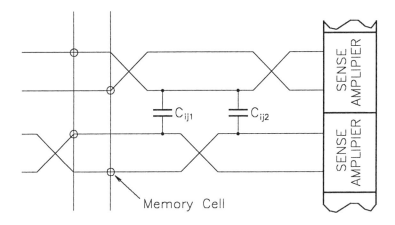

Figure 5.10. Twisted bitline design.

Simultaneous interbitline-capacitance reduction and induced noise signal cancellation can be obtained by the combination of interdigitized and twisted bitline designs (Figure 5.11). Interdigitized designs, furthermore, relax the area-constraints for sense amplifier layouts in memory arrays.

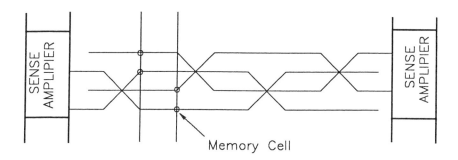

Figure 5.11. Combination of interdigitized and twisted bitline designs.

All types of random access memory arrays can be designed with interdigitized and twisted bitlines so that none of the bitlines float at any time. During the precharge time the bitlines are driven by low-resistance precharge circuits. Just before the time of precharge ends each of the unselected bitlines should be connected either to the bitline loads in static memories, or to a low impedance sense amplifier in dynamic memories.

5.2.4 Power-Line Noises in Arrays

Power-line noises appear every time when a logic state changes in the memory due to the effects of the parasitic resistances, capacitances and inductances which are associated with the ground and supply voltage wires. Although power wires are usually conductive and short enough to develop only negligible voltage drops on their resistances at the maximum DC currents, but when a number of memory constituent subcircuits switches logic states simultaneously and unidirectional, a considerable amount of transient currents combines with the DC currents. The transient currents generate voltage fluctuations on the power-lines, which fluctuations make the supply potentials V_{DD}, V_{SS} and V_{CC} functions of both the time and location of the observation ($V_{DD}(x,t)$, $V_{SS}(x,t)$ and $V_{CC}(x,t)$. Location and time dependency of supply voltage on power lines may result in degraded or diminished operating and noise margins, false precharge and incorrect logic levels in a memory.

Specific to memories are the power-line noises which occur in the memory cell arrays. In an array, the noise-sensitive circuit comprises a power line, a signal line, and either a driver, or a sense amplifier, or both. No sense amplifiers are applied in word-select and decoder circuits, and no drivers are employed in many sense circuits. In all three typical memory circuits, i.e., word select, decoder and sense circuits, power line noises may be analyzed on a single approximation model (Figure 5.12). In this model, the power-line is represented by the resistances R_{P1}, R_{P2} and R_{P3} and inductances L_{P1}, L_{P2} and L_{P3}, the signal-line is simplified in resistances R_{S1}, R_{S2} and R_{S3}, and the noise signal in the power-line is coupled to the signal-line through capacitors C_{C1} and C_{C2} and load impedance Z_L. An arbitrary change in power-line voltage can be imposed by the voltage generator $v_g(t)$ and generator impedance Z_g. A change in $v_g(t)$ causes a change in the loop-current i(t), and from i(t) the noise voltage $v_n(t)$ can be obtained for any increment of the power-line and of the signal-line resistance. Because the power-line and the signal-line are modeled here by means of two T-equivalents, this model allows for approximation of transmission-line delays.

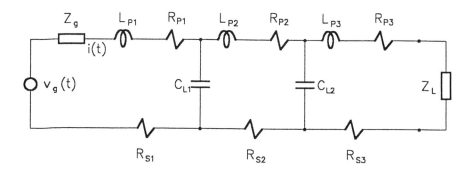

Figure 5.12. Model circuit for power-line noise analysis.

Where transmission-line characteristics are unimportant, i(t) may roughly be approximated by a simplified lump-element model that disregards transmission-line behavior and combines all passive elements in R', L', and C' (Figure 5.13). When the passive elements are driven with

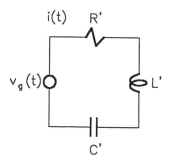

Figure 5.13. Simple power-line circuit model.

a voltage step $v_g(t) = V_g 1(t)$, the Laplace transformed of the loop current $I(p)$ may be expressed as

$$I(p) = \frac{V_0}{p} \frac{pC'}{p^2 L'C' + pR'C' + 1},$$

The reverse-transformed of $I(p)$ to the time domain $i(t)$ yields that

$$i(t) = \frac{V_0}{\beta L'} \left[e^{\left(-\frac{1}{2\tau}-\beta\right)t} - e^{\left(-\frac{1}{2\tau}+\beta\right)t} \right] \text{ if } \frac{1}{2\tau} > \omega_0,$$

$$i(t) = \frac{V_0}{\omega L'} e^{-\frac{1}{2\tau}t} \sin \omega t \text{ if } \frac{1}{2\tau} < \omega_0, i(t) = \frac{V_0}{L'} t e^{-\frac{1}{2\tau}} \text{ if } \frac{1}{2\tau} = \omega_0.$$

Here,

$$\beta = \left(\frac{1}{\tau^2} - \omega_0^2\right)^{\frac{1}{2}}, \omega = \left(\omega_0^2 - \frac{1}{\tau^2}\right)^{\frac{1}{2}}, \omega_0 = \left(\frac{1}{L'C'}\right)^{\frac{1}{2}} \text{ and } \tau = \frac{2L'}{R'}.$$

Depending on the magnitude relationship of $1/\pi$ and ω_0 current $i(t)$ can be either a nonperiodical or a periodical phenomenon (Figure 5.14).

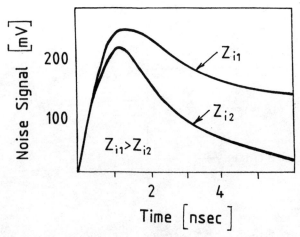

Figure 5.14. Periodical and nonperiodical power-line noise signals.

For both cases the maximum loop-current $\hat{i}$ may be approximated by

$$\hat{i} \approx \frac{V_0}{\omega_0 L} e^{-\frac{t}{\tau}},$$

where

$$t_m = \frac{1}{\beta} \text{arc}\,\beta\tau \quad \text{if} \quad \frac{1}{\tau} \geq \omega_0,$$

and

$$t_m = \frac{1}{\omega} \text{arc}\,\omega\tau \quad \text{if} \quad \frac{1}{\tau} < \omega_0.$$

Knowing i, the maximum voltage drop across any increment of the power wire can be estimated, and, in turn, the changes in operating and noise margins and precharge and logic levels can be calculated (Section 3.1.3). For more accurate computations of the levels, and shapes of noise signals, computer aid with models comprising transmission line losses and reflections should be used (Section 4.1).

The equations of i(t) and i indicate that power-line noises may be decreased to acceptable levels by applying highly conductive materials and by limiting line lengths in the arrays. An increase in line width is constrained not only by packing density decrease, but also by the increase of time constant τ, despite that $L' = f(L_p)$ decreases with increasing line width. Nonetheless, an increased L_p may be compensated by increased $C' = f(C_c)$ because the maximum crosstalk current i is inversely proportional with ω_0. In some CMOS memories, a large C_c between V_{DD} and V_{SS} is implemented as a chip-external capacitor on the printed board or in the package of the integrated circuit.

5.2.5 Thermal Noise

Thermal noises may set the theoretical limit to the minimum signal amplitude that can be sensed, but crosstalk and power-line noises are the dominating noise events in CMOS memories even at very small feature sizes. Thermal noises, nevertheless, add to the effects of the other noises in degrading reliability parameters, e.g., soft error rates SERs and, therefore, the ratio of the data-signal amplitudes to thermal noise-signal amplitudes ρ_{SN} should be large.

Among the variety of memory types, the expected thermal noise contribution to the SERs is the highest in dynamic memories, because they store the data on capacitors. On a capacitor C the average of the thermal-noise induced voltage fluctuations v_C may be determined by using the thermal noise voltage v_R across a resistor R in a simple RC circuit (Figure 5.15). In the depicted RC circuit,

$$v_R = \left(\frac{4KTR}{2\pi}\right)^{\frac{1}{2}}, \quad v_c(\omega) = \frac{\frac{1}{j\omega C}}{R + \frac{1}{j\omega C}} v_R,$$

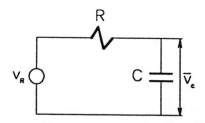

Figure 5.15. Modeling thermal-noise effects on a capacitance.

K is the Boltzmann constant, and T is the temperature in Kelvin grades. The analysis results that the average thermal-noise induced voltage

$$\overline{v_c} = \left(\frac{KT}{C}\right)^{\frac{1}{2}}$$

is independent of R. Applying v_C to a dynamic differential sense circuit that uses a crosscoupled differential amplifier (Figure 5.16), and assuming that all transistors in the amplifier operate in their saturation regions and all passive elements are linear; the signal-noise ratio ρ_{SN} can be approximated [57] as

$$\rho_{SN} \approx V_{PR} V_{DD}^{\frac{5}{4}} \left(\frac{C_S}{C_B}\right)^{\frac{3}{4}} \left(1 - \frac{C_D}{C_B}\right) \frac{(\beta t_{sen} \eta C_s)^{\frac{1}{4}}}{(\gamma KTK_T)^{\frac{1}{2}}}, \quad K_T = \text{Constant.}$$

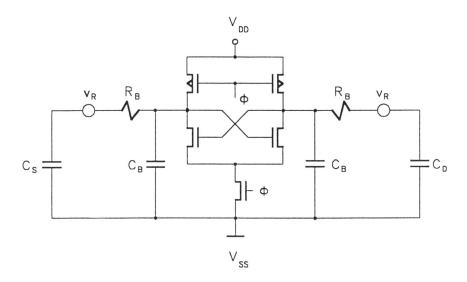

Figure 5.16. Differential sense circuit used in approximation of signal-noise ratio. (Derived from [57].)

The equation for ρ_{SN} leads to the conclusion that higher precharge voltage V_{PR} and supply voltage V_{DD}, large data storage capacitance C_S and dummy cell capacitance C_D, small bitline capacitance C_B, high transistor gain-factor β, long sense time t_{sen}, small storage charge decay η, low transistor noise coefficient γ and low temperature T, improve the data-to-thermal-noise ratio ρ_{SN}. Circuit coefficient K_T and its constituents, the gate-source voltage V_{GS} and the threshold voltage V_T of the crosscoupled transistors MN1 and MN2, are established by the design of the sense circuit and can be considered constants in ρ_{SN} calculations. Improving ρ_{SN}, by adjusting the component parameters of ρ_{SN}, decreases also the sense circuits' sensitivity to crosstalk signals, power line noises and to the effects of atomic particle impacts. Thus, the effects of thermal noises on memory operations decrease with all facets of the general reliability improvements.

5.3 CHARGED ATOMIC PARTICLE IMPACTS

5.3.1 Effects of Charged Atomic Particle Impacts

Impacts of charged atomic particles may result in randomly located and randomly timed anomalies in the memory operation. These anomalies are observed in both terrestrial [58] and space [59] environments, and attributed to ionization in the semiconductor material by alpha particles radiated from the memory's own packaging and by a variety of ions, protons and electrons ever present in cosmic-rays and cosmic events. Alpha particles are emitted by the thorium and uranium contamination of the chip-package and lead-frame materials. Other materials used in the processing may also radiate a trace amount of ionizing particles. In cosmic environments, the impact of heavy ions, including the members of the iron group with atomic numbers of $Z>22$, of the aluminum group with $Z=10-21$, and of the carbon group with $Z=3-9$, as well as the alpha particles (He^+) and protons, cause most frequently errors in memory circuits.

On the Earth, the alpha and other radiations from memory-chip packaging and lead frame may induce soft-errors by upsetting the logic state of memory cells and sense amplifiers and, in a much lower probability, by upsetting the functions of the peripheral logic circuits. Nevertheless, in the effected circuits these radiations do not cause permanent damage.

In cosmic environments structural damages and, thereby, hard-errors may also appear as results of impacts of very high-energy particles, in addition to frequent soft-error occurrences. In the near-Earth atmosphere, e.g., in high-flying airplanes and missiles, the incident ionized particles are of low atomic numbers and, usually, do not have sufficient energy to directly induce errors. However, high-energy charged atomic particles may be generated indirectly by nuclear reactions initiated by the incidents of medium-weight alpha particles, protons and neutrons in the semiconductor crystals and in the oxide materials, and these generated particles may provoke operation errors. In semiconductor memories, which have to operate within the belts of ionizing particles trapped by the Earth's magnetic field, e.g., in satellites orbiting around the Earth, errors occur most often due to MeV-protons. In the regions outside of the effective magnetic field of the Earth, i.e., in the cosmic space, the impacts of heavy cosmic ions are the principal causes of errors. Most of the error-causing particles have very high average energy ranking 10-1000 MeV, and appear in cosmic rays and solar winds. Protective shielding against cosmic events effects are ineffective, because the reduction in incident ion-energy, so obtained, is usually insufficient to appreciably effect the total charge amount induced by the cosmic particle impact.

The ultimate effect of the impact of a charged atomic particle is the creation of free electron-hole pairs along the path a particle travels through the material and around the centers of nuclear bursts initiated by an incident particle in the semiconductor material. In accordance with the generally accepted models for particle impacts [510], those electrons and holes which are raised to the conduction band within the depletion regions and within the gate insulators are separated by the rather high electric field induced by the potentials of the drains, sources and gates of the transistors and by the supply voltage and ground nodes. Electrons are swept to the positive potential, and holes are swept to the negative potential regions. Electrons and holes generated outside the depletion region diffuse through the bulk silicon, and those reaching the boundaries of the depletion region are swept into the storage area. At the data storage nodes, the sense amplifier inputs and various other nodes, the prompt appearance of free electrons and holes generate spurious currents which may upset the stored or the processed data.

The upsets are randomly located in the memory and randomly timed during memory operations. These types of anomalies in memory operations are called single-event-upsets (SEUs) or single-event phenomena (SEP). An SEU rate of a memory is the number of error events caused by SEU per time unit per memory. Since in a well designed CMOS memory, most of the soft-errors are results of particle impacts, the number of soft-errors per time unit per circuit the soft-error-rate (SER), and the number failures per one billion

390 CMOS Memory Circuits

device-hours per memory chip, i.e., the failures-in-time (FIT) are also used, somewhat imprecisely, to indirectly indicate SEU rates. Particle impact induced SEUs have the highest probability to cause the shortest mean-time-between-errors (MTBE) and mean-time-between-failures (MTBF). Thus SEU or SEP rates, SERs and FITs are important indicators of CMOS memory reliabilities.

5.3.2 Error Rate Estimate

For characterization of a memory's susceptibility to atomic particle impacts, mostly the SER is used. The SER is usually provided by the manufacturer as results of accelerated tests. During development and design of a memory, however, test data are scarce or unavailable, therefore the SER should be analytically approximated.

To a memory's SER, the following approach provides an estimate within a factor of two. In this estimate, the circuit-technical base is the introduction of equivalent critical charge Q_C. Here, the critical charge determines the minimum quantity of charge which is able to alter the logic state of a memory cell, and the equivalent critical charge is defined as the charge-quantity that degrades an operation margin to zero in a sense circuit.

The equivalent critical charge Q_C can directly be determined, by using the node capacitance C_n, the width of the particular operation margin V_m, and the time constant $\tau_n = C_n R_L$ of the data leakage in normal operation;

$$Q_C = C_n \int_0^t v(t)dt \approx C_n V_m e^{-\frac{t}{\tau_n}} .$$

In the expression of τ_n, the equivalent resistance R_L can be determined by the node voltage and the leakage current from the node. The time integral of the current pulse $i(t)$ that induced by the impact of a charged particle (Figure 5.17)

$$Q_C = \int_0^t i(t)dt$$

provides connection between the critical charge and the current pulse.

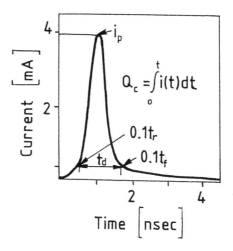

Figure 5.17. Current pulse generated by an atomic particle impact.

To change a datum in a static memory cell, in addition to the critical amount of charge Q_C a minimum peak current i_p also has to be reached (Sections 2.4 and 2.5). With current i_p an equivalent critical charge Q_{Cs} may approximately be calculated as

$$Q_C \approx \frac{1}{2} i_p t_d \quad ,$$

where t_d is the impulse duration from $0.1 i_{pr}$ to $0.1 i_{pf}$, and i_{pr} and i_{pf} are the rise and fall edges of the impulse.

To generate Q_C the incident particle must deposit a sufficient amount of energy in the material. The deposited energy is a function of the energy of the incident particle E, the stopping power dE/dx, and of the length of the ionization track x in the material. The track length x depends on the incident angle α, atomic number Z, mass M, and initial energy E_0 of the incident particle.

An exact calculation of the various energy depositions for the various possible track lengths is a complicated problem and requires extensive use of computers. Moreover, the computations with the incident uncertainties disallow an explicit expression of individual particle energy deposition. First order approach to energy deposition [511], however, may be obtained by taking an average track length S through the sensitive region. Approximating the sensitive region by a parallel-piped of dimensions l, w, and h the average track length S can be expressed as the ratio of the volume V=l x w x h and the average projected area A_p;

$$\overline{S} = \frac{V_n}{\overline{A}_p} \quad \text{where } \overline{A}_p = \frac{lw + lh + wh}{2}.$$

Along a track length S the energy deposited in the sensitive volume E is

$$E = \int_0^{\overline{S}} \left(\frac{dE}{dx}\right) dx \approx \overline{S} \frac{dE}{dx} \geq \check{E}$$

and E should exceed a minimum energy $\check{E}$ to be able to perturb data. Deposited energy E can be related to the equivalent critical charge Q_C of assuming an ionization rate in silicon v_i=3.6 V/carrier pair, electron charge q= 1.60203 x 10^{-13}, electron mass energy equivalent ρ_e= 5.109 x 10^{-1} MeV, and a charge collection efficiency f_n by the equation

$$Q_C[pCb] = \frac{1}{22.5} f_n E[MeV] .$$

Combining the equations of Q_C with the expression of E yields that the minimum stopping power dE/dx required for minimum deposited energy $\check{E}$ to generate an equivalent critical charge Q_C along an average track length S is

$$\left(\frac{d\check{E}}{dx}\right) \approx \frac{\check{E}}{\overline{S}}\left[\frac{MeV}{\mu m}\right] = \frac{22.5 Q_c}{f_n \overline{S}}\left[\frac{pCb}{\mu m}\right].$$

Thus, E and dE/dx restrict the incident particles, which may cause errors, to certain types and energy ranges. Both the particle types and energy ranges can be determined from experimental or calculated stopping power versus energy curves [512] (Figure 5.18).

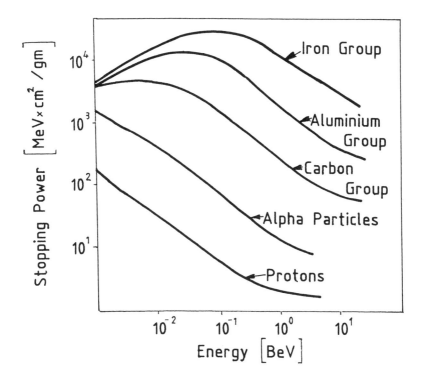

Figure 5.18. Stopping-power versus energy for various atomic particles.

For particles which meet the energy and stopping-power requirements the omnidirectional flux ϕ_E [number of incident particles/cm²day] may be obtained from $\phi_E = f(Q_c, S)$ functions (Figure 5.19). The product of flux ϕ_E and average projected area A_p of the sensitive region approximates the SER. For a memory cell SER_{cell} is

$$SER_{cell} \left[\frac{\text{number of errors}}{\text{day}} \right] = \overline{A}_p [\text{cm}^2 \phi_E] \left[\frac{\text{number of incident particles}}{\text{cm}^2 \text{day}} \right].$$

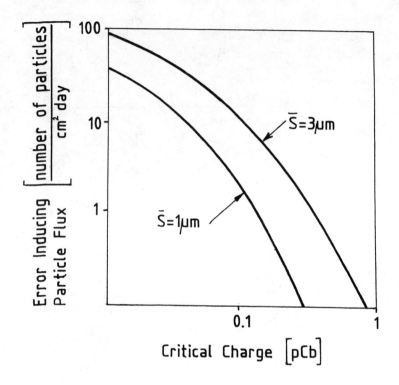

Figure 5.19. Omnidirectional flux versus critical charge at constant lengths. (Source [511].)

The SER of a memory array SER_{array} is the SER_{cell} multiplied by the number of memory cells on the chip N

$$SER_{array} \approx N\overline{A}_p \phi_E = N SER_{cell} \ .$$

This approach to SER chip neglects the soft-errors which may occur on the sense amplifiers and other precharged circuits. Although sense amplifiers often have the most error-prone circuit nodes (Figure 5.20), the probability of read and write errors due to incorrect sensing is by factor of 10^4-10^8 less than the error-probability of data upset in the memory cells.

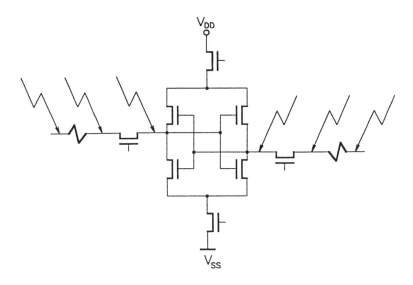

Figure 5.20. Error-prone nodes in a sense amplifier circuit.

This is because in a memory chip sense amplifiers occupy very little silicon area in comparison to the memory cell arrays, and because the state of a sense amplifier can be changed by particle impacts only in a small fraction of the read or write cycle times. Nevertheless, a particle hit on a sense amplifier may cause burst errors, and therefore designs for extreme environments may have to consider the soft error rate of the sense amplifiers SER_{SA} and

$$SER_{SA} \approx M\overline{A'_p}\, \phi_E \; ,$$

where M is the number of the sense amplifiers per chip, and A'_p is the total average projected sensitive area in a single sense amplifier circuit.

Decoder circuits may also be susceptible to atomic particle impacts, especially in designs which allow floating nodes. A particle hit on the floating nodes, or on nodes coupled to others only by high impedances, may result in placing correct data into incorrect addresses. Thus, incorrect addressing by the rate of soft-errors in the decoder SER_{DEC} should also be regarded in the chip total SER

$$SER_{DEC} \approx K\overline{A_p''} \phi_E .$$

Here, K is the number of equivalent decoder subcircuits per chip and A_p'' is the average projected sensitive area per subcircuit.

Other circuits, in a memory chip, may also be susceptible to charged-particle impacts, but experiments proved that their contribution to the soft-error rate of memory chip SER_{chip} is insignificant in most of the designs. Thus, the SER_{chip} may be approximated as

$$SER_{chip} \approx SER_{array} + SER_{SA} + SER_{DEC} \approx SER .$$

In terrestrial environments, experimental SER data on CMOS memories normally yield a simple step-like function in the diagram of the experimental particle-sensitive cross section σ versus linear energy transfer LET (Figure 5.21).

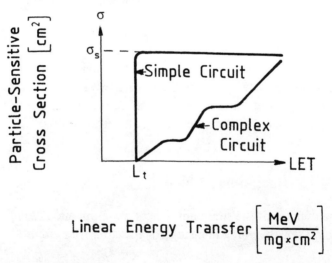

Figure 5.21. Particle-sensitive cross section as a function of linear energy transfer. (After [513].)

LET [eV/mg cm^2] is the linearized equivalent of the stopping power dE/dx, and σ is the experimental counterpart of the theoretical A_P, and at a given LET

$$\sigma = \frac{SER}{\phi_E R_r} \left[cm^2 \right] ,$$

where R_r is rationalization factor for convenient analysis of e.g., R_r = M, N, K. The simple σ = f(LET) curve is usually reduced to a two-point data set that includes the saturated cross section σ_s and the threshold LET L_t. L_t is determining either by linear extrapolation of the σ curve or by projection of $0.1\sigma_s$ to the LET axis. For ions depositing charge above LET = L_t the geometric surface area of the sensitive volume σ increases rapidly to σ_s. At σ_s the collected charge is large enough for every incident particle that strikes the sensitive volume to cause an upset, and σ_s remains approximately constant with further increasing LETs. Experimental σ_s and L_t provides acceptable SER accuracy in estimation of terrestrial package radiation effects, but for complex circuits operating in cosmic environments σ =f(LET) often differs from a step function, and the spectrum of particles, their energy and fluxes vary with the location and time of the orbiting or travelling space object [513].

For both space and terrestrial applications increased accuracy in SER prediction can be obtained by the use of another multiplicative factor, the error probability per particle incidence ε [514]

$$\varepsilon = \int_0^\infty D(Q) dQ ,$$

where D(Q) is the probability that the generated charge Q appears in an error causing zone ΔQ

$$D(Q) \Delta Q = \int_0^\infty dE \int_0^{\frac{\pi}{2}} d\alpha \int_0^{2\pi} d\beta \int_{X_p}^{Y_p} dy_p F(E, \alpha, \beta, X_p, Y_p) ,$$

Here, E is the kinetic energy, α and β are the incident angles, X_P and Y_P are the incident position coordinates and F is a normalized distribution function.

5.3.3 Error Rate Reduction

The first order approach to SER calculations (Section 5.3.2) demonstrate the desirability for large equivalent initial charge, large data-hold current, short track lengths through the sensitive volume and small collection efficiency as means of SER reduction. Both the critical charge and the sensitive track lengths decrease with the evaluation of the CMOS technology. Nevertheless, the SER increase by reduced critical charges is more significant than the SER decrease by shorter track length.

To limit SER, dynamic memory designs enlarge storage capacitors by using three-dimensional (3D) structures and thin insulators with high dielectric constants in dynamic 1T1C memory cells (Section 2.2.4). In static memory technology, the quest for low standby dissipation and high packing density resulted in 4T2R memory cells with extremely high load resistances, low hold currents and small node capacitances, which increased the SER's of the CMOS memories. SER can be improved in 4T2R memory cells by the increase of the current drive capability of the constituent transistors and of the storage node capacitances (Section 6.2.4). Device current and node capacitance enlargements are limited, nevertheless, by the practically deliverable write currents and by the projected size of the memory cell.

Traditionally the best SER characteristics are provided by full-complementary 6T static memory cells due to their high data-hold currents for both log.1 and log.0. In fact, 6T static memory cells with added resistive and capacitive RC elements (Figure 5.22) for Q_C increase, are the prevailing choices to memories operating in space, military and other extreme environments [515], in spite of these cells' rather large size. Since the size of 6T cells can be made the same as that of the 4T2R cells by placing polysilicon thin-film transistors on the top of the traditional transistors (Section 2.4.4) high packing density can be combined with very low SER, very low power dissipation and high operational speed.

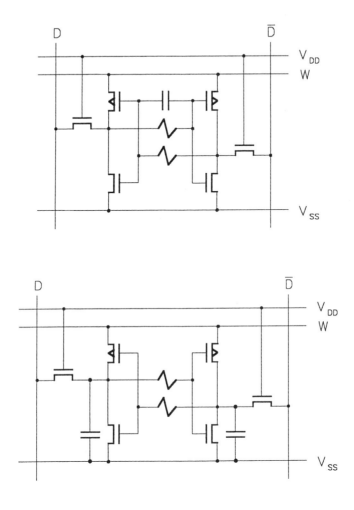

Figure 5.22. Six-transistor static state-retention memory cells.

In CMOS memories SER decrease can also be obtained by using current sense amplifiers because of their low input impedances. In voltage sense amplifiers, a reduction in input impedance results also in receding particle-generated spurious signals [516] (Figure 5.23) and in smaller SERs for the circuit. Reduction in SERs can be obtained, furthermore, by decreased number and, ultimately, by elimination of sense amplifier circuits, e.g., by using orthogonal shuffle and shift register type of memory arrays.

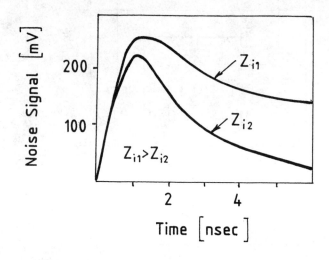

Figure 5.23. Particle-generated spurious signals on a sense amplifier input node at different input-impedances.

In commercially applied memories, the large bit-capacity per chip and cost-effectiveness are the primary concerns and, therefore, most of the commercial memories use the smallest possible dynamic 1T1C and static 4T2R memory cells. To improve SER in commercial memories, the storage capacitances and hold currents have been increased, and the collection efficiencies have been decreased, by process technological approaches, e.g., by three-dimensional structures, doping profile adjustments in the channel areas, epitaxial layers, etc. Because these process technology improvements have been developed simultaneously with feature-size downscaling the failure-in-time (FIT) per bit for terrestrial alpha-particle impacts has been improved with increasing memory bit-capacities per chip (Figure 5.24). With increasing FIT per bit the soft-error-rate SER lessens. Lessening SERs indicate that terrestrially applied memories may not need any special circuit techniques to reduce alpha particle sensitivity. Nevertheless, the need for special circuit techniques within the memory, or in the system outside of the memory, can be determined only after SERs have been thoroughly analyzed on the specific memory design.

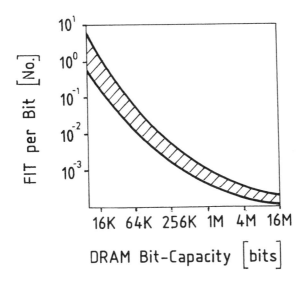

Figure 5.24. Failure-in-time versus DRAM bit-capacity.

For CMOS-bulk memories the epitaxial and other buried layers may be designed so that they divert the free charges induced by incident atomic particles from the data-holding nodes and, thereby, they decrease considerably the charge collection efficiency of the memory circuit nodes.

More improvement in memory SERs can be achieved by the application of CMOS-silicon-on-insulator CMOS-SOI and silicon-on-sapphire CMOS-SOS fabrication technology. CMOS-SOI and CMOS-SOS structures provide short possible track-lengths for an incident particle because they apply very thin silicon films and because the sizes of the transistor islands in the memory cells are small (Section 6.3).

In large bit-capacity memories where the memory cell size minimization is a primary concern, and in memories in which neither circuit nor processing approaches can satisfy environmental requirements; the implementation of error detecting and correcting codes (Section 5.7) and, eventually, of fault-repair (Section 5.6) may be justified. Generally, the implementation of the encoding and decoding of error control codes, the redundant code bits, and the redundant elements for fault-repair, require additional circuits and, in turn, compromises in memory chip size, operational speed and power dissipation. Nonetheless, the improvement in memory reliability, achievable this way, allows to produce memory chips with bit-capacities, environmental tolerances and yields which are difficult or impossible to attain by other approaches.

5.4 YIELD AND REDUNDANCY

5.4.1 Memory Yield

Memory yield is expressed by the percentage of fully functional memory chips from all of the fabricated memory chips or, alternatively, by the ratio between the fully functional memory chips and all fabricated memory chips. Fully functional memory chip means that each and all memory cell and memory circuit perfectly performs each and all designed functions, i.e., access, write, storage, read, input, output and other operations, under the planned conditions. The percentage of fully operational memory chips obtained from all the chips on all the wafers is indicated by the wafer yield, and the yield gained after assembly and packaging is the manufacturing yield.

Memory fabrication yield is limited [517] largely by random photo defects, random oxide pinholes, random leakage defects, gross processing and assembly faults, specific processing faults, misalignments, gross photo defects and other faults and defects (Figure 5.25). The defects and faults result predominantly in random single bit errors and much less frequently, in burst, cluster and double bit errors, and also in totally dysfunctional memories.

The need for memory yield improvement has been instrumental in the development of yield models for all types of semiconductor digital integrated circuits [518]. Yield models approximate the probability that the yield is Y. Y is a function of chip size A, defect density D, wafer size A_w, lot

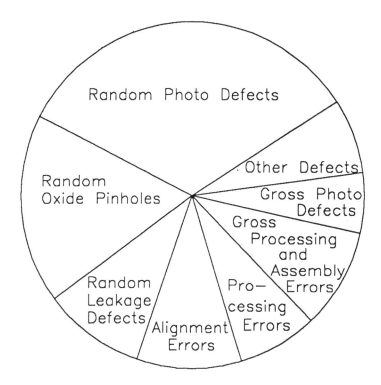

Figure 5.25. Distribution of defect types after fabrication.

size N_L, number of chips on a wafer M, intrawafer fault-clustering factor α_1, interwafer fault-clustering factor α_2, defect size S, geometrical and other variables, and a plethora of sophisticated computer programs have been developed and used for yield approximations.

Simple approximations to Y assume that the yield is determined by the critical area in which point-defects cause functional errors A, and by the mean density of point-defects if they were evenly distributed on the defect sensitive surface D. Further assumptions, in simple models, include that the defect distributions for Y obey either Maxwell-Boltzmann [519], or Poisson [520], or Bose-Einstein [521] statistics. Rudimentary use of Maxwell-Boltzmann and Poisson statistics result too pessimistic expectations, while Bose-Einstein statistics give too optimistic yield estimates. At high yields all models provide good approximations [522], but the usual choice of convenience is the Poisson model. Yield calculations with Poisson model do not require established fabrication data, and computations with Poisson statistics is uncomplicated.

A fabrication yield-loss is usually due to a number of different defect types rather than to one dominant cause. If for a defect type i the critical area is A_i and the defect density is D_i, then for a multiplicity of defect types the assumption of Poisson distribution results a yield

$$Y = \prod_i e^{-A_i D_i} \quad ,$$

while the Bose-Einstein statistic gives

$$Y = \prod_i \frac{1}{1 + A_i D_i} \quad .$$

Individual defect types, nevertheless, tend to cluster into higher and lower defect density area. Dividing the wafer area S to S_i regions, so that $S = \Sigma S_i$, and S_i is the area of the i-th region over which the defect density is reasonably uniform; a good yield estimate for a single wafer may be obtained [523]

$$Y = \sum_i \left(\frac{S_i}{S} \right) e^{-D_i A} \quad .$$

Somewhat more sophisticated models attempt to consider clustering by including probability density functions P(D) or PDF [524]

$$Y = \int_0^\infty P(D) e^{-AD} dD .$$

For P(D) a number functions are introduced and good yield approximations are obtained by the use of triangle distribution

$$Y = \left(\frac{1-e^{-AD_o}}{AD_o}\right) .$$

gamma functions [525]

$$Y = \left(1 + AD_o \frac{\sigma}{D_o}\right)^{-\frac{D_o}{\sigma}} .$$

and Erlang distribution [526]

$$Y = \frac{1}{\left(1 + \frac{AD_o}{K}\right)^K} ,$$

where D_o is the average defect density, σ is the standard deviation and K is the number of processing steps.

Semiconductor processing may cause defects which greatly vary in their sizes. Defect size variations may be considered by the applications of probability density functions PDFs [527]. From the plethora of PDF statistical distribution functions, so far, the Rayleigh distribution

$$f(x,\gamma) = \frac{x}{\gamma^2} e^{-\frac{x}{2\gamma^2}} \quad \text{if } x > 0,$$

$$f(x,\gamma) = 0 \quad \text{if } x \leq 0 ,$$

brought the most acceptable results in modeling single source defects. Here, x is the distance along the wafer diameter, and γ is the defect distribution parameter.

Processing experience indicated that both on-wafer and wafer-to-wafer defect density distributions differ from each other, and most of the chips on the wafer edges are unusable. Assuming within a wafer and among wafers the defect density distribution can be modeled by Poisson and gamma distributions respectively, then for i types of defects the yield [528] is

$$Y = Y_0 \prod_i \left(1 + \frac{\lambda_i}{\alpha_i}\right)^{-\alpha_i},$$

where Y_0 is the yield after gross failures, λ_i is the expected number of defects, and α_i is the clustering coefficient. In large chips the distinction of intrawafer clustering coefficient α_1 from interwafer clustering coefficient α_2 is of increased importance. With the number of on-wafer fault-causing defects λ_w, fault distribution function for λ_w-s wafer-to-wafer variations $g_w(\lambda_w)$, fault density q, fault distribution on a single wafer $P(\lambda_w,q)$ and total number of circuits N; a formula [529] which reasonably estimates semiconductor memory yield Y may be obtained as

$$Y = \int_0^\infty \int_0^\infty e^{-qN} P(\lambda_w,q) g_w(\lambda_w) dq dA_w .$$

Here, by P and g_w are Poisson and gamma distributions, but other distribution functions may also provide results which approach the experientially acquired yield figures.

5.4.2 Yield Improvement by Redundancy Applications

Memory yields can be improved, in addition to stringent fabrication control and statistical worst-case design, by implementation of redundancy. For yield increase redundancy is implemented by on-chip spare elements including spare bits, rows, columns and blocks of memory cells, duplication and triplication of sense, and peripheral circuits, or by repetitions of entire memory circuits. Furthermore, on-chip circuits implementing error detector and corrector codes are also frequently applied in CMOS memories for yield improvements, particularly in large read-only and nonvolatile memories. Redundancy in memories increases access and cycle times, power dissipation chip area, and require modifications in the design. Because of the substantial design tradeoffs, careful analysis should precede the decision whether redundancy should be used, and if it is used, how much redundancy would approximate the maximum achievable yield improvement.

Memory yield improvement by on-chip redundancy applications is justified mostly by requirements for (1) reducing cost-per-bits in large capacity memories, (2) increasing memory bit capacities at immature processing, and (3) providing fully functional parts in very low volume memory production. Cost-per-bit of CMOS memories, with or without redundancy, as of other integrated circuits, can also be decreased by reduction of chip sizes. Thus, CMOS memory chips designed without redundancy application may result considerably lower cost-per-bit than memory chips designed with redundancy do. The use of redundancy may also allow for obtaining fully functional memories fabricated with immature processing technologies when the amount of processing defects would cause near zero manufacturing yield. Moreover, with redundancy some yield may be acquired in such a few processing lots and runs, in which processing parameter variations do not follow the distribution statistics that the memory designer assumed. For correctly designed memories, nonetheless, which are fabricated in large volumes, the very redundancy, which is rendered to improve yield, may become a yield limiting factor as the fabrication technology matures with the time of production (Figure 5.26). To improve yield after the processing matured, the design should provide the option of removal of the inept redundant elements from the memory.

Memory yield improvement by redundancy applications is limited by the amount of redundant elements used on a single chip. On-chip redundancy increases the chip size, and the expansion of the chip size tends to reduce yield. Yield improvement by redundancy implementations may be designed (1) by optimizing the number of on-chip redundant elements to obtain the highest achievable yield, and (2) by using a given number of redundant elements and compute their effects on the yield. In CMOS memory yield improvements, the amount of redundancy for the memory cell

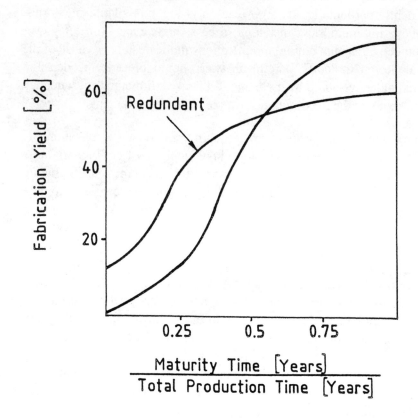

Figure 5.26. Yield increase as a function of fabrication maturity.

arrays are usually optimized, while for the duplication or triplication of elements to peripheral circuits, or functional entities, the yield is designed to be higher than that of the arrays. In memory circuit designs, the yield optimization and yield computation rely on both appropriate statistical models and experimental data.

A model's exactitude in estimating memory yield depends greatly on the parameters of the applied fabrication technology, fabrication facility and circuit design, and what is more, the validity of various models are widely debated and challenged. Inadequacy or misuse of models in yield prediction may lead to financial disasters. For yield optimization, however, not the yield itself but the amount of spare memory cells that maximizes memory yield, i.e., the optimum number of redundant elements, is the most important parameter. To estimate the optimum number of redundant elements for a specific memory design, the evaluation of the effective yield Y_{eff} versus the number of redundant elements N_R at constant defect densities and at given numbers of critical layers (Figure 5.27) may favorably be applied. At a certain defect density, the effective yield is the ratio between the number of

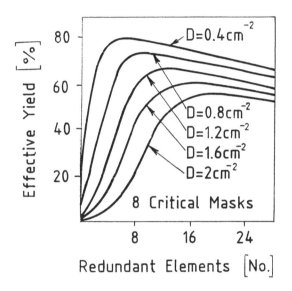

Figure 5.27. Effective yield versus number of redundant elements for a DRAM with 1-Mbit redundant elements.

chips on a wafer with redundancy and the prospective number of chips on the same wafer without redundancy multiplied with the percentage fabrication yield, at a specified number of critical layers. In CMOS memory technology critical is the layer (the mask and the accompanied processing steps) in that a small, but finite size, defect can impair the function of a storage cell. The same defect size, that impairs memory cells, can not necessarily make other memory circuits dysfunctional, because the packing density of peripheral circuits is usually much less than that of memory cell designs. A memory cell as well as complete rows, columns, blocks or arrays of memory cells and any other subcircuits, may be replaced by their on-chip copies which are designated, here, as redundant elements. Above a certain number of redundant elements the effective yields curve optimums. These optimums are rather flat and appears approximately at the same number of redundant elements when computed with a variety of different models. Computation results for effective yields and optimum number of redundant elements indicate that for lower defect densities smaller number of redundant element provide optimum yield, and the effective yield increases with lower defect densities.

Yield increase of large memory chips, e.g., ultra high scale integrated ULSI and wafer scale integrated WSI memories, may require duplication 2X, or triplication 3X of peripheral circuits, whole memory arrays or complete memories. Since in 2X and 3X redundant memory circuits the amount of redundancy is given, the anticipated fabrication yield of the redundant memory Y_R is the parameter to be obtained. For convenient computations, the yield with redundancy Y_R may be viewed as a product of three terms [529] gross failure limited yield Y_o, random defect limited yield without redundancy Y, and random defect limited yield with redundancy Y_r

$$Y_R = Y_o \, Y \, Y_r$$

Y_o is usually obtained from fabrication experience, while Y can either be extracted from fabrication experience, or be approximated by one of the previously discussed analytical yield models. For calculation of Y_r, yield models are often used, because experimental results for Y_r can be scarce at the onsets of the designs. Y_r may be approached by dividing the chips into β number of equal size domains, each domain into two identical prime and unprime elements, and an element into n number of circuits. While the total number of wired up circuits required to function is N = nβ, the total number of circuits per chip in 2X redundancy configuration is $2N = N^1$. Assuming any single fault impairs only one element, the yield Y_r may be expressed as

$$Y_r = \prod_k [1 - (1 - e^{-q_k n})] \; ,$$

where K is the domain identification number and q is the density of the local fault-causing defects. If the q_K-s are Poisson distributed, Y_r may well be approximated by

$$Y_r = e^{-\frac{\lambda_{wo}^2}{\beta}},$$

where λ_{wo} is the expected number of faults in N circuits.

The approach to Y_r of duplicate redundancy may be extended to higher orders of redundancy when one element out of R elements should be functional, e.g., in the special case of Poisson statistics

$$Y_r = e^{-\beta\left(\frac{\lambda_{wo}}{\beta}\right)^R}.$$

R = 3 for simple triplicate redundancy in that a minimum one out of three identical elements should operate. If two out of three identical elements must work, e.g., in a majority decision configuration where the voter is functional, Y_r becomes

$$Y_r = e^{\frac{-3\lambda_{wo}^2}{\beta}}.$$

The expressions of Y_r illustrations the yields' increasing sensitivity to the expected number of faults λ_{wo} with increasing amount of redundancy.

In any real situations, λ_{wo} tend to cluster on a wafer which may be considered by an interwafer clustering parameter α_1, and both λ_{wo} and α_1 vary wafer-to-wafer which may be regarded by an intrawafer clustering parameter α_2. In general, the yield with redundancy Y_R is a complex triple or quad integral expression where

$$Y_R = f(\lambda_{w0}, \beta, \alpha_1, \alpha_2, q_k, ...).$$

Numerical evaluations of Y_R, for the case when both P (Q) and $g_w(\lambda_w)$ are gamma distributions, indicated that yield reduction from intrawafer variations may be compensated for by interwafer variations, and that the yield is more sensitive to interwafer clustering than to intrawafer clustering in large chips and wafers (Figure 5.28).

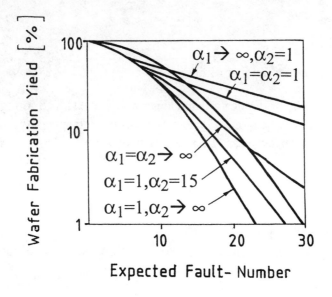

Figure 5.28. Yield sensitivity to defect clustering. (After [529].)

On-chip redundancy that improves yield, does not necessarily improve reliability, although the substitution of marginally operating elements by good ones, increases both reliability and yield. For both yield and reliability improvements the application of circuit redundancy is essentially the same, and can be discussed without discrimination for the intended use.

5.5 FAULT-TOLERANCE IN MEMORY DESIGNS

5.5.1 Faults, Failures, Errors and Fault-Tolerance

In memory technology, as in digital circuit technology the terms fault, failure, and error have distinguished meanings [530].

A fault is an anomalous physical condition. Causes of anomalies include design deficiencies, manufacturing problems, damages, fatigues, deteriorations, extreme ambient temperatures, ionizing radiations, humidity, electromagnetic interference, internal and external noises, misuse, etc. Usually, memory faults are classified by their duration, location, extent and nature (Table 5.2).

Fault			
Duration	**Location**	**Extent**	**Nature**
Transient Intermittent Permanent	Memory Cell Bitline Wordline Sense Amplifier Read Circuit Write Circuit Address Circuit Control Circuit Clock Circuit Supply Network Input Output	Single Multiple Row Column Array Decoder Peripheral Global	Stuck-at Timing Floating

Table 5.2. Fault classification.

A failure is the inability of a memory cell, circuit element, circuit or an entire memory to perform its designed function, which inability is caused by a fault.

An error is a manifestation of a fault in a memory, in that logical state of a memory cell or a peripheral circuit element different from its intended state. Different errors may be categorized by occurrence, direction, pattern, location and orientation (Table 5.3), and all errors can be hard or soft. A hard-error is a result of permanent damage, which may appear either through an abrupt or through a drift failure mode, in a circuit element. A soft error does not persist in the effected circuit after the fault-causing phenomenon disappears, and at reuse the circuit operates as designed without any repair or correction in the memory chip.

Error				
Occurrence	Pattern	Direction	Location	Orientation
Random Systematic Clustering	Single Double Quad Burst Multiple Burst	Bidirectional Unidirectional	Memory cell Row Column Sense Amplifier Logic Element Input Output Global	Symmetrical Biased

Table 5.3. Error categories.

Those memories which feature on-chip fault repair or error-correction circuits are referred, though somewhat inaccurately, as fault-tolerant memories. Fault-tolerance in memories is applied to improve reliability, or yield, or both. Fault tolerance is crucial for reliable operation in space, nuclear, military and other extreme environments, and important to increase yield until the fabrication technology matures. Nonetheless, mature memory products operating in standard environments need rarely on-chip fault-repair or error correction to enhance fault-tolerance.

Fault-tolerance within a memory chip is achieved by redundancy in memory cells, arrays, peripheral circuit-elements and in the redundancy control circuits. The implementation of redundancy may profoundly influence the layout area, speed and power performance and, therefore, must be planned for at the outset of a memory design.

The plan's objective is to establish whether and how much improvements in reliability and yield are required and what techniques are to be used to implement the required improvements. To determine these requirements the circuit design needs to know:

(1) What types and amounts of faults and errors appear and are economical to repair and correct,

(2) What the optimum strategies are to tolerate the faults and errors to be repaired and corrected.

Analyses of faults and errors, which may occur in a particular memory, indicate the reliability and yield parameters of the memory designed without particular features for fault-tolerance. A comparison of the actual reliability and yield parameters to the desired ones show whether any or how much improvement is necessary, while the strategy of improvement depends on the types and amounts of dominating faults and errors.

5.5.2 Faults and Errors to Repair and Correct

For determination of types and amounts of those faults and errors which are necessary and economical to repair or correct for achieving a desired

Error Sources	Hard-Errors	Soft-Errors
Fabrication and Design		
Hot carrier emission	X	
Electromigration	X	
Surface charge spreading	X	X
Ionic contamination	X	
Spurious currents	X	X
Time dependent breakdown	X	
Environmental		
Package alpha radiation		X
Static charge and discharge	X	
Electromechanical corrosion	X	
Electrochemical corrosion	X	
Electromagnetic interference	X	X
Temperature	X	X
Cosmic particle impacts	X	X
Radiation total dose	X	X
Transient Radiation	X	X
Mechanical shock	X	X
Electrical shock	X	

Table 5.4. Hard- and soft-errors influencing reliability.

reliability or yield, analyses of the failure modes faults, and errors are needed in each individual design variation of the memory circuits. The specific goal of the circuit failure mode analysis is to establish the performance requirements for fault-repair and error correction. The performance requirements of reliability improvements may greatly differ from those of yield increase. Namely, reliability is influenced by both hard- and soft-errors arising from faults in memory fabrication, design and environmental sources (Table 5.4), while yield is effected by hard errors only, which may be caused by fabrication and design issues (Table 5.5).

Fabrication	Design
Cleanness	Feature size
Materials	Chip size
Masking	Packing density
Oxides	Cell type
Parameter spread	Layout
Complexity	Parameter variation tolerance
Control	Fault-tolerance
Temperature	Pattern sensitivity
Mechanical shock	

Table 5.5. Fabrication and design issues effecting yield.

In each of the memory circuits, fabrication, design and environmental effects can cause faults through circuit-specific failure mechanisms, and the specific faults are manifested in either symmetrical or unidirectional error types. Circuit faults appear in the data outputs of the memory, most frequently, as discrepancy between write-and-read data and as data stuck at log.0 or log.1 (Table 5.6). A write-read data discrepancy may be caused by a failure either in the addressed memory cell, or in the sense circuit, or in the addressing circuit. Stuck-at faults may occur in any of the circuits, and they may result from three major categories of failure modes in transistor devices and interconnects (1) parameter degradations, (2) short circuits, and (3) open circuits [531]. Parameter degradations and open-circuit faults may result in intermittent levels between log.0 and log.1, but eventually these intermittent levels are amplified to standard log.0 and log.1 levels, while short- and many open-circuit faults cause immediate sticking of logic operations at standard logic levels.

Faulty Circuits	Data Output
Data Input	Stuck at log.0 or log.1
Address Input	Discrepancy between write and read data
Word Decoder	Discrepancy between write and read data Periodic appearance of same set of data
Word Line	Stuck at log.0 or log.1 Discrepancy between write and read data
Storage Cell	Stuck at log.0 or log.1
Bit Line	Stuck at log.0 or log.1
Bit Decoder	Discrepancy between write and read data Periodic appearance of same set of data
Sense Amplifier	Stuck at log.0 or log.1
Read Line	Stuck at log.0 or log.1
Read Amplifier	Stuck at log.0 or log.1
Write Amplifier	Stuck at log.0 or log.1
Data Output	Stuck at log.0 or log.1

Table 5.6. Circuit faults and their most frequent effects on the data output.

Data discrepancy and logic level faults may represent a large variety of physical failures, but they do not cover all possible failure modes, e.g., power-supply line short and break, pattern sensitivity, etc. Nevertheless, the coverage of dominant failure modes is sufficient for establishing what faults and errors should be tolerated in the memory by on-chip repair and correction. By using spare chips in a system, of course, the memory operation can be sustained despite the presence of any number and types of faults and errors.

The frequency of faults and errors depends mostly on the memory circuit and layout design, processing technology, starting material, handling, operating and storage environments. Experience in these issues indicates that in CMOS VLSI memories the single random type of errors (Figure 5.29a), but in CMOS ULSI and WSI memories, beside the dominating single-random errors, the number of double and quad errors in the memory cell (Figure 5.29b) may also be significant in addition, of course, to other miscellaneous error types.

In addition to the determination of the most frequent type of faults and errors, the effects of memory malfunctions on the system operation, has to be weighed. A system operation may absorb certain low amounts of single bit errors, but the appearance of a burst error may have catastrophic consequences. Burst errors in memories are caused most likely by the malfunction of a sense amplifier and, with much less probability, by impairment in some peripheral memory circuits.

Generally, a fault tolerant CMOS memory design has to cope, most likely, with hard and soft errors which are (1) symmetrical and occur randomly in single and double bit patterns in memory cell arrays, and (2) unidirectional and occur as bursts in sense amplifiers.

More prominent appearance of double, quad and burst errors, as well as simultaneous impairments of neighbored rows and columns, are anticipated with the further evolution of the fabrication technology toward smaller feature sizes, and with applications of CMOS memories in increasingly severe environments. Gross errors and global faults impairing entire memory chips may be tolerated in system levels by enabling spare chips. Yield increase by use of spare chips can be considered only in integrated chip systems such as multi-chip-module (MCM) and WSI memories. The analysis of memory failure modes and the experience in fabrication, design and application in specific environments reveals what classes of errors and faults at what frequentness may appear and what is desirable to tolerate. Additionally the implementation of fault-tolerance requires a strategy that is amenable to the

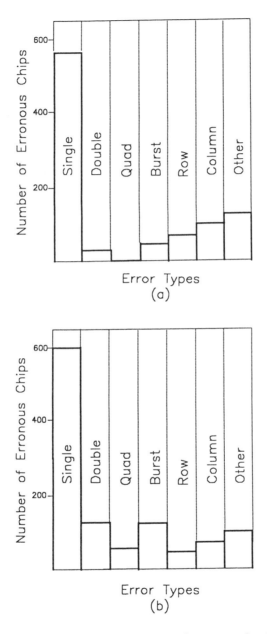

Figure 5.29. Error-type distributions occurring in two different CMOS memories.

repair or to the correction of the faults and errors determined by the failure-mode analysis.

5.5.3 Strategies for Fault-Tolerance

The right fault-tolerance strategy is the approach that provides the required performance in reliability or yield improvement and the optimum implementation efficiency in area, speed and power. To satisfy the range of performance-efficiency requirements in memory designs, a fault-tolerance strategy may include one or more of the following fundamental techniques:

- Repair by reconfiguration separates, bypasses, and replaces the faulty element by an operating spare one,
- Detection indicates error by application of error detection codes,
- Correction reconstructs an acceptable code word from an erroneous data word by application of error detection and correction codes,
- Masking corrects errors by simultaneous use of circuit replicas,
- Containment prevents the propagation of erroneous data out of a faulty circuit,
- Diagnosis identifies the faulty elements,
- Inhibition disallows the use of diagnosed faulty elements,
- Timing and protocol checks compare internally generated clock impulses to each other or to replicas,
- Repetition of rewrites and rereads detects erroneous data,
- Discarding renders erroneous data unusable,
- Maintenance purges errors in the memory operation periodically.

From this variety of techniques, fault-tolerant memory designs incorporate on-chip fault-repair, error detection and correction, and in some instances fault masking, while the other techniques are applied, so far, off-chip on system level. Nevertheless, the on-chip implementation of any approaches may be applied to satisfy the anticipated higher standards in reliability and yield for future large bit-capacity memories.

5.6 FAULT REPAIR

5.6.1 Fault Repair Principles in Memories

Fault repair, here, is a reconfiguration of the memory circuit that inhibits the use of faulty circuit elements and enables the use of operating spare elements. Circuit elements which are repaired in a memory include rows, columns, blocks (clusters), subarrays and arrays of memory cells. Individual memory cells are not, and elements or the whole of peripheral logic circuits are seldom, practical to repair.

The repair procedure consists of three phases: (1) detection and location of faulty elements, (2) assignment of operating spare elements, and (3) disconnection of the faulty elements and integration of the assigned spare elements with the memory operation.

The detection and location of faulty elements or errors are provided either by error control codes, or by an on-chip tester circuit, or by external test equipment. On-chip and external tests allocate also the spare elements which are operational, and allow to assign the spare elements to the circuits which contain the faulty elements. Disconnection and of faulty elements and engagement of spare elements may be implemented externally by laser, fuse and antifuse programming, or by on-chip repair circuit applying electrical fuse, antifuse, EPROM, EEPROM, FRAM, SRAM or other bistable programmable circuit elements.

The operation principle of the memory circuit repair (Figure 5.30) is essentially the same in all types of memories; depending on the content of the fault-address memory either the main memory or a spare memory circuit is selected, and from or to either one of both memories the data are transferred from and to the input-output terminals through an eventual corrector circuit.

The implementation of repair circuits may have the following objectives:

- Improved reliability at minimum or no impact on yield,
- Improved yield at increased or at no change in reliability,

CMOS Memory Circuits

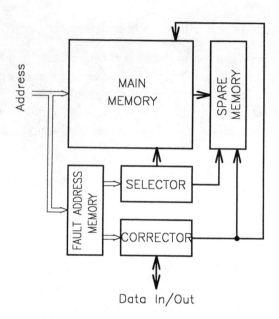

Figure 5.30. Fault-repair principle for memory circuits.

- Minimum increase in area,
- Minimum or no degradation in speed and power performance,
- Reconfiguration transparent to the user,
- Replaceability of both main and spare elements,
- Unchanged processing technology,
- Easy removability of spare elements.

Reliability improvement, by using spare elements, may decrease the yield because the implementation of spare and repair-control circuits increases the layout area. Yield, by varying the number of spare elements, may be optimized, but the use of spare elements for yield improvements may reduce the reliability, e.g., by the debris of fused links, decreased noise margins in electrically programmable memory cells, etc. Nonetheless, both yield and reliability can simultaneously improve at correctly designed and executed repair when not only faulty but marginally operating elements are also replaced.

The implementation of the spare elements and repair-control circuits can be designed so that neither the access time nor the power consumption of the memory increases palpably, and the application of the memory in systems does not need any change in input and output requirements, pin order, and other parameters. A repair method that would require process modification raises questions not only about product cost increase, but it may cause substantial changes in the design of the memory and its testing.

5.6.2 Programming Elements

Programming elements either separate or couple memory and spare circuit elements by creating low- or high-resistances between circuit nodes arbitrarily. Unprogrammed low-resistance wiring elements are used as fuses, while normally high-resistance semiconductor elements are applied as antifuses. Furthermore, both normally high- and normally low-resistance states can be provided by any type of nonvolatile memory cells, and also by volatile memory cells using battery backup.

All the different types of programming elements, which are applicable to memory designs, may be programmed externally or internally, and all the types of fuses and antifuses may be programmed by laser or electrically generated heat. Memory cells applied as programming elements operate under the same write and read conditions as in data storage applications, although they may be designed to prefer an initial log.0 or log.1 state before programming.

For yield improvement the preferred programming method is the laser cut [532] of a polysilicon or metal link, because its implementation requires small layout area that can be fitted to a row or column pitch, influences negligibly the speed and power characteristics of the memory, and provides a large resistance change from the unprogrammed (0.1-10Ω) to the programmed (25-250MΩ) resistance. The layout area of the programming link is determined by the effective diameter of the laser spot D and by the inaccuracy of laser beam positioning α (Figure 5.31).

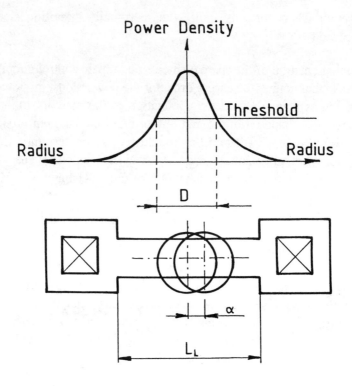

Figure 5.31. Diameter, position accuracy and power density distribution of a laser beam to a fusible link design. (Source [532].)

Decreasing memory feature size require improvements in focusing, position accuracy and shorter laser wave lengths. The widely used yttrium-aluminum-garnet YAG lasers provide $D = 1\text{-}3\mu m$, $\alpha = \pm 0.1\text{-}0.3\mu m$ and operate at a wave length of $\lambda = 1.064\mu m$. For the laser beam, in many designs, holes are left in the passivation layer to promote the removal of contaminants produced by the laser cut. In both polysilicon and metal links, the laser beam destroys much more than its diameter, because a significant amount of power under the fuse-threshold is also absorbed by the link and its surrounding material. Because of this "wicking" the link should be the minimum length of $L_L = D + 0.5\alpha + l_w$, where l_w is the destruction length in the link by wicking.

The cut of a polysilicon link results less debris than the cut of a metal link and, therefore, a polysilicon cut results in more reliable separation. Since the separation does not need any other circuit element other than a short link, the access and cycle times as well as power dissipations are influenced only by the interconnect lines necessary to implement the spare elements and in some instances by the load introduced by the spare elements.

For connection and disconnection of spare and faulty elements laser programmable antifuse links (Figure 5.32) may also be used. When a laser spot on the link covers the p-n junctions the material melts and provides a rather low-ohmic conductance. Both the low and high resistance as well as the maximum operating voltage and current depends on the characteristics of the switchable p-n-p, or n-p-n or amorphous device. Antifuses switch normally high resistances (>10MΩ) to low resistances (50-300Ω), and in their high-resistance unprogrammed states the breakdown voltages are in the range of 3-6V.

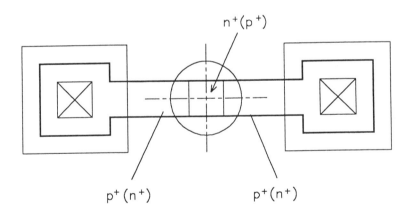

Figure 5.32. A laser programmable antifuse link.

The implementation of laser programmable fuses and antifuses is simple and inexpensive, but the acquisition and maintenance of a fully automatic laser programmer equipment needs substantial capital investment. For yield enhancement when high capital investment into a laser programmer is impractical for any reason, or for repairs of memories which operate in systems, i.e., in-situation or in-situ, the fuses or antifuses may be programmed electrically. Nevertheless, the electrical programming require the application of high current, or voltage, or both, that leads to inflated circuit sizes and penalties in speed and power. Even at external application of high voltage, a part of the on-chip programming circuits has to be able to tolerate the current and voltage stress, and has to contain an interface circuit to separate the programming elements from the operating elements to maintain reliability standards.

Electrically programmable elements may also be implemented in fuse or antifuse devices. Electrically fusible links have similar layout designs as those of laser programmable fuses, but the Joule-heat, rather than the laser generated heat, evaporates the conductive metal or polysilicon material. An insulator material, e.g., Oxygen-Nitrogen-Oxygen [533] or amorphous silicon [534] is melted between two conductive layers in the mostly applied electrically programmable antifuses (Figure 5.33). Electrical antifuses can

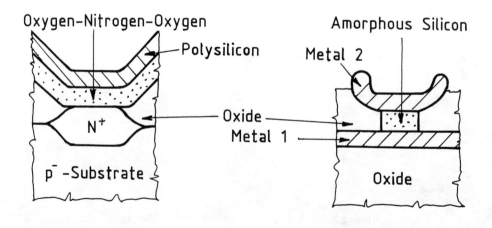

Figure 5.33. Electrically programmable antifuse elements. (Source [534].)

provide very high off-resistances (>1GΩ) and acceptable on-resistances (80-500Ω), and need 3-6mA and 10-20V programming current and voltage. Implementations of electrical programmability may require up to three additional mask-layers and extended layout areas.

The layout area of an electrically programmed element may not be as small as that of a laser programmable element, because its implementation often needs p^+ or n^+ doped guard-rings to avoid destruction in the surrounding circuit elements. Furthermore, electrically programmable elements need at least one extra pad for programming power supply. The extra pad is not bonded out to package and must be held in an adequate potential after testing to prevent any inadvertent programming during the rest of the fabrication process.

Programming of fuses or antifuses may need slight changes in processing to keep open a window in the oxide above the programmable element. Windows allow for better positioning control and for improved gas, debris and heat escape which are generated by programming.

Fuse and antifuse programming with laser assistance are most suitable to enhance yield of memories which apply dynamic and static (1T1C, 4T2R, 6T, etc.) random access memory cells. The reasons for that include the (1) minimum impact of the implementation in circuit area, speed and power, (2) reliability of the programming, (3) reasonability of equipment costs at very high volume production, (4) unchanged processing technology and (5) easy design for transparency in operations and/or for removability of spares. Area, speed and power tradeoffs and design difficulties are compromised more by the application of electrical programming than by the use of laser programming. Nevertheless, the use of electrically programmable fuses and antifuses are well justified for reliability improvements and repairs of memories installed in systems, and to enhance memory yield without substantial investment in laser programmers.

In static and dynamic memories which have battery backup, the applied memory cells can also be used for repair programming, especially where high programming speed, cost effectiveness in implementation, and direct compatibility with the controlling logic circuits, are among the requisites. Nonvolatile [535] and battery-backed volatile [536] programming elements may gain applications also in memories, despite needing additional processing steps for implementation of programming elements, where in-situ memory repairs should combine high reliability operation with high speed performance, low power dissipation, small size and weight, e.g., in memories operating in space, military, radiation hardened and other extreme environments. Programmable elements for most of the applications may be created from the great variety of programmable and static memory cells.

5.6.3 Row and Column Replacement

To provide fault tolerance in CMOS memories, the most widely used technique is the replacement of rows and columns of memory cells [537] by means of laser programmable fuses in NOR circuit based decoders. Programming of a NOR-decoder, that selects a wordline or row, involves disconnecting a wordline from its access circuitry and enabling a spare one by cutting links. The primary difference between a normal (Figure 5.34a) and a spare decoder row (Figure 5.34b) is that a spare decoder row has twice as many transistors as the normal one does, so that a spare row accommodates both the true and its complement of an addressing bit. An unprogrammed spare addressing row need not to be decoupled by blowing a fuse since any address combination keeps an unprogrammed spare row deselected. However, to avoid the selection of a normal row that contains a faulty element, one link, e.g., fuse F, in the access path must be programmed.

Programming of a decoder, that selects a bitline or column, does not need a disconnection of a normal bitline from the write-read data bus (Figure 5.35) because a normal-to-spare N/S switch can avoid a coincident data flow between the normal and spare bitlines. In this schema, the parasitic capacitances for the normal bitlines change insignificantly, but they reduce greatly for the spare columns. The programming of the column-decoder can be the same as that of the row-decoder.

Reliability and Yield Improvement 429

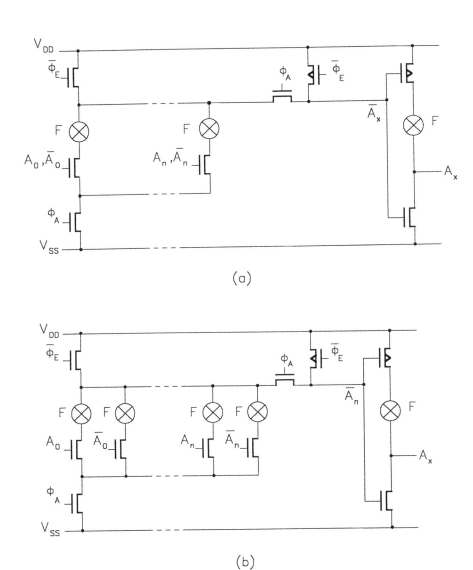

Figure 5.34. Fuse programmable normal (a) and spare (b) rows in a NOR-decoder.

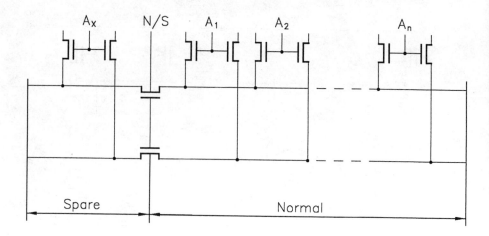

Figure 5.35. Normal-to-spare switching in bitline selection. (Source [537].)

In NAND-gate based decoders (Figure 5.36) the implementation of laser programmable antifuses is the most amenable technique, although the reciprocity of the circuits allows to use fuses and antifuses in NOR- and NAND-gate based decoders arbitrarily.

Fuses and antifuses may also be programmed electrically in the decoders, but each programmable element and each row and column requires at least one additional transistor that is capable to sink high current and to endure the effects of a high-programming voltage V_P (Figure 5.37). The large programmer transistors and the guardrings around the fuse, or the antifuse often disallow to fit electrically programmable links into a row or column pitch. Nevertheless, the in-situ programmability and the cost efficiency of the programming may indicate the application of electrically programmable fuses and antifuse devices despite their inefficiency in layout area. The implementation of electrical melting of antifuse junctions can often be more area efficient than that of fuses.

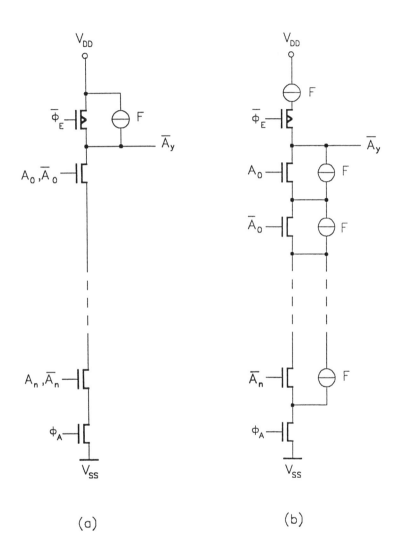

Figure 5.36. Normal (a) and spare (b) columns in an antifuse programmable NAND-decoder.

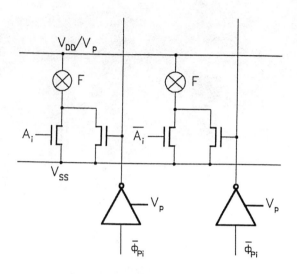

Figure 5.37. Electrical programming in a NOR-decoder.

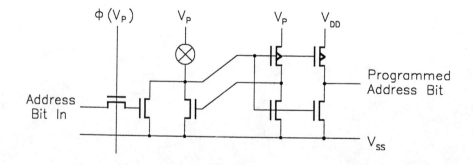

Figure 5.38. Electrical programming of an address bit.

The rather large area requirements for the circuits of electrically programmable fuses or antifuses may conveniently be accommodated if the programming circuits can be placed in the paths of the addressing bits at address inputs, rather than in the decoders. The available space at the address inputs allows to combine a latch with the programming circuit (Figure 5.38). The use of the regenerative latch relaxes the requirements in programming of low and high fuse-resistances to less than 5kΩ and greater than 40kΩ respectively.

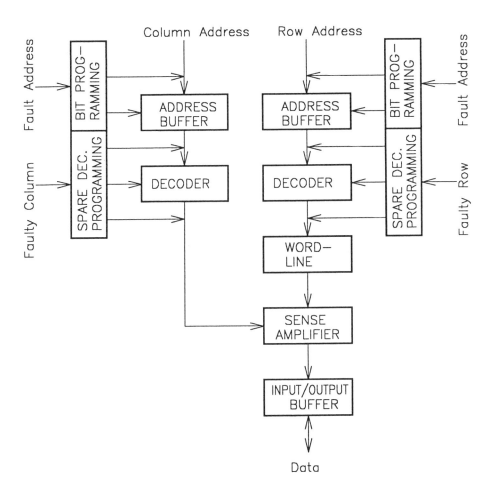

Figure 5.39. Spare decoders in the critical timing paths of a memory.

The implementation of programmability in the word-addressing bits or directly in the column-address decoder changes the memory access time insignificantly, but the bitline programmability of the row-address decoder may introduce considerable delay in the critical time path (Figure 5.39), e.g., in a 64-Mbit static memory the word-decoder's programmability can be hidden but the programmability of the bit-decoder may increase the access time by 4%. The increase in area and power consumption depends on the number of spare rows and columns applied and on the size of the main memory. For large memories the percentage increases are small, e.g., 0.5% for 256-Mbit DRAM, but for small memories the increases may be considerable, e.g., 9% for a 4-Mbit SRAM.

Replacing rows and columns of memory cells is beneficial as long as the defect distribution is uniform. More realistic, however, is to assume clustered defect distribution, which can be repaired by replacement of blocks (subarrays) and arrays of memory cells.

5.6.4 Associative Repair

All types of elements, including arrays, block, rows, columns or cells can be repaired at very little tradeoffs in speed and power by an associative approach [538] that may be iterated within the memory. The approach is associative, because it uses associative memories to store the addresses of the faulty memory locations (Figure 5.40). Chip sizes of CMOS memories, which incorporate associative fault-repair, are enlarged only by the area required to implement an optimized number of redundant elements, e.g.. 5% for a 256-Mbit DRAM fabricated at an average defect density of 1.5 defect/cm^2.

In very large memories the associative approach may be extended to a hierarchical iterative replacement schema, where a large spare element can also be repaired by another, smaller element of a spare submemory, etc. (Figure 5.41). Here, the addresses of the faulty elements which are located in the main memory, in the spare memory and in the spare submemory, are written in CAM1, CAM2 and CAM3 respectively. CAM is a storage medium that can be addressed by searching for data content (Section 1.4). When a set of addressing bit appears on the address inputs $A_0, A_1, \ldots A_n$ of

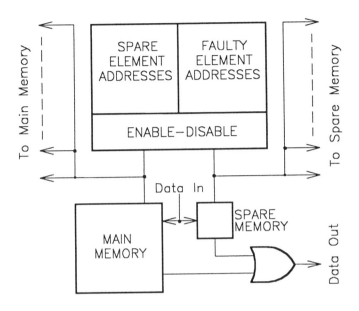

Figure 5.40. Associative repair schema. (After [538].)

the memory, a parallel search determines whether the address is written earlier in CAM1. If no match appears, the address code directly flows into the main memory and the main memory operation is unchanged. If, however, the faulty element memory CAM1 contains the address, a match-flag signal occurs. The flag signal enables the flow of the address code of an operating spare element into CAM2, and deactivates the address input and data output of the main memory. Thereafter, the spare memory itself executes the required read, write or other operation. If a nonoperating element is addressed in the spare memory, the address is deviated again to CAM3 and a spare submemory is activated.

Because the bit capacity and the size of the main memory is much larger than that of the spare memory, e.g., 10:1 or less, the operation time of the spare memory is always much shorter than the access and the cycle time of the main memory. Thus, the operation of the spare memory does not degrade the speed of the main memory and it is entirely transparent for the user. Increases in both memory power consumption and layout area of the are insignificant, because the associative memory has to contain only 4-16 addresses. In large memories the associative hierarchical schema provides a combination of fault repair capability and application efficiency that is difficult to surpass by other approaches.

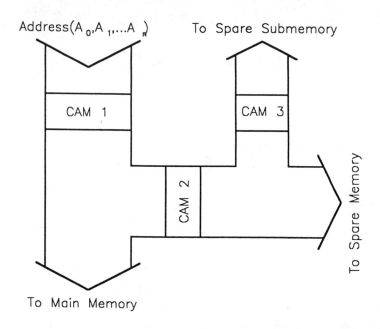

Figure 5.41. Hierarchical iterations of associative repairs.

5.6.5 Fault Masking

To render the peripheral circuits of the memory, and the eventual error detecting, correcting and repair control circuits, insensitive to their own faults fault masking [539] is the most docile approach. Fault masking eliminates the one-to-one correspondence between the failure of a component circuit and that of the entire memory, and each redundant component circuit tolerates the failure of one memory component circuit.

Reliability and Yield Improvement

The redundant component circuits are triplicated and a 2:1 majority vote logic circuit (Figure 5.42) is often applied to decide whether a log.0 or a log.1 is the correct datum when a component circuit fails. If each component of

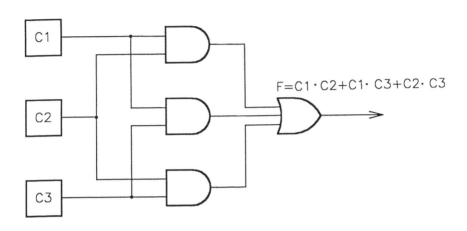

Figure 5.42. A 2:1 majority vote logic circuit.

the C1, C2, C3 has a probability of failure p, the probability that such a one level majority configuration fails P is

$$P = 3p^2 - 2p^3.$$

When n level of circuits are protected by majority vote logic, then the probability of failure of n-level majority configuration P_n is

$$P_n = \{1 - [1 - (3p_n^2 - 2p_n^3)]\}^n \quad \text{and} \quad P_n < p_n \quad \text{if} \quad p_n < 0.5,$$

where p_n is the probability of failure in the circuit n. In the equations of P and P_n the majority vote logic circuits are treated as perfect infallible elements. The effect of failures in the majority logic itself may easily be regarded by the introduction of the error reduction factor F. F is the ratio of failure probability of the nonredundant circuit configuration P and that of the redundant circuit configuration P_n, i.e., $F = P_o/P_n$. F increases with increasing n and with decreasing P_o (Figure 5.43) at the assumption that the voter failure is much less likely than the redundant component failure. Voter logic has the most promising application to sense amplifiers and decoders for masking the effects of charged atomic particle impacts.

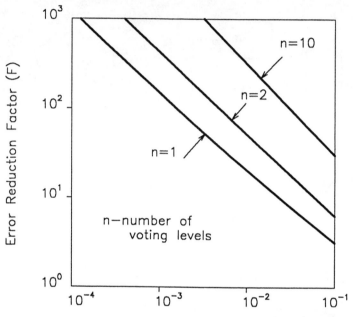

Figure 5.43. Error reduction factor versus failure probability and number of voting levels. (Source [539].)

5.7 ERROR CONTROL CODE APPLICATION IN MEMORIES

5.7.1 Coding Fundamentals

Error control coding (ECC) [540] adds a small number of redundant bits to the write data to form code words. Every code word is a part of a code space. The error control code is designed so that a certain pattern and number of errors transforms a code word into a data word which is out of the code space. Error detection identifies any read word outside the code space, while error correction uniquely associates an out-of-code-space read word with the originally written code word.

Before data would be written into the memory-cell array, an encoder circuit creates a legitimate code word from the original data, and after read a decoder circuit indicates illegitimate words containing errors and, if correction required, determines the location of the erroneous bits in the read word and reverts the erroneous bits to their correct original binary values (Figure 5.44). For data processed in binary form the encoding is simple, but the decoding can be both complicated and complex, requiring a combination of encoder, syndrome-generator, decoder and corrector circuits. The encoding and decoding of the write and read data have to perform an improvement in error probability and have to be efficient in implementation.

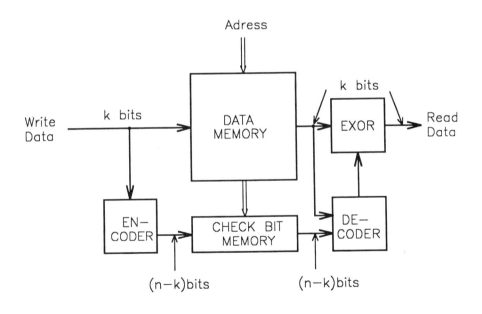

Figure 5.44. Encoding and decoding.

In accordance with Shannon's coding theorem [541] the error probability, i.e., the probability of incorrect decoding P_{ICD}, can be made arbitrarily small by increasing the code length n while holding the code rate

R constant for any information transfer rate less than the channel capacity C at any energy per code bit E_b:

$$P_{ICD} = 2^{-nE_b(R)} \quad \text{where } R < C.$$

k is the number of information bits in a code word, the n is the code length in bits. The theorem, however, provides no means for constructing effective codes and suggests that requirements for very low error probabilities compel the use of very long code words and, in turn, of very complex decoding operations. In memory applications codes required to have short code lengths and simple decoding, and these need to detect and

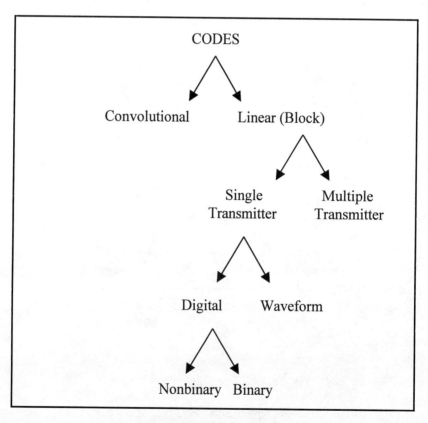

Table 5.7. Family of codes.

correct only few, e.g., random single and burst, error types. To construct error detection and correction codes the design has to find the code that is the most suitable one to the particular memory application. From the family of codes (Table 5.7) the most amenable codes for memory applications occur to be the linear, single transmission, and the digital binary codes.

Linear codes are systematic (unmodified data stream is contained in the encoded data sequence), structured (information bits are separated from redundant check bits), and can be described by mathematical methods which is rather easy to compute. Convolutional codes, tough easy to generate and widely used in communications, are poorly suited to memory application, because they have low code rate and require complex apparatus to decode. Other code families including waveform nonbinary and multiple transmission codes are also unfitting to memory applications, hence memories store data in a digital binary system, and the information is written and read in a single transmission schema.

There is no acceptable single figure of merit upon which a decision can be made for a code application within a code family, and no single class of codes is best for all CMOS memory applications. Nevertheless, in fault-tolerant memory designs, the postdecoding error rate functions for performance, and the percentage increase in layout area, circuit delay and power consumption for efficiency, provide good criteria for code selection. Analyses of performance and efficiency parameters as well as design experience indicate that the following code classes can most likely gain implementations in CMOS memories:

Single parity check,

Berger,

Hamming,

Reed-Solomon,

Bidirectional.

If no code of these classes can satisfy the requirements, of course, other existing code classes should be investigated, or new codes may be devised.

To apply any code to a memory the code's performance and implementation efficiency have to be analyzed.

5.7.2 Code Performance

In memories the only benefit of the use of error control coding is the reduced probability for write, storage and read errors. This reduction can be characterized by comparing the error probability without error control coding p to the error probability with error control coding P.

The objective of error control performance investigation is to find or design a code which is capable to improve a certain p to a required P at the lowest code rate $R = k/n$. A low code rate transforms to little number of redundant bits and to small memory area increase, and usually results in a high efficiency code implementation.

The vast majority of codes applied in fault tolerant memories are in the family of binary linear block codes [542]. Under the assumptions that in the codes binary 0 and 1 occur with equal probability and that the errors are independent from each other, i.e., binary symmetric channel with random error distribution; the most widely used parameters for performance calculations are the probability of correct decoding P_{CD}, the probability of incorrect decoding P_{ICD} and the probability of post-decoding bit error P_b [543].

Error probabilities P_{CD} and P_{ICD} may be expressed as

$$P_{CD} = \sum_{i=0}^{n} \binom{n}{i} p(1-p)^{n-i} \quad ,$$

$$P_{ICD} = \sum_{i=d-1}^{n} \binom{n}{i} p(1-p)^{n-i} \quad ,$$

where i is the Hamming weight i.e., the number of binary 1s in a word, d=2t+1 is the Hamming distance i.e., the number of bits in which binary 0 and 1 values differ, and t≥1 is the guaranteed error correction capability. For reasonable values of p and n the first term dominates, thus

$$P_{ICD} \approx \binom{n}{i} p^i .$$

Applying weight distribution A_i, i.e., the full enumeration of the code word number of every possible Hamming weight, P_{ICD} becomes

$$P_{ICD} = \sum_{i=d}^{n} A_i p^i (1-p)^{n-1} .$$

The upper bound for P_{ICD} with r number of redundant bits in a code word is

$$P_{ICD} = 2^{-r} - 2^{-n} \approx 2^{-r} .$$

P_{ICD} for error detection alone can be approximated by

$$P_{ICD} = 2^{-r} p^d .$$

Assuming bounded distance error detection and correction in which any error pattern i>t causes decoding failure, P_{ICD} for block errors can be approached as

$$P_{ICD} = \frac{(np)^{t+1}}{(t+1)!} .$$

If i errors are present in a read word, and the decoder can insert at most l additional errors, then the post-decoding bit-error probability P_b is

$$P_b \leq \frac{1}{n} \sum_{i=d-l}^{n} (i+l) \binom{n}{i} p^i (1-p)^{n-i} .$$

At bounded distance decoding P_b can be calculated as

$$P_b \approx \frac{d(np)^{t+1}}{n(t+1)!}.$$

Numerical evaluation of P_{CD}, P_{ICD} and P_b with CMOS memory parameters show that P_{CD}, P_{ICD} and P_b strongly depend on the number of check bits r or on the guaranteed error correction capability t, but they are rather weakly influenced by the code length n, by the error probability without error control coding p, and by the weight distribution A_i.

For determination of the optimum code rate of a Bose-Chaudhuri-Hocquenghem (BCH) code the performance curves of $1-P_{CD}$ as a function of p, k, n, and t (Figure 5.45) can most conveniently be used. These performance curves, also called as waterfall curves because of their shape, are obtained under the assumptions that the errors are results of incorrect decoding only, the error correction limits are set to the maximum, the weight distributions are binomial, and the errors are independent events. Diagrams of $1-P_{CD}$, P_{ICD} and P_b as functions of p, indicate that with increasing code rates and with increasing number of redundant bits and with decreasing code lengths the code performance improves. BCH codes, however, exist only to certain check-bit numbers and code lengths [544]. Thus, to fit BCH codes to standard binary data widths or to optimum cell-array size the code lengths should often be shortened. A number of shortened BCH codes and product codes which are composed of BCH codes, nonetheless, do not obey the 2^{-r} performance bound.

An indirect performance parameter the asymptotical minimum distance d_a is also used in code evaluation. For primary binary BCH codes d_a is [545]

$$d_a = 2n \frac{\ln \frac{n}{k}}{\log^2 n} \quad \text{if} \quad n \to \infty,$$

which indicates that these codes are asymptotically bad codes, although up to a length of 1023 bit their distance properties are reasonable. Asymptotically good codes, e.g., Goppa codes [546] approaches the

Gilbert-Varshamov bound [547] i.e., d/n > 0 at given code rate, but BCH codes do not do so.

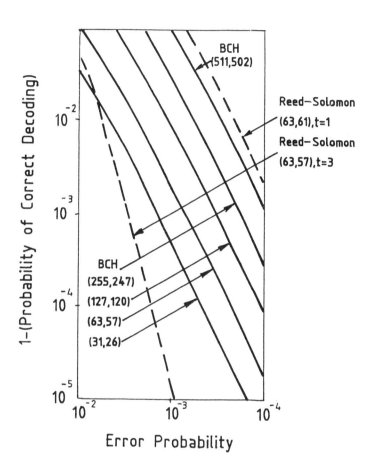

Figure 5.45. Performance curves of some binary Bose-Chaudhuri-Hocquenghem and Reed-Solomon codes. (Source [543].)

The code performance parameters for BCH codes can also be adopted to Reed-Solomon (RS) codes because RS codes can be regarded as a special case of nonbinary BCH codes [548]. When the symbol field and the location field are the same; BCH and RS parameters are also the same. RS codes are particularly suited to correction of burst errors. Burst correction can fail in both ways, either a single burst is longer than the designed length of correctable bursts, or multiple bursts appear within a block. If the error bursts are random events and their starts are distributed according to Poisson statistics the probability of burst correction P_{BC} can be approximated [549] as

$$P_{BC} = \frac{1}{2}\left[\sum_{l=1}^{n} B(l)\right]^2 + \sum_{l=1}^{n} B(l) .$$

Here, l is the burst length in bits, and B(l) is the probability of occurrence of an l length burst. The $1-P_{CD}$ versus character error rate P_{CE} functions at given k, n and t parameters for RS codes are very similar to the $1-P_{CD}$ versus p curves at different k, n and t parameters, and show that the performance improves with increasing code rate and distance. A comparison of $1-P_{CD}$ performance of a BCH code to that of a corresponding RS code, indicates that RS codes in a Galois code field of $GF(2^m)$ [550] outperform all binary BCH codes at the same code rate and length.

In memories code rates and, thereby, the number of redundant bits are limited to a few ones by yield- and cost-considerations, while the code lengths are short for simple decoding and restricted by the widths of the memory-cell arrays, or of the write-read and input-output data streams. For a code design the type of errors to be corrected and the required code performance provide the foundation. Nevertheless, a code selection and design based merely upon error and performance analysis can well be misleading without the investigation of the implementation efficiency, i.e., the impact on the layout area, operational speed and power dissipation.

5.7.3 Code Efficiency

The efficiency of codes in CMOS memory applications is determined by (1) the expansion of the layout area, (2) the degradation in access, cycle and data-transfer times, and (3) the increase in power dissipation, which result from the implementation of the error control circuits. The additional circuits are combined of (1) the number of memory cells for the storage of redundant bits, (2) encoder and (3) decoder for the error control code.

The number of redundant bits depend on the code type, and on the types and number of errors to be detected and corrected. Although with increasing code lengths the percentage of redundant bits decreases, the implementation of a long code may increase the complexity of the encoder and decoder circuits significantly. To investigate and design encoding and decoding schemes algebraic approaches can conveniently be used [551].

The encoding of linear noncyclic codes into a code word C can be described by a vector multiplication of a generator matrix [G] with the message k-tuple vector M, where I_k and P are the identity and parity matrices respectively, and k is the number of information bits in a code word:

$$C = M[G] \text{ and } [G] = [I_K P] .$$

For linear cyclic codes a code word in polynomial form $x^{n-k} + r(x)$ can be constructed by dividing the polynomial x^{n-k} by a generator polynomial g(x)

$$\frac{x^{n-k}}{g(x)} = q(x) + \frac{r(x)}{g(x)} ,$$

and then the code word is

$$c(x) = x^{n-k} m(x) + r(x) = q(x) + g(x) ,$$

and where r(x), q(x) and m(x) are the remainder, quotient and message polynomials, respectively.

The decoding of linear noncyclic codes consist of the following steps:
(1) Computing syndrome S from the received code vector V by means of the transpose of the matrix [H]

$$S = VH^T \text{ and } H = [P^T I_{n-k}] ,$$

where P^T is the transpose of matrix P.

(2) Determining the correctable error pattern e from the syndrome

$$s = e[H]^T ,$$

(3) Adding up vectors V and e to find the message vector

$$c = V + e .$$

Decoding of linear cyclic codes follows the same procedure:

(1) Computing syndrome $s(x)$ from the received vector $v(x)$ with generator $g(x)$, code quotient $q(x)$, and error pattern quotient $q_e(x)$ vectors from

$$\frac{v(x)}{g(x)} = \frac{c(x)}{g(x)} + \frac{e(x)}{g(x)} ,$$

where $c(x)$ and $e(x)$ are the message and error-pattern polynomials.

(2) Determining the error pattern from the syndrome e.g., by (a) table look-up, (b) Megitt decoder, (c) trial and error, (d) majority logic, (e) algebraic procedures, or by other methods,

(3) Finding the information message

$$c(x) = e(x) + v(x) .$$

Clearly, the decoding of information is a complex operation, and as such its implementation is one of the most influential issues to chip size, speed and power performance of memories which feature on-chip error control coding. Many of the traditional algebraic decoding methods for high performance error control codes are unacceptable to CMOS memory designs, because their complexity leads to very large overhead circuit areas, long access times and high power consumptions.

To optimize the efficiency of circuit implementation the redundancy, encoding and decoding requirements of all codes which are candidates for application in a particular memory design have to be investigated. An accepted method of the code-efficiency investigation includes the following steps:

(1) Analyze the mathematical encoding and decoding procedures of the code family,
(2) Select codes that have potential for simple circuit implementation,
(3) Design logic circuits for encoding and decoding of the selected codes,
(4) Approximate the number of transistors needed for the implementation of the logic circuits,
(5) Estimate wiring area,
(6) Approximate total area required for encoder, decoder and redundant circuits,
(7) Simulate and calculate circuit delays added by error control circuits,
(8) Estimate total power dissipation added by error control circuits.

All computation for circuit area, speed and power must apply the same layout-rules, process and design parameters for fair comparison.

Calculated and experimental graphs of normalized area, speed and power parameters versus code lengths [552] for the most prevalently used codes, e.g., bidirectional parity check codes in series-parallel (BD S/P) and in parallel-parallel (BD P/P) configurations, cyclic Hamming code (CH) and noncyclic Hamming code (NCH), etc., may greatly alleviate the computational burden. Experimental designs indicate that the percentage of the overhead area used for code implementation decreases nearly exponentially with linearly increasing code word lengths, and the rates of decreases are different for different code types (Figure 5.46). With longer code words, however, the total overhead area, that is normalized to the memory cell size, expands. Moreover, the projected overhead area expansion of a dynamic memory (Figure 5.47a) differs from that of a static memory (Figure 5.47b). Here, the memory overhead area, that is expanded by code implementations, includes the areas required to implement all the encoder, decoder and redundant memory cells.

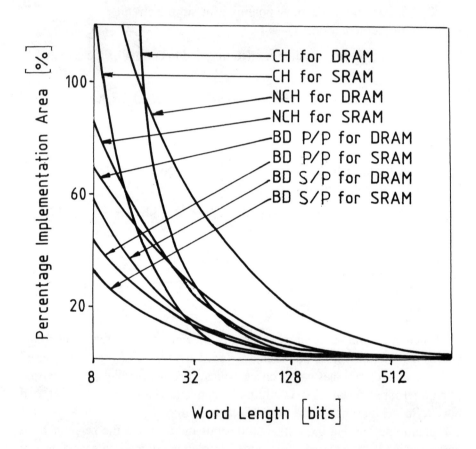

Figure 5.46. Percentage overhead area versus code word length for a variety of codes. (After [552].)

Reliability and Yield Improvement 451

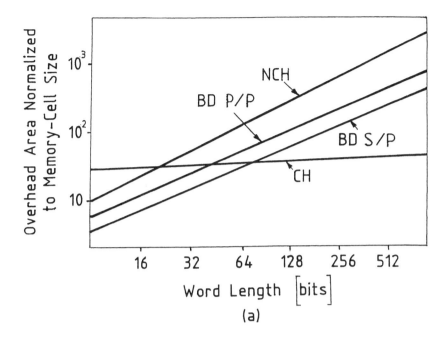

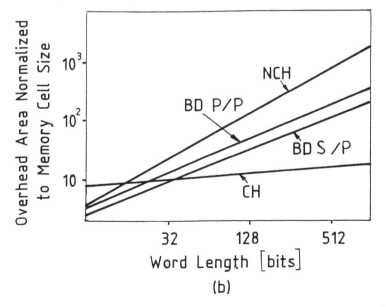

Figure 5.47. Overhead area versus code-word length to a dynamic (a) and to a static (b) memory design. (Source [552].)

The penalty to be paid in memory operational speed may be represented by a diagram of overhead delay versus code lengths (Figure 5.48), where the delays are normalized to the delay of the 2-input NOR gate. The diagram shows that overhead delays, which result from the encoding and decoding of codes, increase rapidly with increasing code-word lengths. Furthermore, it also occurs that for the codes BD S/P and BD P/P the overhead delays may be the same, because the circuit complexity of the parallel-parallel code implementation may cause as much delay as the sequentially operating simple circuits in the serial-parallel code implementation do.

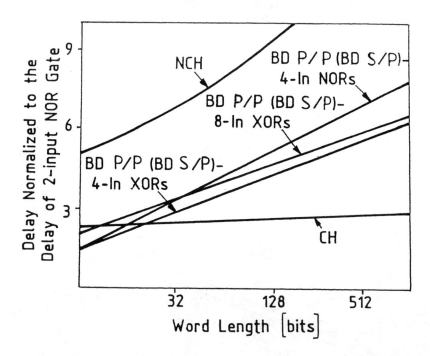

Figure 5.48. Delay versus code word length for a static memory design.

In most implementations the power-dissipation versus code-length curves, traditionally, follow the tendency of the overhead area versus code-length functions. This is because the overhead circuit's power dissipation increases with the expanding number of logic gates and with the increasing wiring lengths at a constant operational speed.

Normalized area, speed and power diagram parameters illustrate general behavior tendencies among candidate codes for a particular memory design only. To the determination of code efficiencies in each individual memory design, the potential degradations in memory chip area, access time and power consumption should be analyzed. For large bit-capacity memory chips, the implementation of error control coding results much smaller percentage degradations in important characteristics than those in medium and small size memories. Thus, in large bit-capacity memories the implementation of error control coding has good potential and importance. For error control in CMOS memories linear codes appear to be the most suitable family of codes.

5.7.4 Linear Systematic Codes

5.7.4.1 Description

A linear or block code of length n is a set of n-tuples which forms a vector space over the Galois-field of q elements GF(q). For binary codes q = 2, for j symbols q = j. A linear code is systematic if each word consists of k unaltered information bits, followed by n-k = r linear combinations of these bits. Many linear systematic codes are easy to implement in memories because their mathematical structure makes the coding circuits easy to fit to memory architectures and layouts.

From the family of binary linear systematic codes, CMOS memories may apply single-parity-check, Berger and Bose-Chaudhuri-Hocquengem (BCH) codes, and from the BCH codes the Hamming and Reed-Solomon (RS) codes, most advantageously. The characterization, encoding and decoding of these codes are summarized in the next sections.

5.7.4.2 Single Parity Check Code

The simplest linear code is the single error detecting parity, or imparity, check code, called shortly as single-parity-check-code. This code uses only one check bit appended to a block of k information bits. Because k=n-1, the

454   CMOS Memory Circuits

single-parity-check-code is an (n, n-1) code. In both parity and imparity schemes the code can detect 1,3,5...erroneous bits per word.

The encoder creates a code word by adding a 0 or 1 bit to the data. The appended bit makes a binary modulo-2 sum of each k information bit, which result either uniformly 0 for parity, or uniformly 1 for imparity. The decoder repeats the modulo-2 summing, checks whether each word results the uniform 0 or 1 sum, and indicates error when the sum is 1 at parity detection or 0 at imparity detection.

Single parity and imparity check codes are easy to encode and decode. The series encoder and decoder circuits require only a flip-flop and a logic gate (Figure 5.49), while the parallel alternatives need an n-input XOR logic circuit for implementation (Figure 5.50).

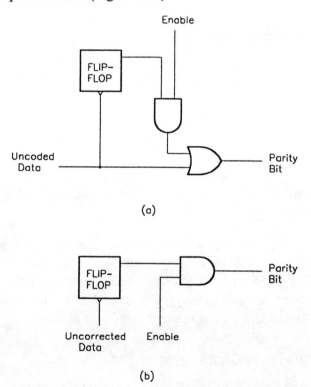

Figure 5.49. Series encoder (a) and decoder (b) for single parity check code.

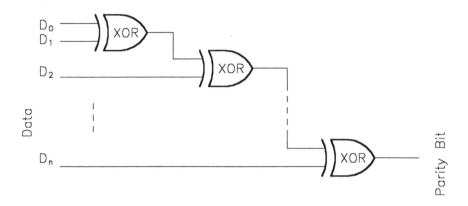

Figure 5.50. Parallel encoder/decoder for single parity check code.

5.7.4.3 Berger Codes

For fault protection of memories, in which only unidirectional binary errors occur, Berger codes [553] can provide excellent performance and efficiency. These codes detect any number of 0 or 1 errors, if 0 or 1 errors never mix in a code word. In Berger codes, k information bits are augmented with $r = 1+(\log^2 k)$ check bits to form an (n, n-r) code word.

The error-detecting capability of these codes rests on the fact that a binary number representation is a weighted-digit representation. Thus, any loss of 1-s reduces the binary weight of the word and increases the number of 0-s in the word, and vice versa. Discrepancies between the number of 1-s or 0-s at write and read of a word indicate errors.

A plain circuit implementation of a unidirectional error correcting series encoder and decoder requires only a shifting binary counter and a few logic gates (Figure 5.51). The shifting binary counter operates as a binary counter for the first k digits and then as a shift register for the next r bits. Inputs A and B switch the circuit between counter and shifting modes.

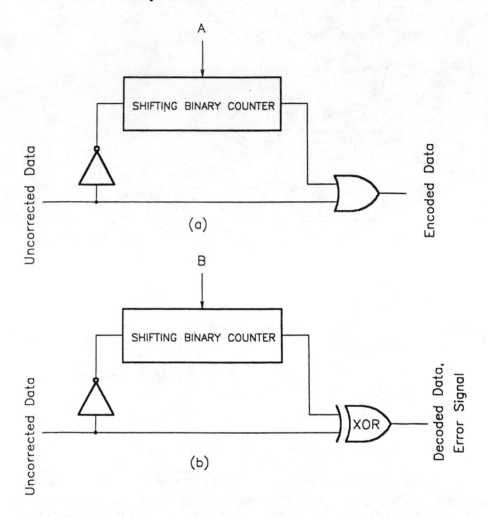

Figure 5.51. Series encoder (a) and decoder (b) for Berger codes.

A parallel encoder circuit implementation needs a k-input XOR combination, while the decoder can be constructed of a gate-complex that comprises a 2-input XOR and two 2-input AND/NOR gates, and a flip-flop for each bit of an n-bit word.

5.7.4.4 BCH Codes

The Bose-Chaudhuri-Hocquenghem (BCH) code family [554] is the best known large class of codes which can correct random errors. These codes perform as good as possible in the range of parameters of memory applications, and for many BCH codes the decoding is reasonably simple. BCH codes exist for all integers m and t such that

$$n = 2^m - 1, \quad n-k = mt \quad \text{and} \quad d = 2t+1,$$

where d is the code distance required for correct t number of errors.

In the family of BCH codes, those subclasses which are most amenable to CMOS memory applications, comprise the Hamming and Reed-Solomon (RS) codes.

5.7.4.5 Binary Hamming Codes

Hamming codes [555] are defined for any integer m as

$$n = 2^m - 1, \; n-k = m, \; \text{and} \; \check{d} = 3.$$

At the minimum distance $\check{d} = 3$ the number of errors that can be corrected is $t = 1$ if the code length is $n = 2^r - 1$. Here, $r = n-k$ is the number of redundant bits, k is the number of information bits and n is the code length. Hamming codes are the longest single-error-correcting binary linear codes which can be constructed with r check bits and, in turn, provide the least percentage area-increase that results from the application of redundant memory cells.

The number of memory cells in a row and column, number of data inputs, outputs and address input, generally, do not match the natural length n of Hamming codes. To fit n = k+r to a memory design, n can be shortened by s bits to n-s = (k-s)+r so that the number of redundant bits r and, thereby, the code's error correcting capability t is unaffected. Shortened Hamming codes are most commonly applied in the data outputs for single error correction and double error detection (SECDED), e.g., for 16 output memory a (22,16) SECDED code is used.

In CMOS memories, most of the encoders use the traditional XOR gate complex. Layout area may be reduced by applying k-to-(k+r) encoding tables or shift register circuits where the delay of the XOR encoder circuits are comparable with the delay of table look-up or binary shift encoders.

The decoding of Hamming codes is significantly less complex than that of other BCH codes since a simple r-to-2^r conversion provides the error vector. Thus, in the general decoding schema of Hamming codes (Figure 5.52) a SECDED code implementation requires (1) 2^r of 3-input XOR gates for syndrome generation (2) 2^r circuits consisting of two 3-input XOR and on 2-input AND gates for error location and (3) 2^r of 2-input XOR gates for correcting the erroneous bits.

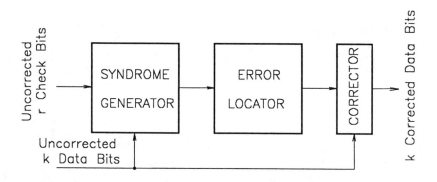

Figure 5.52. General decoding schema for Hamming codes.

To memory-internal error correction the table look-up decoding schema (Figure 5.53) may be the most amenable approach, but it is practical only for single and double error correcting codes of moderate lengths. The applicability of a table look-up decoder is limited by the access time and by the chip size and, in turn, by the bit-capacity C_B of the look-up ROM in a CMOS memory chip. A large C_B is required for decoding a code that has extensive code length and multiple error correcting capability.

Reliability and Yield Improvement   459

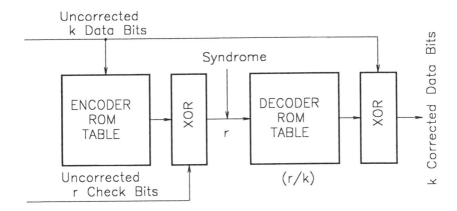

Figure 5.53. Table look-up decoding schema.

A variety of encoding and decoding techniques, which are applicable to memories, can be devised by taking advantage of some of the properties of cyclic Hamming codes. The most widely exploited property of a cyclic code is that the shifting of a code word cyclically produces another code word. Every code word in an (n,k) code is associated with a polynomial of degree n-1 or less, and no polynomial of degree greater or equal n corresponds to a code word. If a polynomial g(x) of degree n-k divides x^n+1, then the set of polynomials which are divisible by g(x) forms a cyclic (n,k) code. A large variety of shifting and other encoding and decoding techniques of Hamming codes are described in the literature, e.g., [556], but only the most economical ones can be used in memory chips.

For a cyclic Hamming code that corrects single random errors the following properties may be useful in the design:

(1) The parity matrix columns can be ordered so that each row is a cyclically shifted version of the previous row.

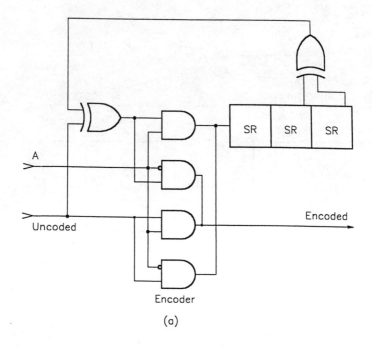
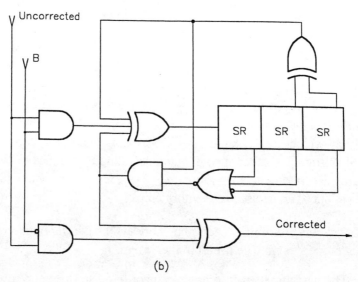

Figure 5.54. Encoder (a) and decoder (b) for a cyclic Hamming code.

(2) The succession of digits in rows obeys a simple third order recurrence relation; each digit is the XOR (Modulo 2) sum of the digits two and three positions prior to it.

(3) The last three columns correspond to three check digits.

With the applications of these properties, a three-digit shift register and some logic are sufficient to form a serial single error correcting encoder or decoder circuit (Figure 5.54). The encoder circuit operates in two phases: (1) when A = 1, n-k = 4 bits flow into the r = 3 bit feedback shift register and to the output; and (2) when A = 0, the r = 3 bit content of the shift-register is flushed to the output through the feedback logic constructing an n = r+k = 7 bit long word. The decoder logic has also two operational phases: (1) when B = 1, the n = 7 bit word is sequenced into the r = 3 bit shift register; and (2) when B = 0, the shift-register cycles the data through the feedback n = 7 times and flushes the data bit-by-bit to the output XOR gate. If and when the register content is 001 a 1 is sent to the output XOR gate which flips the output bit to the correct value.

5.7.4.6 Reed-Solomon (RS) Codes

Reed Solomon (RS) codes [557] may be considered as a subclass of cyclic BCH codes whose symbols are binary m-tuples of bits rather than bits. The code lengths is $n = 2^m-1$ symbols which makes $n = m(2^m-1)$ bits. The use of r = n-k check symbols, or r = m(n-k) check bits, can correct t = (n-k)/2 symbol errors and all together t = m(n-k)/2 bit errors which are randomly located in the word. A t-error-correcting RS code can also correct either one burst of a total length of (t-1)m+1 or i bursts of total length of (t-i)m+i.

RS codes on $GF(2^m)$ outperform binary codes with the same rate and length at small error rates. The reason for the superiority of RS codes is that their distance features are much better than the distance properties of binary BCH codes. Encoder and decoder circuits for RS codes are similar to those of BCH codes, but the use of symbols, in place of bits, significantly increases the complexity of the decoder circuits.

To reduce the circuit complexity of decoding, a variety of methods [558], e.g., fast Fourier transform, Chien search, Kasami procedure, Massey-Berlekamp algorithm, etc., are devised and applied. The optimum choice for application depends on the specific performance and implementation efficiency requirements.

In CMOS memories, the inefficiency of RS code implementations are so significant that only a few designs for very large memories, which have to operate in extreme environments, could justify RS code use. Nevertheless, RS codes in integrated memory systems, e.g., in wafer-scale-integrated (WSI) and multi-chip-module (MCM) systems can provide both excellent performance and efficiency as outer codes in concatenated code combinations. In WSI and MCM systems namely, for the outer codes the encoder and decoder circuits can economically be implemented in extra control chips.

5.7.4.7 Bidirectional Codes

Bidirectional codes, also known as iterative and product codes [559], can be constructed from two or more error detecting or correcting linear block codes by forming a rectangular structure. This structure is inherently amenable to applications in memories, because one code (n_1, k_1) can be applied for row check, and the other code (n_2, k_2) can be used for column check for a $k_1 \times k_2$ array of memory cells (Figure 5.55). The two codes may be the same or different. Because both codes are linear, the bidirectional code can be handled as a single (n,k) code where $n = n_1 n_2$ and $k = k_1 k_2$. Product codes may also be generated to higher orders, e.g., for tri-directional codes $n = n_1 n_2 n_3$ and $k = k_1 k_2 k_3$.

The performance of bidirectional codes depends on the error detection and correction capability of the component codes. Two single-parity check codes arranged in a bidirectional schedule with a code length of $n = (n_1, n_1-1) \times (n_2, n_2-1)$ have a minimum distance of $d = d_1 d_2$, a code rate of $R = (n_1-1)(n_2-1)/n_1 n_2$, and a guaranteed capability to correct all single and detect all double errors as well as a variety of multiple error patterns. The corner error check bit is consistent with that row or column checkbits of which it is a part. To correct multiple errors in a row or column one of the component codes must have multiple error detecting capability.

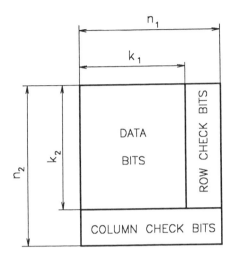

Figure 5.55. Bidirectional code structure for a memory cell array.

Error control by bidirectional codes is most advantageous when the encoding and decoding circuits may operate in serial mode for one of both codes and in parallel mode for the other code, so that the code digits are dispersed in time and space. Since the timing in random access memories is such that first a word line and, thereafter, a bit in the word is addressed, the word selection with parallel error-detecting and the bit selection with serial error-detecting can be associated. Parallel-parallel error-detection is difficult to apply for bidirectional coding, because the encoder and decoder circuits consume excessive power and their implementation require large silicon area; while the error-detection by all-serial working circuits is overly time consuming.

The area and power efficiencies of the encoder and decoder circuits, which implement bi- or multi-directional codes depend on the properties of the component codes. For multiple error corrections many one-directional codes can be implemented at substantially better efficiencies than bi- or multidirectional codes.

5.8 COMBINATION OF ERROR CONTROL CODING AND FAULT-REPAIR

The combination of error control coding (ECC) and fault repair in a memory chip produces a synergistic effect that may result in an immense improvement in both reliability and yield. Although a reliability of over 0.5×10^6 hours MTBE for a CMOS WSI memory operating in radiation hardened space environments [560], and a yield near to 100% for over 3,000 failing cells of a 16-Mbit CMOS SRAM [561] were reported, the on-chip combination of ECC and repair is seldom practical. Reasons for the impracticality include that the silicon surface-area requirements for the implementations are large, and that the vast majority of requirements in reliability and yield can readily be satisfied by other, less complex, means. The area-need and circuit complexity are high, because the on-chip combination of ECC and fault repair requires the addition of (1) test or self-test circuits, (2) decision making logic, and (3) executor circuits to the memory (Figure 5.56).

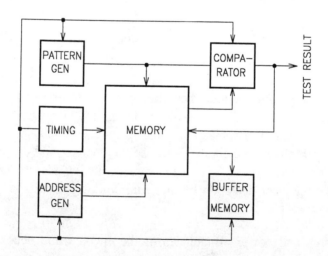

Figure 5.56. Major circuit blocks in a memory combining ECC and fault-repair implementation.

Selftests within CMOS memory chips can efficiently be provided by applications of (a) error detecting codes and (b) write-read data comparison. Linear systematic codes (Section 5.7.4) are particularly suitable to error detection in large memory chips, because their encoder and decoder circuits are simple, and they can conveniently be appended to the original memory design at minimum area increase. Furthermore, designs with linear error detecting codes degrade performance and power parameters only by small percentages. Yet, the rather small variety of error patterns, e.g., single-random, double-random, quad-burst, etc., that can economically be corrected by linear error detecting codes, limits their application area. Theoretically, unlimited number of error patterns can be detected, at no degradation in memory access and cycle times, by write-read data comparison. A memory self-test circuit applying write-read data comparison includes an (1) address generator, (2) data pattern generator, (3) buffer memory, (4) data comparator and (5) timing circuits (Figure 5.57). Before testing a memory bit, the data content of a small array or a block of memory cells is transferred to a buffer memory to save the originally stored data. At test, test-data are provided by the pattern

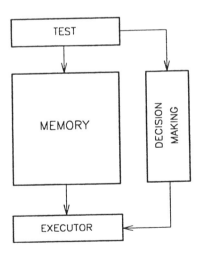

Figure 5.57. Memory selftest schema.

generator, and this data are sent to the addresses determined by the address generator. On the selected address, either a log.0 or a log.1 datum is written into a memory cell, then the read datum is compared with the write datum obtained directly from the pattern generator. The operation of the pattern generators, the comparison of write and read data, and the transfer between the memory and the buffers are timed so that they do not interfere with the normal memory operation.

Memory self-tests detect the errors and indicate the addresses of the errors, but give no information whether the error is a soft- or a hard-one, what the error pattern is, in what circuit and subcircuit, the error appears. More information about the errors, interpretation of test results and determination of appropriate actions can be provided by a decision making circuit, which may range in complexity from a few logic gates to complex implementations of artificial intelligence. Generally, an intelligent circuit structure includes a (1) knowledge base, (2) rule base, (3) information collector and (4) operation and repair control (Figure 5.58). The knowledge base keeps book of the addresses of faulty and spare elements, contains characterization of errors and faults, and descriptions of the repair

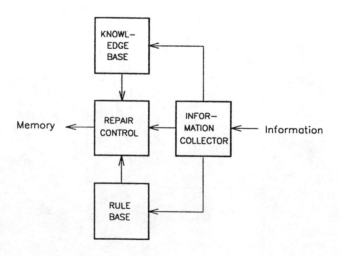

Figure 5.58. Intelligent decision making structure.

methods. The rule base for repair and correction contains the criteria and magnitude regions necessary to decide which method, fault-repair or ECC, in which one of the circuits are to be applied. The information collector is applied to provide updated information of read, write and addressing operation and errors for the control circuits. The operation and repair control associates the obtained data with the knowledge base, evaluates the associated data by means of the rules, compares options and decides and generates control signals for the subcircuits of the memory. The eventual decision is carried out by executor circuits. As executors CMOS memories may apply ECC and fault-repair circuits. ECC circuits can correct certain errors and error patterns (Section 5.7), but do not eliminate the reason of the errors. Therefore, the appearance of too many errors or of error patterns which are undetectable by the code, may overburden the error-correcting capabilities of the applied code.

To prevent an overload in error-correcting capabilities of a code, a certain number of the error-causing faults should periodically be repaired. In CMOS memories, fault-repairs (Section 5.6) disconnect the faulty elements from the operating circuits, and connect some operating spare elements to the working parts of the memory chip. A periodic maintenance of memory operations determines those faults which are necessary and can be repaired, executes the repair of certain faulty elements, and rewrites the correct data to those memory cells which contain soft errors. Usually an initial fault repair and error purging after fabrication is used to improve the yield (Section 5.4), and periodic maintenance procedures during operation are performed to increase the reliability (Section 5.1) of the memory.

To CMOS memories the on-chip implementation of both ECC and fault-repair circuits seems to be overly complex and inefficient in layout designs. Both the complexity and layout area may be reduced, however, by limiting the capability of the error correction and fault repair to a few dominant type of errors and faults, and by using chip-external test-pattern generation, address generation, timing and buffer memories. By these compromises the increase in memory area, access and cycle time, and power dissipation can be kept low, e.g., between 4% and 11%, in designs of large bit-capacity memories.

6

Radiation Effects and Circuit Hardening

The advent of extraterrestrial space utilization, and the requirements to operate many advanced military and some commercial systems in radioactive environments, brought the radiation hardening of semiconductor integrated circuits to the mainstream of the technological developments. Among the various CMOS integrated circuit types, the CMOS memory circuits manifest the highest susceptibility to the effects of radioactive radiation events and, usually, their hardness limits the applicability of the system in radiation environments. To improve the radiation hardness of CMOS memories special processing, device and circuit techniques can be used. This final chapter introduces those radiation effects on CMOS-bulk and CMOS SOI (SOS) transistor devices and circuits which are important to memory designs, and discloses the circuit and design techniques which may be applied to enhance the radiation hardness of CMOS-bulk and CMOS SOI (SOS) memories. The presentation of CMOS SOI (SOS) supports both hardened and nonhardened designs and includes the effects of floating substrates, side- and back-channels, and diode-like nonlinear elements.

6.1 Radiation Effects

6.2 Radiation Hardening

6.3 Designing Memories in CMOS SOI (SOS)

6.1 RADIATION EFFECTS

6.1.1 Radiation Environments

In extreme environments, such as the space, high atmospheric altitudes, nuclear weapons, nuclear propulsions, particle accelerators, colliders and nuclear power plants, CMOS memories may be exposed to ionizing radiations of energetic atomic particles and photons. Ionizing radiations effect the hardware base of satellite and space telecommunications, space defense and surveillance, high altitude flying, intelligent missiles, on-board rocket controls, electronic equipment in space stations, controls of nuclear weapons, nuclear energy production, robots in nuclear fabrication, and all military equipment which are required to operate during and after nuclear attacks.

The effects of nuclear ionizing radiations in MOS devices and circuits have been comprehensively analyzed in the literature, e.g., [61], but very small amount of information have been disclosed about circuit technological approaches to improve the radiation sensitivity of CMOS integrated circuit, and specifically, of CMOS memory devices. This chapter briefly describes the radiation effects on CMOS devices and focuses on the main aspects of radiation hardening of CMOS memories by circuit technological approaches.

A memory operating in radiation environments may have to cope with the effects of (1) permanent ionization due to environmental radioactive radiation, (2) transientionization caused by short-pulse environmental radiation events, (3) semiconductor fabrication induced ionizing radiations, (4) neutron fluence in war environments and (5) combined radiation events. One specific single-event phenomenon, the charged atomic particle impact, is detailed previously (Section 5.3), because it occurs in both standard and radiation environments.

In general, ionizing radiation generates mobile electrons and holes in both the insulator and the silicon substrate and in some other materials of CMOS devices and, by that, causes a variety of errors and faults in CMOS memories. The characteristics of these radiation induced errors and faults must be known prior to the start of the design work, because the design has to incorporate specific circuit techniques for radiation hardening.

6.1.2 Permanent Ionization Total-Dose Effects

Exposure of CMOS memories to radioactive radiation results in permanent and accumulative degradations in their constituent MOS transistor devices and, in turn, in memory characteristics. The rate of the degradation in MOS device characteristics is a function of the absorbed total dose of radioactive irradiation, the voltage bias applied on the MOS devices, the temperature and the time of post-radiation annealing and, in a much lesser degree, of other parameters. Radiation total-dose absorption in CMOS is expressed in the units of rad (Si), and changes in individual MOS device characteristics depend on the voltages on the drain, source, gate and substrate during the irradiation and the annealing. Since voltage biases of the circuit-constituent transistors vary, the radiation caused changes in the transistors are nonuniform within a CMOS memory chip.

Profound total-dose induced changes appear in the threshold voltages (Figure 6.1) due to the substantial build-up of positive charges in the gate-oxide and due to the large increase in oxide/silicon interface traps. Postradiation threshold voltage changes, in general, are larger at higher absorbed total doses, and greatly changes with the voltage bias on the particular device [62]. Positive voltage on the gate electrode causes worst-case shift in threshold voltages, because under this bias condition a very large amount of the electrical charges are trapped near the oxide-silicon interface. The amount of charge trapped, and, thereby, the threshold voltage change, depends also on a variety of parameters other than total-dose and voltage bias, including the frequency and duty-cycle of the switching between log.0 and log.1 states during radiation, dose rate and temperature during radiation, and temperature and time of the annealing after radiation [63]. Radiation sensitivity of threshold voltages is determined chiefly by the thickness of the gate-oxide; the thinner the gate-oxide is the less radiation dependent the threshold voltage becomes. Although a slight radiation dependency of lateral transistor sizes is observed, no regularity of threshold voltage shift as a function of lateral size and radiation dose has been demonstrated so far.

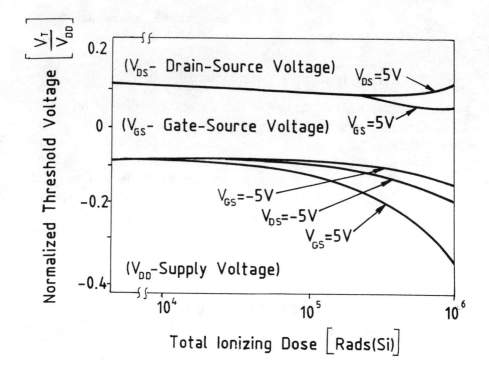

Figure 6.1. Threshold voltage variations as functions of radiation total dose and voltage bias. (After [412].)

Radiation induced interface traps between the oxide and semiconductor decrease the slope of the subthreshold current-voltage characteristics and increase subthreshold drain-source leakage currents [64]. Bias dependent drain-source leakage current increase is substantial in all types of CMOS transistors, but particularly in devices fabricated with some CMOS SOI and many CMOS SOS processing technologies, because in CMOS SOI and SOS devices the radiation may lower the threshold voltages and increase the subthreshold currents in the parasitic side- and back-channels. The total amount of subthreshold leakage currents (Figure 6.2) through the unaddressed memory cells may exceed the current generated on the bitline by an accessed memory cell, and thus may make memory designs impractical in some cases.

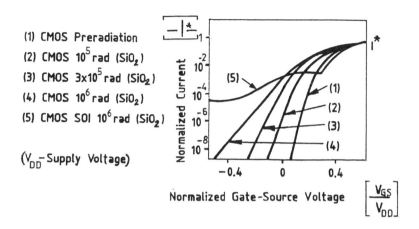

Figure 6.2. Subthreshold drain-source currents versus gate voltage and radiation total dose. (After [64].)

Designs for operation in total dose environments also have to take in account substantial decrease in the mobility of carriers μ in the channels of MOS transistors (Figure 6.3) which decrease is also correlated with the buildup of radiation induced interface traps [65].

Radiation total dose, furthermore, increases junction, leakage currents, transistor-to-transistor leakage currents, and the transistor's sensitivity to hot carrier emission effects. Moreover, total dose effects may result in oxide and junction breakdowns, damages in the in- and output protective devices and, sometimes, latchups in bipolar structures inherent in CMOS implementations. Degradations and damages in CMOS devices may partly recover and rebound, especially at elevated temperatures; but without the use of radiation hardening techniques in processing, memory designs may require either tradeoffs in packing density, speed and power performances, or the designs may result in memory circuits which malfunction in radiation environments.

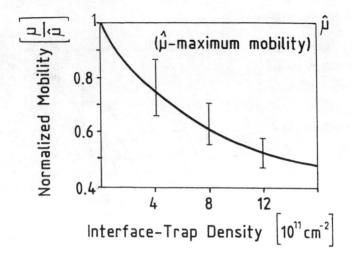

Figure 6.3. Electron mobility degradations at increasing interface-trap density. (After [65].)

Appropriate memory fabrication techniques have to minimize the build-ups of oxide-trapped charges at high dose-rates absorbed in short times, and of the interface-trapped charges at low dose-rates absorbed during long exposure times. As an outcome of the fabrication technique, the total-dose absorption caused spread of design parameters, e.g., the radiation induced variations in threshold voltages, leakage currents, gain factors, etc., must be brought within a range that allows to design a memory that works in the specified environments. The ranges of parameter changes caused by the effects of semiconductor processing, temperature, body (substrate) biasing and ionizing radiation are cumulative and provoke serious limitations or, sometimes, unsatisfiable conditions for the circuit design.

Careful circuit design may reduce the probability of oxide and junction breakdowns, protective device failures and latchups by application of specific guidelines provided by the processing technology. Because the analysis of damage mechanism and the guideline development are process technological tasks, nonspecific to memories, and well described in the literature [66]; the avoidance of breakdowns, protection failures and latch-ups are not discussed in this work.

To accommodate memory designs, the radiation induced variations in threshold voltages and leakage currents are greatly reduced by the emergence of radiation hardened CMOS-bulk, CMOS SOI and CMOS SOS processes [67]. Depending on the processing technology, nevertheless, circuit designs for a radiation dose of 10^6 rads (Si) still may have to allow for large changes in n-channel threshold voltage V_{TN} e.g., from 0.6 V_{TN} to 1.3V_{TN}, in p-channel threshold voltage V_{TP} e.g., from V_{TP} to 2.5V_{TP}, for increases in drain-source leakage current I_{LD} up to 200I_{LD} and for decreases in gain factor β down to 0.5β. Here, V_{TN}, V_{TP}, I_{LD} and β are the preradiation parameter values. These parameter changes are voltage-bias, temperature and annealing dependent and are nonuniform within the memory chip.

6.1.3 Transient Ionization Dose-Rate Effects

Transient ionization can be caused by both cosmic particle impacts and short-pulse high-dose-rate nuclear radiation events, and conventionally classified as single-event and dose-rate phenomena, respectively. Single events can coerce erroneous data into the memory cells and peripheral circuits locally, while dose-rate events can threaten the information stored and processed both locally and globally on the memory chip [68]. Both local and global faults can manifest themselves in either soft or hard errors.

The failure mechanisms and error analyses, and the circuit technologies to tolerate the effects of cosmic and package-emitted particle impacts on memory circuits, are detailed previously (Section 5.3).

Short-pulse high-dose rate events affect the memory by generating large temporary photocurrents and shifts in MOS device parameters, and also by causing permanent damages in device material and latchups. After some recovery time, the photocurrents and the device parameters return approximately to their predisturbance values. Nevertheless the disturbance can cause a local loss of information in CMOS SOI (SOS) memories, and global data scrambling in CMOS-bulk memories. Local photocurrents appear across each semiconductor junction in the circuit, but on those source-nodes of MOS devices which are connected to a power supply pole the photocurrents have insignificant effects on circuit operations (Figure 6.4). Photocurrent simulations show that in a full-complementary storage cell fabricated on n/n+ epitaxial silicon substrate the well-photocurrent [69] is much larger than the other photocurrents in the cell (Figure 6.5).

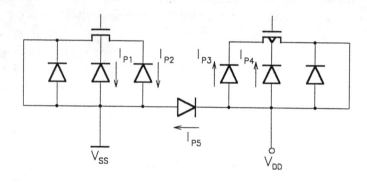

Figure 6.4. Local photocurrent paths in MOS devices.

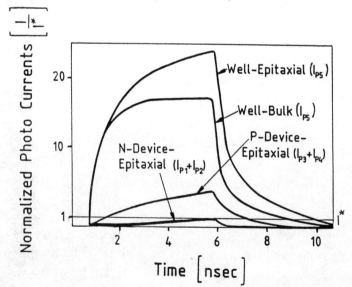

Figure 6.5. Well photocurrents in a 6-transistor CMOS memory cell. (Source [69].)

At low levels of expected photocurrents the information loss can be avoided by the use of state-retention memory cells (Section 9.23). However, beyond some specific dose rate, e.g., 10^9-10^{10} rads (Si)/sec, photocurrents can be as much as 4mA/mm, transient threshold-voltage shifts can exceed the supply voltage and also other parameters may change enormously. At this high rate of parameter changes the scrambling of data is seemingly unavoidable, but techniques for post-event data recovery are conceived and restricted to military use.

Operation during high-level transient events is usually unnecessary, because an external radiation detector may disconnect the memory from the system for the time of the exposure and recovery. An exposure to high dose-rates, however, may result in such thermodynamical stress that fuses interconnects open or short, breaks loose bond wires, burns out semiconductor junctions and destroys other circuit components.

6.1.4 Fabrication-Induced Radiations and Neutron Fluence

Fabrication of CMOS memories may involve the use of very energetic charged particles and photons introduced by electron-beam and X-ray lithographies, reactive ion, plasma and sputter etching, ion implantations, electron-gun deposition, and other radiation techniques. While these techniques satisfy stringent requirements in control of dimensional and material characteristics, they can also cause radiation damages to the memory circuits while being fabricated.

The two most common effects of fabrication by radiation techniques are the buildup of positive charges in the oxides and the increase in the interface traps. The charge buildup in the oxide influences most significantly the threshold voltage and its long-term stability, and the interface trap increase can modify leakage currents, mobilities and other parameters. The fabrication-induced radiation effects are generally alleviated by thermal, hydrogen-assisted thermal and plasma annealing processes. Beside processing adjustments, the memory design does not have to use particular circuit techniques to alleviate the effects of fabrication-induced ionizing radiations.

The effects of neutron radiation in CMOS memory circuits, up to a fluence of $10^{15}/cm^2$, are generally insignificant. Neutron fluence, however, degrades the lifetime of electrons and holes and, thereby, can cause failures in bipolar elements and circuits.

6.1.5 Combined Radiation Effects

A division of radiation-effect analysis into the areas of total dose, single events, dose-rate, fabrication-induced and neutron-fluence episodes is arbitrary. In reality, the effects of the various radiation types are combined, because each of the radiation episodes leave an imprint in the memory and the radiation events join each to the other in certain extreme environments.

Typically, a satellite that applies memories may be affected by radiation total dose [611] in the Van Allen Belt, by cosmic particle impacts [612] during orbiting in space, and to transient radiation [613] at an eventual hostile nuclear detonation. The radiation total dose effects decrease the operation margins (Sections 3.12 and 3.13) and, in turn, the decreased margins make the memory circuits more susceptible to cosmic particle impacts. Similarly, cosmic particles invariably deliver total dose, and so do the dose-rate phenomena that associated with nuclear detonations, and each of these events leaves less margin for total-dose tolerance (Section 6.2.2). During nuclear events the variety of phenomena affect the memory chips simultaneously.

The concept of combined phenomena may be examined through the effects of an experimental irradiation with a single monoenergetic, homogenous beam. Experiments with proton and electron beams show that proton-impact caused pulses degrade the data-upset threshold a lot more than electron beams do at a given duty-cycle and an impulse length [614] (Figure 6.6). Short proton beam impulses, which deliver about 10^3 protons can upset the data stored in four-transistor-two-resistor 4T2R memory cells. Nevertheless, full-complementary CMOS six-transistor 6T memory cells show much less sensitivity to direct-ionizations from protons than 4T2R memory cells do, because, in most of the practical cases, CMOS 6T cells can recover much faster than the mean-time-between-events MTBE in a proton-radiation environment.

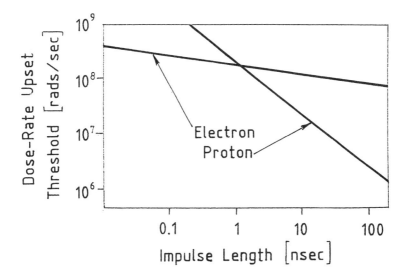

Figure 6.6. Simulated data-upset thresholds of a 4T2R memory cells versus pulse lengths of proton and electron beams. (After [614].)

Combined radiation events readily result in latchup and snapback. Possible latchup is inherent to CMOS structures, where the pnp and npn parasitic bipolar transistors form a pnpn semiconductor controlled rectifier SCR [615]. Normally, the SCR stays in its high-impedance state, but when (1) the loop gain, i.e., the product of the gains by npn and pnp transistors, exceeds unity, (2) the loop current achieves the SCR's exceed a certain turn-on (latch) limit, and (3) the voltage across the SCR is high enough to sustain the latch-situation, the SCR turns and stays latched in a low-resistance state (Figure 6.7) [616]. Exposures of the parasitic SCRs to certain dose-rate ranges do, while to other dose-rate levels do not, result in latchups in CMOS circuits. The latchup-free regions are called latchup windows. Between the latchup windows the operation of CMOS bulk memory can either entirely or partially impaired by switching the whole or

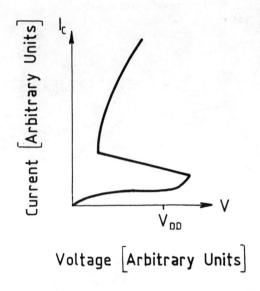

Figure 6.7. Current-voltage characteristics of a latchup sensitive CMOS structure. (After [616].)

an isolated well of the memory to permanent low-resistance conditions [617].

In a snapback phenomenon [618] the avalanche voltage V_A is lowered into the operating region of an n-channel MOS device characteristics (Figure 6.8) without any latchup effects. Unlike latchup, no positive feedback loop exist to make the snapped back situation permanent [619]. Thus, a snapback state can be reversed, e.g., by bringing the drain-source voltage to zero.

By process technological approaches both the radiation induced latchup and snapback phenomena can be controlled, and the design has to assume a latchup and snapback free operation in the memory circuits.

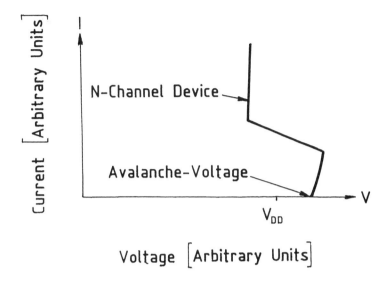

Figure 6.8. Current-voltage characteristics of a snapback-sensitive CMOS structure. (After [618].)

6.2 RADIATION HARDENING

6.2.1 Requirements and Hardening Methods

Radiation hardening is the implementation of those of circuit-design and processing techniques which make a circuit able to operate or survive in a particular, or in a combination of ionizing radiation environments. In space and other nonmilitary electronics, rather arbitrarily, radiation hardened, radiation tolerant and commercial grade memory classes are differentiated to characterize the applied design and processing techniques and the radiation environments in which the memory is able to operate (Table 6.1). Radiation hardened military memories have to satisfy stringent requirements not only in total dose and immunity against the effects

Applications	Radiation Hardened	Radiation Tolerant		Commercial Grade
Design for Radiation Environments	Specific	Specific	Nonspecific	Nonspecific
Process for Radiation Environments	Specific	Nonspecific	Specific	Nonspecific
Total Dose [krad(Si)]	>200	10-200		<10
Threshold LET [MeV/mg/cm^2]	>80	50-80		<5
SEU Error Rate [errors/bit/day]	<10^{-10}	10^{-10}-10^{-5}		>10^{-5}

Table 6.1. Radiation hardened, tolerant and commercial grade memory characteristics.

of atomic particle impacts, but also in dose rate, neutron fluence and, eventually, in specific other parameters which relate to combined radioactive events (Table 6.2). Given parameters, i.e., single-event-upset (SEU), single-event-error-rate (SER), threshold linear-energy-transfer (LET), etc. (Sections 5.3.1 and 5.3.2), impose general requirements to reliable operation in radiation environments and, together with all the other requisites, indicate whether and what specific hardening technique should be used.

Some degree of radiation hardness is inherent to all CMOS memory circuits, but many applications in radiation environments call for the addition of specific hardening techniques in either or in both processing and circuit design techniques. Thus, CMOS memories which can operate in radiation hardened environments may be obtained by the use of four general methods:

(1) Radiation tests and screening of commercial-off-the-shelf (COTS) available memories,

(2) Process technological radiation hardening,

(3) Circuit technological radiation hardening,

(4) Combined process and circuit technological radiation hardening.

Military Radiation Hardness	
Design	Radiation Hard
Processing	Radiation Hard
Total Dose	>1 Mrad
Dose Rate	>10^9 rad(Si)/sec
Neutron Fluence	>10^{15} neutron/cm^2
SEU Error Rate	<10^{-12} errors/bit/day
Threshold LET	>150 MeV/mg/cm^2

Table 6.2. Military radiation hardness.

The application of COTS circuits in radiation environments spurred the development and use of sophisticated test and screening techniques [620]. Wide range of tests have shown that the down-scaling of feature sizes in CMOS fabrication technology increases the amount of total radiation dose at which CMOS memories still can work. This improvement in radiation hardness is attributed mainly to the reduction in gate-oxide thickness and, in turn, to the decreased radiation induced threshold voltage variations. Although with down-scaling field-threshold voltages decrease and the effects of parasitic capacitances increase; the resulting augmentation in leakage currents and crosstalk-signals can reasonably be controlled by processing improvements.

Radiation hardened CMOS processing techniques [621] are vitally important to keep radiation induced threshold voltage fluctuations, leakage currents, gain-factor degradations, crosstalk-signals and other parameter variations in small magnitudes, and to avoid radiation caused field-threshold voltage lowerings, latchups, snapbacks, thermo-electric and thermo-mechanic breakdowns, and other harmful effects. Furthermore, the effects of incident atomic particles and transient dose-rate phenomena on memory reliability, can be greatly reduced by the application of a hardened CMOS SOI or SOS technology in place of a traditional CMOS-bulk technology. Radiation hardening of CMOS processings has emerged as a significant and elaborated area of semiconductor fabrication, and it has evolved as a driving force to SOI and SOS process developments. Process hardening often involves (1) the development of high-quality very thin channel-oxides to reduce transistor threshold voltage variations, (2) doping profile control for transistor and field-threshold voltage adjustments and for leakage current minimization, (3) low temperature fabrication for better parameter control and for decrease of process-caused damages, (4) a variety of annealing techniques to decrease parameter spreads and to improve device reliability, and (5) other technological methods. Additionally, SOI and SOS processing techniques provide very small active device area and allow for full-oxide isolations among transistors and wirings. Moreover, the formation of transistors in SOI and SOS technologies do not require the use of p- and n-wells; and the lack of wells reduces the path-lengths of incident atomic particles in silicon, and decrease the probability of local latchups which may be caused by high-energy transient radiations.

Radiation hardening through circuit design techniques [622] create circuits which are able to accommodate the total amounts of parameter variations resulting from the effects of various radioactive events and from the fluctuations of CMOS processing, power supply, device bias, temperature, hot-carrier emission, and other parameters. The very large parameter-spread that has to be tolerated by a radiation hardened memory circuit, and the parameter-spread tolerance requirements for nonhardened operations, may be exemplified in a comparison of threshold voltage ranges for hardened and nonhardened designs (Figure 6.9). For hardened designs, in addition to the capability of operation with greatly extended threshold voltage fluctuations, the targets and spreads of both threshold voltages should be adjusted to obtain reasonable ranges for operation and

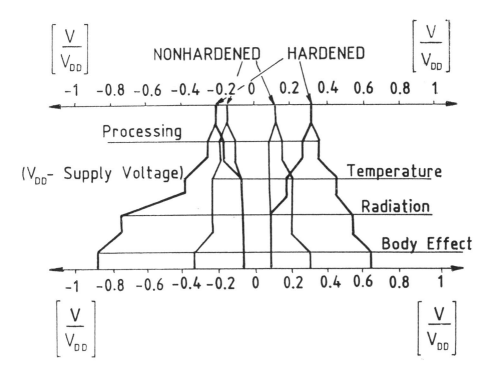

Figure 6.9. Threshold voltage ranges for hardened and nonhardened designs. (Source [622].)

noise margins. Reduced margins and shifted margin-ranges decrease and limit the levels of total doses at which the circuit is able to operate, and increase the susceptibility of the memory to both atomic particle impacts and dose-rate type of transient radiation events. To improve radiation hardness CMOS memory designs apply (1) full-complementary static digital circuits to implement Boolean- and sequential-logic requirements and (2) full-complementary static six-transistor (6T) memory cells for data storage and in flip-flops. The large operation and noise margins of full-complementary logic and storage elements are unmatched by the sense amplifiers. To improve the operation and noise margins of irradiated sense circuits, the design may feature (1) self-compensation, (2) voltage-bias limitation and (3) parameter tracking and eventual other radiation hardening measures. For mitigation of the effects of transient radiations and atomic particle impacts the use of (1) state-retention memory cells (Section 5.3.3), and (2) CMOS SOI (SOS) process and circuit technologies, are preferred. If the peripheral logic circuits also need improved radiation hardness, logic gates which adjust their operation characteristics to radiation induced parameter changes, may be employed. For general radiation hardness improvements however, the implemen-tation of fault-tolerance (1) by memory self-repair (Section 5.6) and (2) by error control coding (Section 5.7) have emerged as the most effective and most economical circuit techniques, so far.

Because processing techniques are not subjects of this book, and the properties of full-complementary digital logic circuits are widely published; the following sections focus on the radiation hardening of sense circuits and memory cells, reviews the radiation hardness enhancement of logic gates, and briefs the use of fault-tolerance techniques (Sections 5.5-5.8) to radiation hardened CMOS memories. Additionally, the most important issues of the application of CMOS SOI (SOS) processing technologies to memory designs are described in the final sections of this chapter.

6.2.2 Self-Compensation and Voltage Limitation in Sense Circuits

The sense circuit has, historically, been the most susceptible memory circuit to both uniform and nonuniform parameter changes, and, thus, to the effects of radioactive irradiations. Memories, which have to operate in radiation hardened environments, use symmetrical differential amplifiers (Sections 3.2-3.5) to sense data generated by full-complementary static memory cells. Postradiation internal operating margins of sense circuits (Figure 6.10) in typical static memories, even which are fabricated with radiation hardened processing, may disappear already at low radiation doses [623]. This is chiefly because of the effects of radiation-induced increases in circuit imbalances, leakage currents, and charge transfers. The imbalance increase is a result of nonuniform bias dependent changes in threshold voltages and carrier mobilities; the high leakage-currents are caused mainly by the enlarged subthreshold drain-source and device-to-device parasitic currents; and the charge transfer increase is an effect of

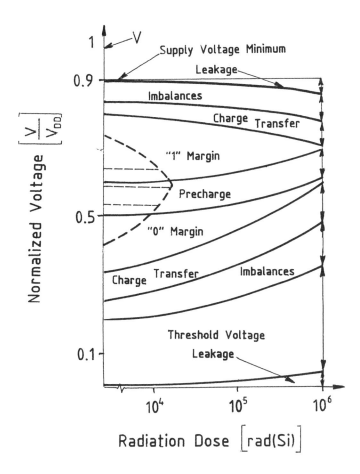

Figure 6.10. Operation margin degradations in an SRAM sense circuit with p-channel memory-cell access devices. (Source [623].)

variations in MOS capacitances. To reduce the radiation sensitivity of CMOS symmetrical differential sense amplifiers, self-compensation and voltage-bias limitation can be used in the sense circuit.

Imbalances in a symmetrical sense circuit appear as sense amplifier offsets. Thus, all of the offset reduction techniques (Section 3.5) can also be applied for radiation hardening of sense amplifiers. Radiation induced parameter changes can most effectively be compensated by negative feedback, sample-and-feedback and by other sense amplifiers which have inherent offset compensation capability.

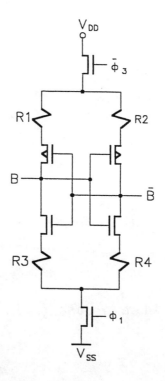

Figure 6.11. Negative feedbacks in sense amplifier for offset reduction. (Source [319].)

From the variety of sense amplifiers which use negative feedback for offset reduction (Section 3.5) the most suitable ones for radiation hardening are those in which the constituent MOS devices function under small bias voltages (Figure 6.11). At small drain-source, drain-gate and source-gate voltage differences, namely, the radiation induced parameter changes are significantly reduced. In these amplifiers voltage biases are reduced because (1) the drain-source and gate-source voltage drops on the individual MOS transistors are fractions of the supply voltage, (2) the signal amplitudes on the gates of the input and output transistors are small, and (3) the clock-impulse amplitudes may also be small. Because of the small output signal amplitudes these feedback circuits are applied often as presense amplifiers and followed by a low-sensitivity large-signal sense amplifier. The susceptibility of these differential sense amplifiers to parameter variations may greatly be decreased by the application of negative feedback (Section 3.5.3). Negative feedback in the pre-sense amplifiers are provided by resistors R1-R4 or transistor devices MP5, MP6, MN7, and MN8. These feedback devices must be designed so that the negative feedback effectively compensates certain limited offset, but the circuit remains still capable to amplify the signal that is generated by the spatially most remote cell to an acceptable amplitude to the follow-up amplifier.

A sample-and-feedback amplifier (Figure 6.12) is virtually free from the tradeoff between offset-reduction and signal-amplification, but it is rather complex and the sampling requires extra time. This amplifier takes sample voltages from the precharged pair of bitlines or from the output nodes, applies these voltages or the output voltages to the gates of regulator devices MP5, MP6, MN7, MN8 through coupling transistors MP9, MP10, MN11, MN12, and stores the sample voltages on the gate capacitances of the regular devices. Because the sample voltages on the gates of the regulator devices change the individual drain-source resistances of these devices so that they act against an eventual output voltage difference, the sense amplifier compensates, in some degree, the effects of its initial imbalances. The accuracy of compensation depends mainly on the symmetry of the regulator device pairs. To reduce non-uniform device parameter changes caused by radioactive irradiations, the sample-and-feedback amplifier designs may also use voltage limitations in the amplitudes of the input and output signals.

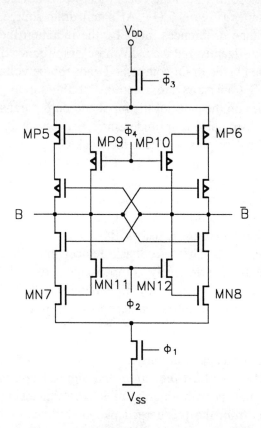

Figure 6.12. A sample and feedback sense amplifier. (After [538].)

Voltage amplification combined with negative feedback or sample-and-feedback have proved to be adequate to satisfy many radiation hardening requirements. Beyond the usual requirements in combining high packing density, operational speed and radiation hardness, the use of the current sense amplifiers, with inherent capabilities for imbalance compensation (e.g., Section 3.4.4), becomes increasingly attractive.

6.2.3 Parameter Tracking in Reference Circuits

Parameter tracking is the capability of a circuit to adjust its operating region in accordance to changes in one or more MOS device parameters. A tracking of the average change of threshold voltages in the access devices of the memory cells can considerably improve the radiation hardness of a memory by making the precharge voltage a function of the total dose in sense circuits (Figure 6.13).

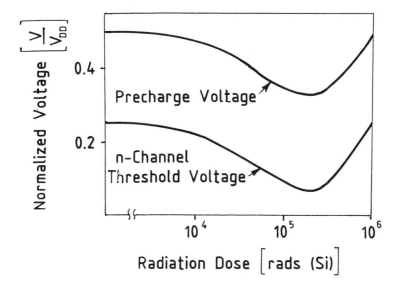

Figure 6.13. Precharge voltage tracks radiation induced threshold voltage variations.

In sense circuits, to which the precharge voltage V_{PR} is provided by a division of the supply voltage (Sections 3.1.3.7 and 4.2.2) and V_{PR} does not track threshold voltage changes, the variation range of the precharge voltage, i.e.,

$$\Delta V_{PR} \approx \rho_d (V_{DD} - V_{SS}) \pm \Delta \rho_d (V_{DD} - V_{SS}) ,$$

is determined by the division factor ρ_d that has very little fluctuation $\Delta\rho_d$ as the total dose absorption changes, and $\Delta\rho_d$ follows the variations of the supply potential difference (V_{DD}-V_{SS}). Namely, the division of (V_{DD}-V_{SS}) is implemented in nonhardened designs as a resistive or capacitive divider circuit to provide a stable V_{PR} that is minimally influenced by device parameter changes. In sense circuits, this stability of V_{PR} may lead to dramatic decrease of the 0 or 1 operating margin at low radiation doses, because the operating margins bend as the threshold voltages of the memory cells access devices vary with the absorbed total dose.

Precharge voltage can also be obtained by deducting one or more threshold voltage V_T from the supply potential (Section 4.2.2). In this case, the precharge voltage and its range follow the threshold voltage variations

$$\Delta V_{PR} \approx (V_{DD} - V_{SS}) - |V_T| \pm \Delta V_T \quad .$$

Since threshold voltage variations ΔV_T-s are greatly radiation dependent, V_{PR} can be designed to track the radiation induced changes in V_T, so that the decrease in operating margins at increasing doses is markedly less than that provided by divider circuits.

For the adjustment of operating margins the output voltage of a precharge generator circuit $V_{PR}(t)$ (Figure 6.14) should follow approximately the radiation-induced variations in the threshold voltage of transistor MN1 and of the access transistors of the memory cells, which are under the same voltage bias conditions during data storage. For MN1, the worst appearing bias combination in radiation environments, i.e., $V_S = V_{SS}$, $V_{DS} = V_{GS} = V_{DD}$, is provided when clock ϕ_{PR} turns devices MP2 and MN3 on and, thereby, the drain and the gate of MN1 are coupled to the positive supply voltage V_{DD} and the source of MN1 is connected to V_{SS}. Here, V_S, V_{DS} and V_{GS} are the source, drain-source and gate-source voltages for the transistor MN1. At precharge, transistor MN4 is brought to highly conductive state, MP2 and MN3 are turned off, and after a transient time, a precharge voltage

$$V_{PR} = v(t_\infty) = (V_{DD} - V_{SS})R_1/(R_1+R_2+r_{d4}) - V_{TN}(V_{BG})$$

occurs. Here, R_1 and R_2 are the voltage-divider resistances, r_{d4} is the drain-source resistance of MN4, V_{TN} is the n-channel threshold voltage and V_{BG} is the backgate bias of MN1. Since the bias of MN1 mimics the worst case bias of the access devices of memory cells, the output voltage of the generator and the precharge voltage follow the worst-case radiation-induced changes in the threshold voltage. Similar tracking of threshold voltage changes can be designed to enhance the operating margins of sense circuits, which contains p-channel access devices to the memory cells (Section 4.2.2).

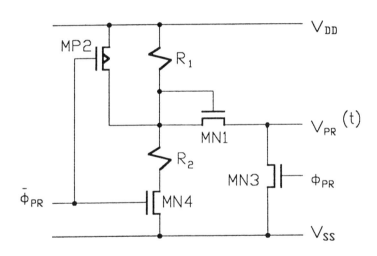

Figure 6.14. Threshold-voltage-tracking precharge generator circuit.

6.2.4 State Retention in Memory Cells

For the design of radiation hardened memories the choices are the full-complementary static six-transistor 6T types of memory cells (Section 2.4). These 6T memory cells can tolerate the largest amount of device parameter changes, and provide the largest operating and noise margins among all types of memory cells when applied in arrays. Furthermore, their application in memories is compatible with fast write and read operations, they retain data despite the appearance of high perturbing currents, their sizes are acceptably small and they are readily available from mainline nonhardened memory designs.

Total dose hardness of a large memory cell array may be extended by using p-channel access devices in the 6T memory cell (Figure 6.15). When radiated, threshold voltages in p-channel devices tend to be more negative under any bias condition, while n-channel threshold voltages may increase or decrease (Section 6.1.2). As an n-channel transistor's threshold voltage decreases, the subthreshold drain-source current of the memory-cell access-devices may increase to an amount that can perturb the datum stored in the memory cell (Section 3.1). The chances of subthreshold-current-generated perturbations are much less at the use of p-channel access devices than those at the application of n-channel access transistors.

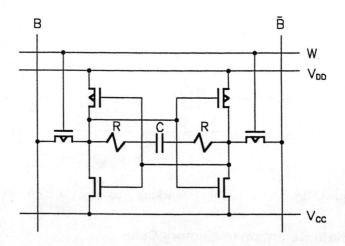

Figure 6.15. RC elements and p-channel access devices in a six-transistor state-retention memory cell.

Dose-rate and atomic-particle-impact caused data loss or scrambling may be reduced or avoided by applying an additional capacitor C and two resistors R to the 6T memory cell (Section 5.3.3). For C the dielectric material is a high quality thin oxide, and R is formed of doped polysilicon. Preferably, both the plates of C and all surfaces of R should interface with dielectrics to keep the time constant $\tau = RC$ large when radiation-induced spurious currents appear. A sufficiently large τ disallows the complete charge or discharge of C during transient radiation events, and the charge remaining on C should be capable to return the memory cell to its preradiation stable state after the effects of the transient radiation abate.

Transient radiation hardness may be aggrandized by fabricating the memory that combines RC enhanced 6T state-retention memory cells with SOI or SOS processing. CMOS SOI and SOS structures greatly decrease the charge collection paths of incident radioactive particles and eliminate device-to-device currents through pn junctions. Nevertheless, side and back channel parasitic currents and uncontrolled charge buildups in the channel substrates may cause significant difficulties in designs with SOI and SOS devices (Section 6.3). Despite the superior radiation hardness of CMOS SOI and SOS six-transistor 6T static memory cells, there is a tendency to use one-transistor-one-capacitor 1T1C dynamic memory cells (Section 2.2) wherever it is possible, due to the small size of the 1T1C memory cell implementations. To improve radiation hardness in a 1T1C memory cell, the cell-internal storage capacitor may be expanded and a p-channel access device may be applied.

State retention in memory devices may be provided also by the application of circuit elements other than capacitors and resistors, e.g., by EPROM, EEPROM, etc., components. These electrically programmable components are radiation sensitive. Radiation hard, however, are the FRAM (Section 2.2.4.2) components, which have great potentials for use in future radiation hardened memory devices.

6.2.5 Self-Adjusting Logic Gates

In CMOS memories the full-complementary static peripheral digital circuits, usually, do not require specific circuit techniques for the increase of radiation hardness, because they are less sensitive to total dose effects than the sense circuits, because they do not have to store data during radiation events, and because they do not require to operate during transient dose-rate episodes. Nevertheless, when operating at high total doses, their input-output voltage characteristics may be modified and their noise margins may considerably be reduced (Figure 6.16) as results of the radiation-induced threshold voltage changes and of the drain-source leakage-current increase in the n-channel devices (Section 6.1.2).

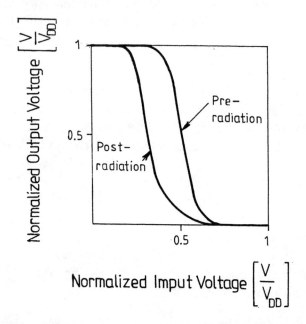

Figure 6.16. Pre- and postradiation input-output characteristics of a CMOS inverter.

This radiation induced noise margin reduction can be compensated. The compensation does not have to retain the circuit's preradiation transfer curve, but it should keep the noise margins larger than a predetermined extent through the entire total dose region in which the memory should operate.

In peripheral digital circuits of memories, a single resistor R or transistor device MN5 that is added to an inverter or logic gate (Figure 6.17) may be sufficient to counteract the noise-margin modifying effects of the changes in threshold voltages and in leakage currents. In these circuits the threshold voltages V_T (V_{BG}) of the n-channel transistor MN1 get elevated when the substrate or backgate bias V_{BG} increases due to a current increase. When the drain-source current i_{dl} of the n-channel device MN1 increases due to a radiation induced threshold voltage lowering $\Delta V'_T$, then node potential V_R, substrate bias V_{BG} and, thereby, the postradiation $V'_T(V_{BG})$ also tend to increase. Thus, the bias-dependent threshold voltage increase ΔV_T can compensate the radiation induced reduction $\Delta V'_T$ through an appropriately designed resistor R coupled to the source of the n-channel devices. Resistor R may be implemented in a polysilicon resistor, or in an n-channel transistor device MN5 that operates in its triode region. To MN5, triode-operation may be obtained by the use of a fixed gate voltage, e.g., V_{DD}, or by the application of gate voltage V_{PR} which tracks the radiation-induced threshold voltage variations (Section 6.2.3).

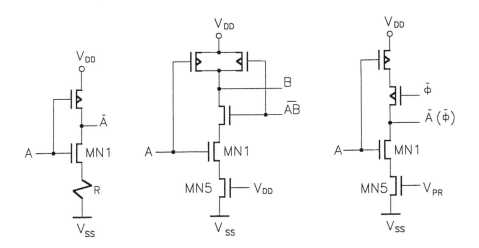

Figure 6.17. Counteracting radiation-induced shifts in input-output characteristics of inverters and logic gates.

More complex circuit variations for input-output characteristics and noise margin adjustment, e.g., requiring three additional transistors per logic gate plus radiation tracking $\Delta V'_T$, and preradiation V_T generators (Figure 6.18) [624] may also be considered. In this circuit, a compensation in postradiation input-output voltage characteristics and in noise margins

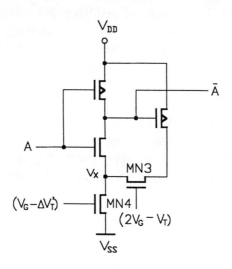

Figure 6.18. Noise margin adjustment in an inverter. (After [624].)

can be provided, if the gain-factor ratio $\beta_4/\beta_3 = 4$. The desirability of this ratio may be substantiated by the equation of the currents through device MN3 and MN4 when A is a log.0:

$$\frac{1}{2}\beta_3(2V_G - V_T - V_x - V_T') = \frac{1}{2}\beta_4(V_G - \Delta V_T - V_T') ,$$

where β_3 and β_4 are the gain-factors of MN3 and MN4, V_G is the gate voltage, V_x is the voltage on node X, and postradiation threshold voltage V_T is assumed to be approximately the same for MN3 and MN4. Because in radiation environments V_T-s, ΔV_T-s and β-s change nonuniformly and a preradiation V_T is difficult to maintain, the effectiveness of this circuit is very limited in improving radiation hardness.

6.2.6 Global Fault-Tolerance for Radiation Hardening

Radiation hardness of CMOS memories can greatly be enhanced by designing fault-tolerant features globally in the memory chips. In radiation hardened memories fault-tolerance is achieved by either one or by the combination of the following approaches:

(1) Error control coding,

(2) Fault repair,

(3) Fault masking.

Error control coding (Section 5.7) is used to detect and correct many of the random soft-errors induced by cosmic particle impacts and dose-rate events, and some of the hard errors resulting from the effect of total-dose and dose-rate type of radioactive radiations. Effects of radiations most commonly detected and corrected by

- Single error correcting bidirectional parity (imparity) check code,
- Two to four error correcting Hamming codes,
- Burst correcting Berger codes.

Implementations of other codes which provide higher performance, e.g., Reed-Solomon, Viterbi codes, require such excessive layout areas that their on-chip applications appear to be uneconomical beyond about $0.2\mu m$ processing feature sizes.

Fault repair techniques (Section 5.6) replace the permanently damaged rows, columns and clusters of memory cells by operating circuits. Replacements of faulty circuits are usually applied to avoid the overrun of the capabilities of an error correcting code and to repair origins of errors which are uncorrectable by codes. The most successful repair techniques are implementation of the associative approach (Section 5.6.4).

500   CMOS Memory Circuits

Fault masking (Section 5.6.5) is a convenient approach to improve the reliability of the peripheral circuits beyond the reliability of the memory cell array, and to render the control circuits, which provide fault-tolerance, insensitive to their own faults. For fault-masking the triplicate majority logic is the nearly exclusive choice, but for repair of the input and output buffers duplication seems to be the practical approach.

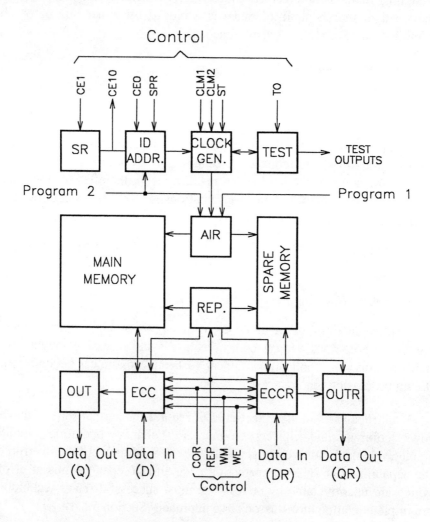

Figure 6.19. Architecture of a radiation hardened CMOS static memory. (Source [625].)

The combination of all approaches fault-tolerance, hardened sub-circuits and hardened processing is required to provide a high level of radiation hardness in memories. A heavily radiation hardened memory architecture (Figure 6.19) [625] comprises a main memory cell array, a spare memory cell array, duplicated error control coding ECC and ECCR, duplicated output circuit OUT and OUTR, a control circuit of repairs for yield improvement REP, a control circuit to provide associative iterative repair for reliability increase AIR, built in test generator and control circuit TEST and other circuits. In this memory, all circuits except the memory cell arrays and except the output buffers, are implemented in triplicated majority-voting logic circuits. In spite of circuit triplications and duplications and the accommodation of fault tolerance and self-test, the peripheral circuits take only about 4%, while the spare memory occupies approximately 7% of the memory chip. This memory architecture is developed for applications in military satellites.

6.3 DESIGNING MEMORIES IN CMOS SOI (SOS)

6.3.1 Basic Considerations

6.3.1.1 Devices

CMOS SOI (SOS) processing and device technologies are the most widely used derivatives of the basic CMOS-bulk technology. In addition to improved radiation hardness the feature sizes of CMOS SOI (SOS) transistor devices are readily scalable to well below $0.1\mu m$, while a feature size of $0.12\mu m$ seems to be the lower practical limit for the down-scaling of CMOS-bulk transistor devices. This amenability of CMOS SOI (SOS) transistors to down-scaling greatly increases the future potentials of CMOS SOI (SOS) technology applications not only in radiation hardening, but also in general integrated circuit processing and manufacturing. In some aspects, CMOS SOI (SOS) processing and device technologies deviate from CMOS-bulk technologies, and the deviations effect the characteristics of both the active and passive circuit elements. In circuit designs, the structures and properties of the elements which are effected by the use of nonstandard CMOS technologies, must be taken into consideration.

The technology that implements complementary metal oxide semiconductor (CMOS) transistor devices in semiconductive silicon islands on an insulating layer, rather than directly on a semiconductive silicon bulk, is called CMOS silicon-on-insulator (CMOS SOI) technology [626]. The early CMOS SOI technologies used sapphire as insulating basis and, therefore, the technology was named CMOS silicon-on-sapphire (CMOS SOS) technology [627]. Somewhat imprecisely, the term CMOS SOI is used also to distinguish the technology that applies an insulator layer on a bulk silicon from the CMOS SOS technology. Commonly, CMOS SOI (Figure 6.20a) applies a SiO_2 layer between the MOSFET device and the carrier substrate-semiconductor. The semiconductor carrier

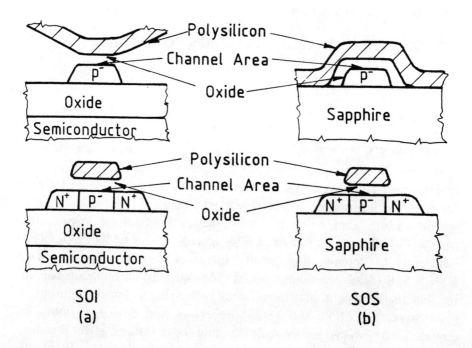

Figure 6.20. Cross-sectional views of an n-channel transistor fabricated with SOI (a) and SOS (b) technologies.

substrate readily allows for creation of oxide isolation among the semiconductor MOSFETs and other devices. In CMOS SOS, the device-to-device isolation in CMOS SOS (Figure 6.20b) is provided by the sapphire substrate and by the oxide developed on the surfaces of the semiconductor islands.

In semiconductor islands the CMOS transistors can be implemented to operate in a variety of modes, e.g., in partially depleted (PD), fully depleted (FD), dynamic threshold (DT), and other operation modes, which allow to meet diverse technical requirements. In a PD device, the electric field of gate depletes the silicon body only to a depth that is less than the thickness of the silicon film. Because of the limited depth of the depletion region PD devices operate similarly to traditional bulk devices. Thus, for memory designs with PD devices, the circuit and design techniques of the mainline CMOS-bulk technologies can be adopted. Although CMOS-bulk transistor models have to be somewhat modified, mainline design tools and methods are also applicable to PD CMOS SOI (SOS) memory cells. PD devices combine high drain-source currents and radiation hardness, and make possible to produce high-performance memory devices which can operate in severe environments. In FD devices the total depletion charge, including front, back and lateral depletion charges, exceeds the possible depletion charge in the silicon body. The total depletion results in low subthreshold leakage currents and large drain-source resistances in the saturation region. Low leakage currents are important for the operation of the access transistors in memory cell arrays (Section 3.1.3.3), and to achieve low standby power dissipations, while high resistances in the saturation regions facilitate high gains in sense amplifiers (Sections 3.3, 3.4 and 3.6) and in analog devices (e.g., Section 3.3.6.4). In DT devices, the gate is connected to the body and, thereby, activate the parasitic bipolar transistor. The operation of the bipolar transistor increases the current between the drain and source and reduces the threshold voltage of the CMOS SOI (SOS) device. The threshold voltage decreases due to forward biasing the body-source or the body-drain diode. Currents through the diodes contribute to the standby current, limit the number of memory cells coupled to a bitline, and increase sense amplifier offsets. Furthermore, the increased body potential may enlarge the diodes junction capacitances.

In CMOS SOI (SOS) memory technology, the high current drive capabilities of PD devices, the low subthreshold leakage currents and high saturation resistance of FD devices, an be combined by switching between PD and FD operation modes. For the implementation of PD-FD switching in dual-mode transistors the polysilicon backgate technique (Figure 6.21a) has high potentials, because the implementation of the PD-FD switch effects the packing density very little, its integration into the processing is not difficult, and it can be used also for threshold voltage switching and control. Dual threshold voltage application may be necessary in the designs of the memory cell arrays and sense circuits, especially in designs for low voltage operations. Array designs may have to apply bootstrapped wordline drivers, the boosting capacitor may also be implemented in CMOS SOI (SOS) at small modification in the polysilicon backgate technique [628] (Figure 6.21b).

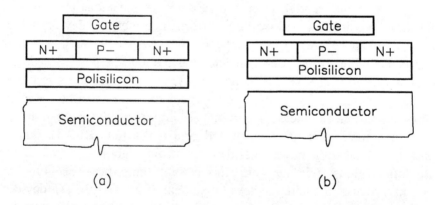

Figure 6.21. Backgate (a) and boosting capacitor (b) implementations.

In the following, the CMOS SOI (SOS) transistors are understood to operate in the partially depleted mode unless the text designates the operation mode of CMOS SOI (SOS) transistor device otherwise.

6.3.1.2 Features

The interest in making CMOS SOI (SOS) memories has been spurred by the anticipation of significant improvements in (1) radiation hardness, (2) operational speed, (3) power dissipation, and (4) packing density, in comparison to those provided by CMOS-bulk memories.

Radiation hardened memory designs use CMOS SOI (SOS) technologies to greatly reduce the number of single event errors and the probability of latchups which may be caused by the impacts of ionizing atomic particles, by transient dose-rate events and by permanent total-dose ionization. Incident ionizing particles find much shorter charge collection paths S-s and produce much lower single event error rates in CMOS SOI (SOS) than in CMOS-bulk designs. Taking the sensitive regions as parallelepipeds, then the approximate maximum $\hat{S}$ in a CMOS SOI (SOS) transistor is

$$\hat{S} \approx [(W+2d)^2 + d^2 + t_s]^{1/2}$$

while that in a CMOS-bulk transistor is

$$\hat{S}^1 \approx [(W+2d)^2 + d^2 + (L+2d)^2]^{1/2}$$

Here, W and L are the width and length of the transistor drain- or source-area, d is the depth of the depletion region, and t_s is the thickness of the semiconductor film. In practice, $t_s \ll L+2d$, thus $\hat{S} \ll \hat{S}^1$. A parasitic bipolar transistor amplifies the charge Q_i that is induced by the incident atomic particle along S, by a factor of $(1+\beta^x)$, where β^x is the effective charge gain. Thus the charge collected by the drain Q_d may be approximated by [629]

$$Q_d \approx (1+\beta^x)Q_i$$

and

$$Q_i = K \cdot S \cdot LET ,$$

where K is a constant, and LET is the linear energy transfer. To alter a datum in a memory cell LET = LET_C and $Q_d = Q_c$ required, where LET_C is the critical LET, and Q_C is the critical equivalent charge. Applying the equation for Q_i the LET_C may be given as

$$LET_c = \frac{1}{K} \frac{Q_C}{(1+\beta^x)S}.$$

The expression of LET_C indicates that for large LET_C and, in turn, for high immunity against the effects of incident atomic particles, large Q_c, small S and small β^x are required. Since the charge collection paths S-s are significantly smaller in CMOS SOI (SOS) memories than those in CMOS-bulk memories, CMOS SOI (SOS) memories can operate at much lower SEU rates than CMOS-bulk memories do in radiation environments. Furthermore, in CMOS SOI (SOS) circuits the complete insulation of semiconductor transistor devices and the use of highly doped semiconductor interconnects, eliminate the inadvertent creation of large parasitic bipolar transistor and thyristor devices and, thereby, avoid the possibility of global latch-ups which can be induced by transient dose-rate events. Although latch-ups may occur locally in those silicon islands which include a pnpn structure, many of the local failures can be tolerated in memories by the application of error detecting and correcting codes and fault repair circuits. Since radiation hardened memory circuits have been indispensable in satellites, missiles, rockets, high-altitude aircraft and a variety of space, high atmospheric and military equipment, the evolution of CMOS SOI (SOS) technologies have been well supported.

The anticipation of high speed in CMOS SOI (SOS) memory operations is based on the effects of the rather small parasitic transistor- and field-capacitances, and of the low effective threshold voltages. Parasitic capacitances in the CMOS SOI (SOS) transistors are quite small because the drain-body and source-body junction areas are very little, and because the substrate is isolated from the transistors. Field, i.e., substrate-wiring, capacitances may also be small because the thickness of the insulator between the carrier substrate and the silicon film adds to the thickness of the oxide between the wires and the insulator surface. CMOS SOI (SOS) transistors, which are placed on silicon island have little threshold voltage increases during operations because the appearing backgate bias voltages are mostly such that they reduce rather than increase the effective threshold voltages. Low threshold voltages result in high effective gate-source voltages and, in turn, in high drain currents, which are particularly important in output drivers and NAND type of decoder and logic circuit designs. In practical circuits, however, the speed-increasing effects of the small parasitic capacitances and threshold voltages are upset by the speed-reducing effects of the floating substrates, charge-carrier mobility degradations, eventual side- and back-channel currents, threshold voltage fluctuations caused by film thickness variations, reduced field oxide thickness, and other CMOS SOI (SOS) phenomena. Thus, the speed of CMOS SOI (SOS) memories are only 25-35% faster than the speed of CMOS-bulk memories. This speed gap is anticipated to increase with decreasing feature sizes in the deep submicrometer region, where the parasitic capacitances in CMOS SOI (SOS) reduce in greater percentage than those in CMOS-bulk devices do.

The expectation for low power consumption in CMOS SOI (SOS) is indicated also by the small parasitic capacitances and by the low effective threshold voltages. Small capacitances need small energy to charge and discharge, and low effective threshold voltages allow the use of low supply voltages. Capacitances proportionally and supply voltages quadratically effect the dynamic power dissipation, thus their reduction can substantially decrease the total power dissipation. Nevertheless, in CMOS SOI (SOS) memories the reduction of the total power consumption is limited by the substantial wire-island capacitances, threshold voltage variations and subthreshold currents. The anticipated improvement in control of threshold voltages and subthreshold currents may make the use of CMOS SOI (SOS) technology more attractive also for memory designs.

Device packing densities in CMOS SOI (SOS) memories can significantly be higher than those in CMOS-bulk memories. CMOS SOI (SOS) circuit implementations, namely, do not need wells and well separations, and CMOS SOI (SOS) processing are very amenable to fabricate stacked device structures.

Generally, the use of CMOS SOI (SOS) processing technology definitely results in high radiation hardness and packing density, but the anticipated advantages in operational speed and power dissipation may not be significant enough to justify the higher production costs. CMOS SOI

(SOS) production costs are high, not only because of the expensive starting material, but also because the processing equipment, and the engineering and design tools deviate from the CMOS-bulk standards in a number of aspects.

Memory circuits applied in CMOS SOI (SOS) designs are nearly identical with those used in standard CMOS-bulk technologies. CMOS SOI (SOS) memory designs, nevertheless, have to overcome the effects of the (1) floating substrates, (2) side- and back-channels, and (3) diode-like parasitic elements and others, which should be taken in account in the designs despite improvements in CMOS SOI (SOS) processing technologies, e.g., [630].

Additionally, designs of certain CMOS SOI (SOS) circuits may be very complex due to the self-heating of the transistor devices on insulating substrates [631]. Unlike the standard CMOS transistor devices which are placed on a common semiconductor substrate in a chip, the individual CMOS SOI (SOS) transistors are isolated dielectrically and thermodynamically from a common substrate and from their adjacent elements. The small thermal-conductance of the insulating material in the vicinity of a transistor device may allow for substantial temperature elevation when the transistor operates at a high drain current. A temperature-change influences important device parameters, e.g., threshold voltage, gain factor, etc., which, in turn, vary the drain-current. Because of the interdependency between the drain-current and temperature, the operating temperature of numerous transistor devices should individually be calculated. For computations and mappings of device temperatures on a CMOS SOI (SOS) chip, computer programs are available, and allow for combining temperature variations in individual transistors with electric circuit analysis, e.g., [632]. High operating temperatures of transistors may impose substantial constraints to the layout designs, and may require the design of thermo-specific structures and the combination of thermal-conductors and thermal-shields with electric circuit elements. Nonetheless, in memory-internal circuits, i.e., memory cell arrays, sense circuits, decoders, input interface and peripheral logic circuits, the self-heating effect causes only a small, e.g., 8°C, temperature raise. Thus, thermal effects may significantly influence the design of the output buffers and direct-current reference circuits, but may result only in insignificant timing and performance variations in all other memory circuits.

6.3.2 Floating Substrate Effects

6.3.2.1 History Dependency, Kinks, and Passgate Leakages

Floating substrate, or floating body, means that the channel region of a MOS transistor has no low-resistance connection to any fixed potential source such as ground V_{SS}, power supply V_{DD} or V_{CC}. The lack of wiring of the channel regions to V_{SS}, V_{DD} or V_{CC} results high packing density, but also makes the potential of the channel region of each individual transistor different from each other, depending on their drain-, source- and gate-potentials as well as on the time of observation and the potential variations before the observation. Effects of the floating substrate and of the consequent substrate potential changes in a CMOS SOI (SOS) transistor may result in the occurrence of (1) history or time dependency of threshold voltages $V_T(t)$-s and (2) kinks and premature breakdowns in the direct current DC drain-current I_D versus drain-source voltage V_{DS} and gate-source voltage V_{GS} characteristics $I_D = f(V_{DS}, V_{GS})$, and (3) transmission or passgate leakage currents [633].

The history or time dependent characteristic of $V_T(t)$ may be taken in account by the variations of the backgate bias as a function of time $V_{BG}(t)$ (Figure 6.22), because $V_T[V_{GB}(t)]$. At a particular time t, $V_{BG}(t)$ is determined by the signals which appeared prior to the observation time on the device's drain, gate- and source-terminals, by the device-internal charge and discharge times and by the generation and recombination mechanisms of the electrons and holes. Within a CMOS SOI (SOS) transistor device (Figure 6.23) capacitances C_{DB}, C_{GB}, C_{CB} and C_{SB} are charged and discharged through capacitances C_{GD}, C_{DS}, C_{GC} and C_{GS}, diodes D_{DB} and D_{SB}, resistances R_D, R_S, R_{DB}, R_{SB}, R_{BD} and R_{BS}, and bipolar junction transistor (BJT) T_{DBS}. Here, for a capacitance C, diode D and transistor T the indices D, S, G, B and C designate drain, source, gate, body and channel of a CMOS SOI (SOS) device, respectively. Signal

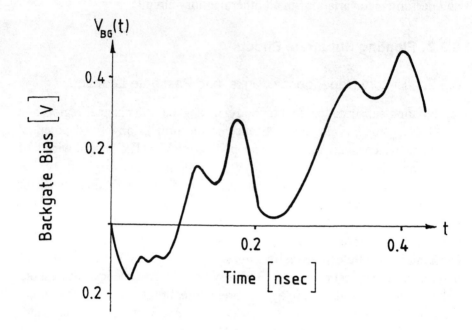

Figure 6.22. Simulated backgate bias as a function of time in an n-channel SOI transistor

variations, on the terminals of a CMOS SOI (SOS) transistor, change the device-internal voltages and currents, which changes induce differing device-intrinsic charge and discharge times and differing carrier generation and recombination mechanisms. These differing events result in a hysteretic behavior in $V_T[V_{BG}(t)]$ when the device is controlled by an impulse that has symmetrical rise and fall transients. The hysteresis in $V_T[V_{BG}(t)]$ affects the signal delays at small control signal amplitudes V_o-s and at low supply voltages V_{DD}-s significantly more than at large V_o-s and high V_{DD}-s, because the switching times are functions of the terms $V_o - V_T[V_{BG}(t)]$ and $V_{DD} - V_T[V_{BG}(t)]$, where $V_o \approx V_{DD} > V_T[V_{BG}(t)]$.

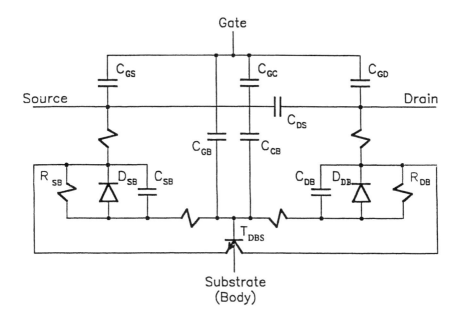

Figure 6.23. A model of parasitic elements present in a CMOS SOI (SOS) transistor device.

Time dependency and hysteresis in $V_T[V_{BG}(t)]$-s can lead to incorrect timing and undesirable race conditions for clock, control and general logic signals, operation and noise margin degradations, read and write pattern sensitivities, increased sense and small-signal amplifier offsets, and to other type of inadequacies in circuit operations. The timing of memory subcircuits and the propagation delays of the clock signals, in addition to others, are influenced also by the $V_T[V_{BG}(t)]$ variations of the individual transistors in the drivers and in the receivers. Increased variations in the signal delays may result in incorrect activation of subcircuits and in erroneous output signals in logic gates. In sense circuits $V_T[V_{BG}(t)]$ decreases may particularly be harmful because low $V_T[V_{BG}(t)$-s are associated with large subthreshold leakage currents in the access devices. The enlargement of leakage currents degrades the operation and noise margins, may cause false reads or writes, and may make the read and write operations pattern sensitive in memory cell arrays (3.1.3.3). Read and write operations, furthermore, may significantly be slowed or impaired by the nonuniform $V_T[V_{BG}(t)]$ variations caused offset augmentations in differential sense amplifiers (Sections 3.1.3.5 and 3.5). The possible overall effects of $V_T[V_{BG}(t)]$ variations include increased read and write error rates, degraded radiation hardness, longer access- and cycle-times, excessive power dissipation, unreliable memory operation and impaired functionality.

Memory operations may adversely be influenced by the eventual occurrence of kinks and premature breakdowns in the n-channel transistors' DC $I_D = f(V_{DS}, V_{GS})$ characteristics (Figure 6.24). Commonly, kinks and

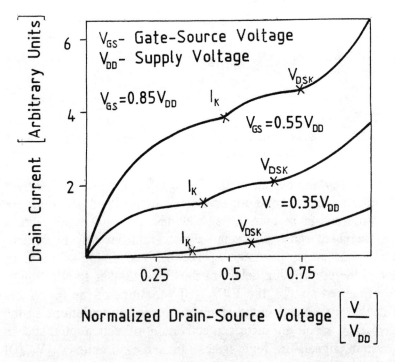

Figure 6.24. Kinks and premature breakdowns in the DC current-voltage characteristics of an n-channel SOI transistor

premature breakdowns are attributed to the effects of floating substrates and parasitic side-channels (Section 6.3.3) where side-channels exist. Both the floating substrate and the side-channel effects may reduce $V_T[V_{BG}(t)]$, the lower $V_T[V_{BG}(t)]$ increases the drain-current I_D, the large I_D onsets an expanding rate of impact ionizations, and this further raises I_D and reduces $V_T[V_{BG}(t)]$. $V_T[V_{BG}(t)]$ may be lowered, furthermore, by the activation of the parasitic BJT that forward biases the diodes D_{DB} or D_{SB} or both, and by the presence of the quasi-neutral and heavily doped region in the body. To the body potential and $V_T[V_{BG}(t)]$ changes other mechanisms, i.e., thermal carrier generation and tunneling, contribute also. When the contributing phenomena create a positive feedback between an increasing I_D and a decreasing $V_T[V_{BG}(t)]$ beyond a certain current I_K a pronounced raise, a kink, occurs. The rate of the I_D increase is intrinsically controlled until a breakdown voltage V_{BR} (V_{GS}), and above V_{BR} (V_{GS}) the combined effects of the floating substrate, side-channels, thermal feedback, avalanche impact ionization, etc. leads to premature appearance of breakdown in the $I_D = f(V_{DS}, V_{GS})$ characteristics.

Kinks and premature breakdowns in the $I_D = f(V_{DS}, V_{GS})$ characteristics of CMOS SOI (SOS) transistors may make very difficult the design of effectively operating sense and other small signal amplifiers. To achieve acceptable amplification A (Sections 3.2-3.6) for the practical ranges of input signal amplitudes Δv_i or Δi_i in the vicinities of the quiescent operation points, the amplifier circuits should be designed of constituent transistor devices which have adequately extended flat current saturation regions in their $I_D = f(V_{DS}, V_{GS})$ characteristics. Current saturation regions, namely, may be small or nonexisting due to the floating substrate induced kinks and premature breakdowns. Kinks and premature breakdowns in the $I_D = f(V_{DS}, V_{GS})$ characteristics may also cause thermal instabilities, reduced radiation hardness, excessive active power dissipation and unreliable circuit operation.

Memory circuits may also be plagued by the effects of the transmission gate, or passgate, leakage currents. Passgate leakage currents are profound manifestations of the operation of the parasitic bipolar junction transistor BJT that is inherently present in each CMOS SOI (SOS) transistor device. The drain current of a pnp or an npn BJT may be significant when the history dependent body voltage forward biases the base-emitter diode. Such forward bias may frequently occur in passgate devices and in dynamic NOR gates even when the CMOS SOI (SOS) transistor devices are turned off. In an n-channel device that is in a low-conductance state, when its drain and source are on the supply voltage V_{DD} and its gate is on the ground potential V_{SS}, a pulldown of the source potential can temporarily turn the BJT on, and can generate a transient drain current (Figure 6.25). BJT-s may be activated not only by particular combination of device terminal voltages, but also by the impacts of charged atomic particles, and by transient radiation events when the memory operates in radioactive environments.

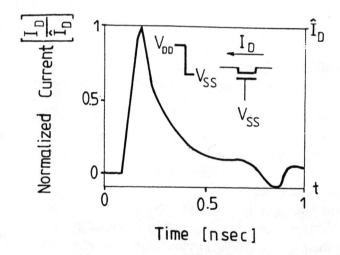

Figure 6.25. Simulated BJT current in a turned off passgate device.

In memories, the passgate leakage currents are critical in the operation of memory cell arrays, sense amplifiers, and NOR decoders. The access devices of the unselected memory cells may generate such large accumulated parasitic bitline currents that the selected cells read or write current can only partly, or can not at all, counteract the parasitic currents (Section 3.1.3.3), and these may result in slow or impaired read or write operations. Furthermore, BJT currents can alter the stored data in DRAM cells and also in many SRAM and other memory cells. False data reading can also occur by sense amplifier offsets which may be enlarged by the nonuniform changes in the BJT currents (Sections 3.1.3.5 and 3.5). The total of BJT currents may pull down the high output nodes of the NOR decoder circuit and, thereby, the BJT currents may induce incorrect addressing and multiple access of memory cells. The general effects of passgate leakages on memory circuits comprises aggrandized number of read and write errors, lengthier access and cycle times, decreased radiation hardness, higher power dissipation, unreliable or impaired memory operation.

Clearly, CMOS SOI (SOS) memory designs are heavily challenged by the effects of the floating bodies. These effects may be summarized as

- timing failures in subcircuit activations and logic gate functions,
- operation and noise margin degradations,
- read and write data pattern sensitivities in arrays,
- offset increases in sense amplifiers,
- gain reductions in sense and other amplifying circuits,
- data losses in memory cells,
- false addressing by NOR-type of decoders,
- other malfunctions.

Memory circuit malfunctions caused floating-body effects may be temporary or permanent and, in general, they may substantially degrade the reliability, speed, power and radiation hardness and may render the entire memory dysfunctional. Floating body induced dysfunctions are also environment, e.g., humidity, temperature, radiation, etc., dependent. Since environmental parameters may change with time on a given place, the floating-body effects may or may not result in substantial degradations in operating characteristics or in dysfunctions. Therefore, conventional operation tests may need revisions. Revised CMOS SOI (SOS) specific tests revealed that the floating-body effects dramatically reduced the fabrication yield of CMOS SOI (SOS) memory products.

6.3.2.2 Relieves

Floating body effects on CMOS SOI (SOS) memory circuits are much more severe than on CMOS SOI (SOS) logic circuits (Section 6.4.2.1). While CMOS SOI (SOS) logic circuits, e.g., central computing units, data processors, and others, may require only circuit modifications [634] CMOS SOI (SOS) memories need to combine process, transistor device and circuit design approaches to alleviate the floating-body effects.

Specific processing techniques are developed, predominantly, to enhance the recombination properties of the source-body and drain-body function, e.g., by implanting Ar or Ge into the drains and sources of the transistor devices to decrease carrier-life times and energy bandgaps, respectively. Although Ar and Ge implantations and other processing improvements reduce the floating substrate effects, the improvements achievable by purely process-technological means are usually insufficient to obtain reliably operating memories and reasonable fabrication yields.

Operational characteristics and yields of CMOS SOI (SOS) memories, however, can be brought to acceptable levels by specific transistor device designs. The mostly applied specific transistor designs use body ties, very short channels and full depletions to mitigate the effects of the floating substrates.

Floating substrate effects can nearly be eliminated by low-resistance body-ground ties in the n-channel transistors, and, if it is needed, also by body-supply ties in the p-channel transistors. In most of the CMOS SOI (SOS) memories, only the n-channel transistors of the access devices of the memory cells, the n-channel transmission devices, and the n-channel transistors in the sense- and small-signal amplifiers require the use of body ties. Low resistance body ties, however, call for high doping concentrations and special doping profiles in the body as well as for significant extension of the transistor area to accommodate the highly doped area and the contact on the body. Body ties may greatly expand memory cell and transmission device areas, complicate processing, increase parasitic device capacitances in the circuit, and, in combination with the connected circuit elements, may form parasitic thyristor structures in CMOS SOI (SOS) circuits. Thus, body-tie applications in memory circuits may greatly compromise the number of memory cells which can be accommodated in a single chip, and increase fabrication, development and design costs, and may magnify latch-up probabilities in certain circuit configurations. Nevertheless, body-to-wafer bonds vertically to the wafer-surface under the channels can be implemented [635] as well and, thus, the full potential of CMOS SOI technology may be exploited also in memory designs.

CMOS SOI (SOS) memory designs often apply body-ties not only to improve reliability and yield, but also to raise sensing speed, and to reduce supply voltage and power dissipation. Benefits in speed, supply and power characteristics are consequences of the body-tie-assisted reductions in nonuniform fluctuations of body potentials $V_{BG}(t)$-s and of threshold voltages $V_T[V_{BG}(t)]$-s. In a simplified sense circuit (Figure 6.26), the $V_{BG}(t)$-s of the individual transistors are controlled by clocks ϕ_{BP}, ϕ_{BA}, ϕ_{BS}

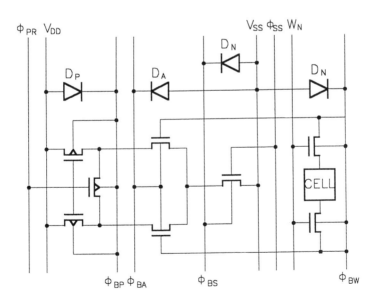

Figure 6.26. Body-tie applications in a CMOS SOI (SOS) sense circuit.

and ϕ_{BW}. These clocks provide $V_{BG}(t)$-s which result $V_T[V_{BG}(t)] \approx 0$ for the controlled transistor devices during the times of precharge, high signal amplification and memory-cell access and $V_T[V_{BG}(t)] > 0$ during the other times. During the times when the $V_T[V_{BG}(t)] \approx 0$, high substrate currents may occur, which may promote hot carrier emissions, and can generate ground-supply and other noise signals. Decreases in hot carrier emissions and noise signal amplitudes are obtained, here, by adding shunt diodes D_P, D_A, D_S and D_N to the sense circuit.

Needs for body-ties in memory circuits, may be alleviated by the use of down-sized deep-submicrometer CMOS processing technologies which, as a byproduct, reduce the effects of substrate-bias on threshold voltages. Furthermore, the use of deep-submicrometer CMOS technologies greatly decreases the radiation induced variations in threshold voltages and in drain-source leakage currents and, thereby, extends the radiation hardness of both CMOS-bulk and CMOS SOI (SOS) memories.

Floating substrate effects in CMOS SOI (SOS) memories may also be mitigated by the use of fully depleted FD transistor devices. FD devices, namely, have much less subthreshold leakage currents, and their drain currents are much less effected by kinks, than the traditionally applied partially depleted PD devices do. To combine the high currents and other benefits of PD devices with the advantages of the FD devices, FD and PD operation modes can be switched by using backgates (Section 6.3.1) or other methods in memory cells, sense amplifiers and NOR decoder circuits.

Circuit technical approaches attempt to enhance the memory circuits' tolerance of the floating body effects, and the applicable techniques vary circuit to circuit. The most significant effects of the floating substrates (Section 6.3.2.1) and their most prevalent circuit technical relieves are concisely described next.

In memory subcircuit activations and in logic gate operations, the floating substrate caused delay variations are rather small, e.g., 5%, in comparison to the total delay times. These delay variations should be taken into account in the computation of worst case clock signal delays and, in the rare cases where selftiming is used, also in the interface designs. Logic gate designs, although they need no structural change, should be timed so that the worst case race conditions cause no erroneous logic operation.

Operation and noise margins degradations (Section 3.1.3) are the most significant in the memory cell arrays, where the floating body induces large leakage currents, and the large wire-to-wire capacitances and the power supply lines couple great noise signals into the sense circuit. CMOS SOI (SOS) circuits, in contrast to bulk circuits (Sections 4.1.1 and 4.1.2), have little wire-to-substrate capacitances which decouple a part of the noise signals. The large leakage currents opposing the read or write currents (Section 3.1.3.3) may also cause pattern sensitive read or write operations. Operation and noise margin degradations as well as pattern sensitivities may be alleviated by reducing the number of memory cells which are connected to a single bitline and to a single wordline, by decreasing the wordline-bitline and the wire-to-wire capacitances, by using high current write amplifiers and wordline buffers, by increasing the input/output conductance of the sense amplifiers, by increasing the threshold voltage of the access devices in the memory cells, by boosting the wordline voltage well beyond the supply voltage, by increasing the minimum high level and by decreasing the maximum low level of data stored in the memory cells, by decreasing precharge voltage variations, by modifying the detection thresholds in the sense amplifier circuits, and by others. In peripheral logic circuits the operation and noise margins may be extended by avoiding the use of transmission gates and dynamic logic circuits, by increasing drive currents, by applying noise filters to the power lines, by reducing line resistances and capacitances, by adding circuit elements to backward bias the parasitic base-emitter diodes, by reducing the number of parallel-coupled transistors in the circuits, and by others.

In sense circuits, the floating substrate effects aggrandize the imbalances and the sense amplifier offsets. Offset reductions (Section 3.5) may most economically be provided by the applications of negative feedback (e.g., Sections 3.5.3 and 6.2.2) sample-and-feedback (e.g., Sections 3.5.4 and 6.2.2) and positive feedback current (e.g., Sections 3.4.4, 3.4.6 and 3.4.9) sense amplifier circuits. Sense amplifier gains may be reduced by the floating-body induced anomalies in the transistors' saturation regions. The quiescent operation points of the sense amplifiers should be placed to the low gate-source voltage and low drain-source voltage regions where the kink and early breakdown have little or no effects on the saturation currents.

In CMOS SOI (SOS) memory cells, the leakage currents through the access transistors may be so large that it can alter the stored data when the cells are unselected (Section 3.9.3.3). In dynamic memory cells (Sections 2.2.2.3, 2.7.2, 2.8.2 and 2.9), this type of data loss can be counteracted by the increase of the refresh frequency, and in static memory cells (Sections 2.4, 2.5, 2.7.3, 2.8.2 and 2.9), the data loss may be avoided by the use of low load resistances and of high-current constituent transistor devices.

NOR-type of decoders (Section 4.3) and wide NOR logic gates may dysfunction, because the cumulative floating-body generated leakage currents can exceed the current of a turned-on device or the current of the load device, when all parallel devices are turned off. To outweigh the effects of the leakage currents wide load transistors, long parallel transistors and periodical charge and discharge of the NOR circuit-internal nodes may be applied. In dynamic NOR gates special techniques such as input data setup during precharge, precharge of circuit-internal nodes, crossconnected input pairs, and others [636] may improve functionality and performance.

Generally, the functionality, speed, environmental tolerance, reliability and yield of CMOS SOI (SOS) memories which may be hampered by the effects of the floating substrates, can be improved to exceed the characteristics of CMOS SOI (SOS) memories. For the improvements however, packing density, process and circuit complexity, and power dissipation may have to be compromised.

6.3.3 Side- and Back-Channel Effects

6.3.3.1 Side-Channel Leakages, Kinks and Breakdowns

In CMOS SOI (SOS) transistors, conductive channels may be induced not only on the top of the semiconductor island but also on the sides and the bottom of the island [637]. The top, side and bottom surfaces of n-channel CMOS SOI (SOS) transistors have numerous physical differences. Significant differences may exist in the crystal orientations between the top and the sides of an island. Silicon islands, may have crystal orientations <100> on the top and nearly <111> orientations on the sides. Generally, the crystal orientation on the sides depends on the etched side-top angle. This angle may change greatly with the use of different isotropic and anisotropic etching procedures, and may vary somewhat due to nonuniformities in the photoresist and etching processes. Nonuniformities appear also in the crystal structure since the initial silicon growth on the surface of insulator is highly disordered and only with the further growth becomes regular semiconductive silicon crystal. Variation in the crystalline structure results in unavoidable variations in surface dopings of the island sides. The surfaces of the island sides are subject to further changes by the thermal oxidation that follows the ion implantation of the doping material. Moreover, during thermal oxidation "V" shaped grooves may be formed in the silicon along the edges of the silicon-insulator junction which grooves ultimately reduce reliability.

Side-channel effects on circuit operation and reliability can greatly be reduced by process technological approaches such as oxide backfill, highly doped side-channel stops and edgeless configurations. Furthermore, side-channel effects may be subdued by routinely using field-oxide among perpendicularly cut islands. Although improvements in the CMOS SOI (SOS) processing can significantly subdue side-channel operations, many circuit designs may have to take into account the (1) side-channel caused leakage currents, and (2) side-channel induced kinks and modified breakdown features in the DC $I_D = f(V_{DS}, V_{GS})$ characteristics.

Drain-source leakage currents I_{LDS} in n-channel transistors may considerably be increased as an effect of the side-channel threshold voltage V_{TS} which can be markedly smaller than the threshold voltage of the top-channel V_{TT} (Figure 6.27). This difference in threshold voltages $\Delta V_T = V_{TT} - V_{TS} > 0$, is a result of the distinctive crystal orientation between the top and the side channels as well as of the variations in the crystalline structure, surface dopings and oxide thickness. Because $V_{TS} < V_{TT}$, the side-channels turn on at a gate-source voltage V_{GS} that is smaller than V_{TT}, and the drain currents in the side-channels appear as subthreshold leakage currents for the transistor device on the top. These subthreshold leakage currents reduce the number of memory cells connectable to a sense amplifier, decrease the achievable operational speed, can make the array of memory cells pattern sensitive and may impair circuit operations.

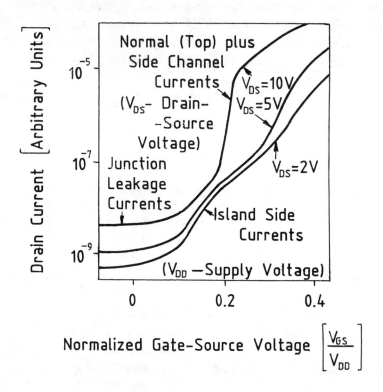

Figure 6.27. Side-channel effects on drain-source leakage currents. (After [637].)

Sense amplifier operations may be compromised by the occurrence of side channel induced kinks in the n-channel transistors $I_D = f(V_{DS}, V_{GS})$ characteristics. Side channel operation caused kinks are similar to the floating substrate induced kinks (Section 6.3.2). As a result of side channel operations the drain current increases, and the threshold voltage gets lower, because the impact ionization from the side-channel currents produce body-source biases that can reduce top-channel threshold voltages. Lessening threshold voltages increase both the drain currents and the impact ionizations, and the resulting strong current-increases occur as kinks in the saturation regions of the $I_D = f(V_{DS}, V_{GS})$ characteristics of an n-channel transistor. Breakdown features of n-channel CMOS SOI (SOS) devices may also be modified (Figure 6.28) by the interaction of the parasitic side-transistors with the basic top-transistor, and by the effects of the forward biases in the drain and source junctions of the devices [638].

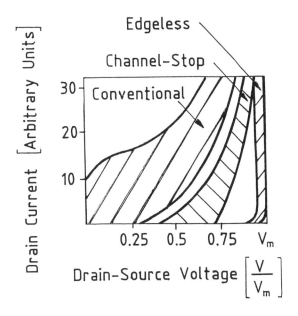

Figure 6.28. Breakdown features of n-channel SOI devices fabricated with conventional, channel-stop and edgeless technologies. (After [637].)

Due to side-channel operations, conventional CMOS SOI (SOS) transistor devices may exhibit soft breakdown characteristics, but transistors implemented with channel stops and in edgeless configurations show sharp breakdown characteristics. Soft breakdown features, in mild forms, reduce achievable gains and operational speed in sense and in other amplifiers, and in emphasized forms soft breakdowns can make circuits unstable and circuit designs impractical. Impacts of kinks and modified breakdowns on memory circuit operations and designs are described generally under floating substrate effects (Section 6.3.2).

6.3.3.2 Back-Channel- and Photocurrents

In addition to side-channel effects, circuit designs may have to cope with back-channel generated currents [639]. On the back side of a CMOS SOI (SOS) transistor island, near the silicon-insulator interface, in the insulator, charges may be trapped (Figure 6.29). Charges in the insulator

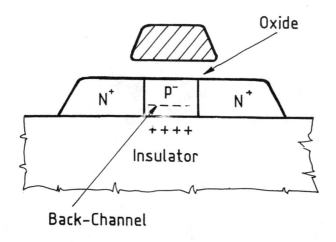

Figure 6.29. Parasitic back-channel MOS device.

may appear for a variety of reasons, e.g., for hot-carrier generation by the electric field of the drain, electrostatic charge, etc., but the most remarkable effects can be induced by radioactive radiations. Ionizing radiations may generate positive charges in the insulator, which attract electrons to the silicon surface in an amount that makes the substrate material slightly conductive. The increases in conductance and in the associated drain current are functions of the absorbed radiation dose and of the gate-source and drain-source voltage biases (Figure 6.30). Subthreshold drain current increase by back-channel conductivity is an important restriction in the design of radiation hardened memory cell arrays and sense circuits. Power dissipation of the memory circuits may also be increased by the occurrence of back-channel currents.

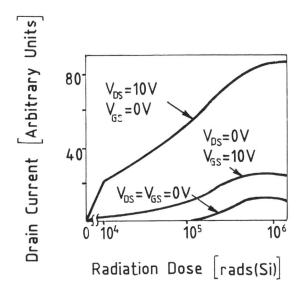

Figure 6.30. Drain current as a function of radiation dose and voltage bias influenced by a back-channel parasitic device. (Derived from [639].)

During short high-energy transient radiation events, on the back side of the CMOS SOI (SOS) transistor devices, in the insulator material, photocurrents appear in addition to the other parasitic currents. For a single transistor device this SOI (SOS)-specific photocurrent I_{IP} may be calculated by the exponential approach [640]

$$I_{IP} = W\dot{\gamma} A(1-e^{B(V_{RE}+0.9)}) \ ,$$

where W is the junction width, e.g., in mils, γ is the dose-rate in rads (Si)/sec, A and B are material dependent constants, e.g., $A = 4.5 \times 10^{-15}$ and $B = -0.044$ V for sapphire, and V_{RE} is the reverse voltage bias on the junction, e.g., $V_{RE}=V_{DD}=3V$. Large insulator photocurrents alter the biases of the individual transistors, cause upsets in the data stored in the memory (Section 6.1.3) and impair memory operations (Section 6.2.2). Unlike the CMOS-bulk memories, however, high dose-rates and high photocurrents can not result global latchups in CMOS SOI (SOS) memories, and proper designs can minimize the probability of local latchups.

6.3.3.3 Allays

Simulations of memory circuit operations, which apply the models of short-channel CMOS SOI transistor devices [641] and involve the effects of floating substrates, side- and back-channel and photocurrents indicate that the full-complementary 6-transistor (6T) memory cells (Section 6.2.4) with body-ties, self-compensating sense amplifiers (Section 6.2.2) with body ties, and full-complementary static logic gates without transmission gates (Section 6.2.5) are the most amenable subcircuits to CMOS SOI (SOS) radiation hardened memory designs.

In CMOS SOS (SOI) memories fully-depleted, rather than traditional partially depleted, transistor devices may be used, to minimize the side- and back-channel as well as the other subthreshold currents. The use of fully-depleted transistors in memory designs, however, results in significantly slower operation than the speed that can be achieved by applications of partially-depleted transistors. An exchange between fully- and partially-depleted operating modes by alterations of the individual transistors' backgate biases, can combine the advantages of both operation modes (Section 6.3.1). The switch between operation modes may be designed without the use of a backgate bias plane in the vicinity of a certain backgate bias V_{BG} and of a certain gate-source voltage V_{GS}, e.g., at $V_{BG} = 0.5V$ and $V_{GS} = 0V$ [642]. Voltages V_{BG} and V_{GS} can be set by design of the doping concentration and profile in the channel-space. The part of the channel-space near to the silicon-insulator interface is partially depleted and contains repelled charges with opposing polarity of the charges in the channel. These repelled charges can be used as virtual backgate electrodes and make possible to form low-resistance contacts at the sides of the islands. Through the low-resistant contacts the body biases and, in turn, the exchange between fully- and partially-depleted operation modes can be controlled.

6.3.4 Diode-Like Nonlinear Parasitic Elements

The sizes of CMOS SOI (SOS) static full-complementary memory cells, and in a lesser degree, also the sizes of other memory circuits, may be reduced by the application of heavily doped polysilicon as short-distance interconnects. The polysilicon material is doped either P^+ or N^+, which can form low resistance contacts only with a similarly doped P^+ or N^+ semiconductor material.

In static memory cells polysilicon-semiconductor contacts can be made in smaller area than metal-semiconductor contacts, and the use of two polysilicon layers allows for combining the crosscoupling and a state-retention capacitance C_{SR} in a small silicon-surface region (Figure 6.31).

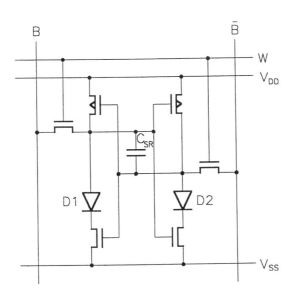

Figure 6.31. Static memory cell circuits applying P^+ and N^+ doped polysilicon crosscouplings.

Because the decreased silicon-surface area is achieved by the exclusive use of either one P^+ or N^+-doped polysilicon, P^+N^+ junctions appear between the drains of the joining p- and n-channel transistors. These P^+N^+ junctions constitute parasitic nonlinear elements which are designated as diodes D1 and D2 in the circuit. Rather than diode-like behavior these parasitic devices have current-voltage characteristics (Figure 6.32) which are similar to those of nonlinear resistors. The nonlinear characteristics of the parasitic P^+N^+ junctions modify the write and read properties of the memory cells and, therefore, they should be considered in the circuit analysis, simulation and design.

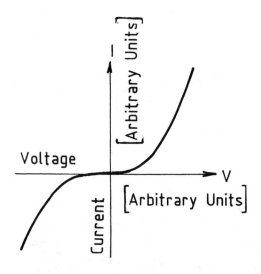

Figure 6.32. Nonlinear current-voltage characteristics of a parasitic P^+N^+ junction.

Generally, the size of all other memory subcircuits may also be decreased by applications of polysilicon-to-semiconductor contacts and by the elimination of the short circuits between P^+ and N^+ drain and source electrodes. In digital logic circuits the effects of the parasitic nonlinear elements on operation and performance are mostly insignificant, but in parasitic elements may result in deviations from the planned characteristics.

Memory circuit characteristics, in addition to the floating body, side-channel, back-channel, and parasitic-diode effects, may also be influenced by a number of other phenomena which are specific to the use of a CMOS SOI (SOS) technology. These other phenomena, usually, have about the same impacts on memory circuits as on the widely applied digital and analog circuits, and they are comprehensively investigated and described along with the domineering CMOS SOI (SOS) events, e.g., [643]. Cumulatively, the other phenomena may reduce the operation margins of a sense circuit (Sections 3.1.2 and 3.1.3) by a term as much as $0.08V_{DD}$. As comparisons, the additive margin degradation caused by the effects of the floating bodies side-channels can be about $0.2V_{DD}$, the parasitic diode may abate the margins by 0.5V, and with properly designed and processed devices, the back-channel and self-heating effects may be neglected. Here, V_{DD} is the supply voltage. Apart from voltage-margin reductions, the other specific phenomena may also unfavorably influence the speed and power features of the memory circuits.

The prevalent features, i.e., radiation hardness, fast operation, small size and convenient down-scaling of CMOS SOI transistor devices, are very attractive not only for radiation hardened memory designs but also to satisfy requirements for high performance, low voltage and low power memory operations, e.g., [644]. To produce dependable CMOS SOI memories in high volumes, the circuit design and device engineering have to overcome the difficulties associated with the effects of floating substrates, side- and back-channels, nonlinear parasitic elements, and other specific phenomena. Alleviation of these adverse effects, and the anticipated advancements in processing technology, device modeling and circuit simulation, may place the CMOS SOI technology to the mainline of the microelectronic industry and CMOS SOI memories may gain universal applications in computing and data processing systems.

References

[11] G. Luecke, J. P. Mize and W. N. Carr, "Semiconductor Memory Design and Application," McGraw-Hill Book Company, 1973.
[12] W. D. Brown and J. E. Brewer, "Nonvolatile Semiconductor Memory Technology," IEEE Press, 1997.
[13] S. Przybylski, "The New DRAM Architectures," IEEE ISSCC '97, Tutorial, pp. 9-13, February 1997.
[14] J. L. Henessy and D. A. Patterson, "Computer Architecture: A Quantitative Approach," Morgan Kaufman, 1990.
[15] B. Prince, "High Performance Memories," John Wiley and Sons, 1996.
[16] J. M. Rabaey, M. Pedram and P. Landman, "Low Power Design Methodologies," Kluwer Academic, pp. 201-251, 1996.
[17] H.-J. Yoo, "A Study of Pipeline Architectures for High-Speed Synchronous DRAMs," IEEE Journal of Solid-State Circuits, Vol. 32, No. 10, pp. 1597-1603, October 1997.
[18] T. P. Haraszti, "High Performance, Fault Tolerant Orthogonal Shuffle Memory and Method," United States Patent Number 5,612,964, March 1997.
[19] T. Kohonen, "Content-Addressable Memories," Springer-Verlag, 1980.
[110] Goodyear Aerospace "STARAN", Company Document GER-15636B, GER-15637B, GER-15643A, GER-15644A, GER-16139, 1974.
[111] J. Handy, "The Cache Memory Book," Academic Press, 1993.
[112] S. A. Przybilsky, "Cache and Memory Hierarchy Design," Morgen Kaufmann, 1990.
[113] R. Myravaagens, "Rambus Memories Shipping from Toshiba," Electronics Products, November 1993.
[114] Ramlink Committee, Electronic Industries Association, "Standard for Semiconductor Memories," JEDEC No. JESD216, IEEE Computer Society No. P1596.4, 1996.
[115] H. Ikeda and H. Inukai, "High Speed DRAM Architecture Development," IEEE Journal of Solid State Circuits, Vol. 34, No. 5, pp. 685-692, May 1998.
[21] W. P. Noble and W. W. Walker, "Fundamental Limitations on DRAM Storage Capacitors," IEEE Circuits and Devices Magazine, Vol. 1, No. 1, pp. 45-51, January 1985.
[22] K. W. Kwon et al., "Ta_2O_5 Capacitors for 1 Gbit DRAM and Beyond," IEEE International Electron Devices Meeting, Technical Digest, pp. 34.2.1-34.2.4, December 1994.

[23] L. H. Parker and A. F. Tasch, "Ferroelectric Materials for 64 Mb and 256 Mb DRAMs," IEEE Circuits and Devices Magazine, Vol. 6, No. 1, pp. 17-26, January 1990.
[24] J. C Burfoot, "Ferroelectrics: An Introduction to the Physical Principles," D. Van Nostrand, 1967.
[25] D. Bursky, "Memory and Logic Structures Are Getting Faster and Denser, "Electronic Design, pp. 40-46, December 1, 1997.
[26] A. Goetzberger and E. Nicollian, "Transient Voltage Breakdown Due to Avalanche in MIS Capacitors," Applied Physics Letters, Vol. 9, December 1966.
[27] P. Chatterjee et al., "Trench and Compact Structures for dDRAMs," IEEE International Electron Devices Meeting, Technical Digest, pp. 128-131, December 1986.
[28] H. Arima, "A Novel Stacked Capacitor Cell with Dual Cell Plate for 64 Mb DRAMs," IEEE International Electron Devices Meeting, Technical Digest, pp. 27.2.1-27.2.4, December 1990.
[29] P. C. Fazan and A. Ditali, "Electrical Characterization of Textured Interpoly Capacitors for Advanced Stacked DRAMs," IEEE International Electron Devices Meeting, Technical Digest, pp. 27.5.1-27.5.4, December 1990.
[210] S. Watanabe et al., "A Novel Circuit Technology with Surrounding Gate Transistors (SGTs) for Ultra High Density DRAMs," IEEE Journal of Solid-State Circuits, Vol. 30, No. 9, pp. 960-971, September 1995.
[211] K. V. Rao et al., "Trench Capacitor Design Issues in VLSI DRAM Cells," IEEE International Electron Devices Meeting, Technical Digest, pp. 140-143, December 1986.
[212] W. M. Regitz and J. Karp, "A Three-transistor Cell, 1.024-bit, 500-ns MOS RAM," International Solid-State Circuit Conference, Digest of Technical Papers, Vol. 13, pp. 42-43, February 1970.
[213] R. G. Middleton, "Designing Electronic Circuits," Prentice-Hall, pp. 221-226, October 1986.
[214] C. F. Hill, "Noise Margin and Noise Immunity in Logic Circuits," Microelectron, Vol. 1, pp. 16-21, April 1968.
[215] A. Bryant, W. Hansch and T. Mii, "Characteristics of CMOS Device Isolation for the ULSI Age," IEEE International Electron Devices Meeting, Technical Digest, pp. 28.1.1-28.1.4, December 1994.
[216] C. E. Chen et al., "Stacked CMOS SRAM Cell," IEEE Electron Device Letters, Vol. EDL-4, No. 8, pp. 272-274, August 1983.
[217] W. Dunn and T. P. Haraszti, "1-Mbit and 4-Mbit RAMs with Static Polysilicon-Load Memory Cells," Semi Inc., Technical Report, October 1974.
[218] T. P. Haraszti, "Novel Circuits for High Speed ROMs," IEEE Journal of Solid State Circuits, Vol. SC-19, No. 2, pp. 180-186, April 1984.
[219] P. R. Gray, D. A. Hodges and R. W. Brodersen, "Analog MOS Integrated Circuits," IEEE Press, 1980.
[220] T. P. Haraszti, "Circuit-Techniques and Applications of MOS LSI," (Schaltungstechniken und Anwendungen von MOS-Grosschaltungen), Semiconductor Seminar, Telefunken, Digest (Kurzfassungen), 1969.
[221] J. T. Koo, "Integrated-Circuit Content-Addressable Memories, "IEEE Journal of Solid-State Circuits, Vol. SC-5, pp. 208-215, October 1970.
[222] R. M. Lea, "Low-Cost High-Speed Associative Memory," IEEE Journal of Solid-Sttae Circuits, Vol. SC-10, pp. 179-181, June 1975.
[223] S.M.S. Jalaleddine and L. G. Johnson, "Associative IC Memories with Relational Search and Nearest-Match Capabilities, Vol. 27, No. 6, pp. 892-900, June 1992.
[224] T. P. Haraszti, "Flip-Flop Circuits with Tunnel Diodes," (Billenokorok Alagutdiodakkal), Radiotechnika, Vol. XV, No. 12, pp. 444-447, December 1965.

[225] G. Frazier, et al., "Nanoelectric Circuits Using Resonant Tunelling Transistors and Diodes," IEEE International Solid-State Circuit Conference, Digest of Technical Papers, pp. 174-175, February 1993.

[226] W. S. Boyle and C. E. Smith, "Charge Coupled Semiconductor Devices," Bell System Technical Jounral, No. 47(4), pp. 587-593, April 1970.

[227] E. R. Hnatek, "A User's Handbook of Semiconductor Memories," John Wiley and Sons, pp. 609-646, 1977.

[228] F. Lai, Y. L. Chuang and S. J. Chen, "A New Design Methodology for Multiport SRAM Cell," IEEE Transactions on Circuits and Systems - I: Fundamental Theory and Applications, Vol. 41, No. 11, pp. 677-685, November 1994.

[31] B. J. Sheu et al., "BSIM: Berkeley Short Channel IGFET Model for MOS Transistors," IEEE Journal of Solid-State Circuits, Vol. SC-22, No. 4, pp. 558-566, August 1987.

[32] B. Song and P. Gray, "Threshold-Voltage Temperature Drift in Ion-Implanted MOS Transistors," IEEE Journal of Solid-State Circuits, Vol. SC-17, No. 2, pp. 291-298, April 1982.

[33] T. P. Haraszti, "CMOS/SOS Memory Circuits for Radiation Environments," IEEE Journal of Solid-State Circuits, Vol. SC-13, No. 5, October 1978.

[34] R. R. Troutman and S. N. Chakvarti, "Subthreshold Characteristics of Insulated-Gate Field Effect Transistors," IEEE Transactions on Circuit Theory, Vol. CT-20, pp. 659-665 November 1973.

[35] W. N. Carr and J. P. Mize, "MOS/LSI Design and Application," McGraw-Hill Books, pp. 19-24, 1972.

[36] P. S. Winokur et al., "Total-Dose Failure Mechanisms of Integrated Circuits in Laboratory and Space Environments," IEEE Transactions on Nuclear Science, Vol. NS-34(6), pp. 1448-1454 (1987).

[37] P. K. Chatterjee et al., "Leakage Studiers in High-Density Dynamic MOS Memory Devicefs," IEEE Journal of Solid-Stte Circuits, Vol. SC-14, No. 2, pp. 486-498, April 1979.

[38] T. P. Haraszti and R. K. Pancholy, "Modern Digital CMOS VLSI," Seminar Manuscript, University of California Berkeley, Continuing Education in Engineering, Universit Extension, pp. 59-60, March 1987.

[39] N. N. Wang, "Digital Integrated Circuits," Prentice-Hall, pp. 257-263, 1989.

[310] F. F. Offner, "Push-Pull Resistance Coupled Amplifiers," Review of Scientific Instruments, Vol. 8, pp. 20-21, January 1937.

[311] K. Y. Toh, P. K. Ko, and R. G. Meyer, "An Engineering Model for Short-Channel MOS Devices," IEEE Journal of Solid-Stte Circuits, Vol. 23, No. 4, pp. 950-958, August 1988.

[312] R. H. Crawford, "MOSFET" in Circuit Design," McGraw-Hill Book, pp. 31-37, 1967.

[313] R. A. Colclaser, D. A. Neamen and C. F. Hawkins, "Electronic Circuit Analysis, "John Wiley and Sons, pp. 408-411, 1984.

[314] J. P. Uyemura, "Circuit Design for CMOS VLSI," Kluwer Academic, pp. 405-408, 1992.

[315] C. G. Sodini, P. K. Ko and J. L. Moll, "The Effects of High Fields on MOS Device and Circuit Performance, IEEE Transactions on Electron Devices Vol. ED-31, No. 10, pp. 1386-1393 October 1984.

[316] T. Doishi, et al., "A Well-Synchronized Sensing/Equalizing Method for Sub-1.0-V Operating Advanced DRAMs," IEEE Journal of Solid-Stte Circuits, Vol. 29, No. 4, pp. 432-440, April 1994.

[317] J. J. Barnes and J. Y. Chan, "A High Performance Sense Amplifier for a 5V Dynamic RAM," IEEE Journal of Solid State Circuits, Vol. SC-15 No. 5, October 1980.

[318] N. N. Wang, "On the Design of MOS Dynamic Sense Amplifiers," IEEE Transactions on Circuits and Systems, Vol. CAS-29, No. 7, pp. 467-477, July 1982.

[319] H. Walker, "A 4-kbit Four-Transistor Dynamic RAM," Carnegie-Mellon University, Research Report CMU-CS-83-140, June 1983.
[320] T. P. Haraszti, "High Performance CMOS Sense Amplifiers," United States Patent No. 4,169,233, Sep. 1979.
[321] T. Seki, et al., "A 6-ns 1-Mb CMOS SRAM with Latched Sense Amplifier," IEEEE Journal of Solid-State Circuits, Vol. 28, No. 4, April 1993.
[322] P. R. Gray, "Basic MOS Operational Amplifier Design¾An Overview," University of California Berkeley, Electronic Engineering and Computer Sciences Department, Tutorial Manuscript, March 1980.
[323] T. N. Blalock and R. C. Jaeger, "A High Speed Scheme for 1T Dynamic RAMs Utilizing the Clamped Bit-Line Sense Amplifier," IEEE Journal of Solid States, Vol. 27, No. 4, pp. 618-625, April 1992.
[324] E. Seevinck, P. J. van Beers and H. Ontrop, "Current Mode Techniques for High Speed VLSI Circuits with Application to Current Sense Amplifier for CMOS SRAMs," Vol. 26, No. 4, pp. 525-536, April 1991.
[325] J. Fisher and B. Gatland, "Electronics From Theory Into Practice," Vol. 2, Oxford Pergamon, 1976.
[326] J. P. Uyemura, "Circuit Design for CMOS VLSI," Kluwer Academic, pp. 405-408, 1992.
[327] K. Seno, "A 9ns 16-Mb CMOS SRAM with Offset-Compensated Current Sense Amplifier," IEEE Journal of Solid-State Circuits, Vol. 28, No. 11, pp. 11198-1124, November 1993.
[328] K. Ishibashi, et al., "A 6-ns 4-Mb CMOS SRAM with Offset-Voltage-Insensitive Current Sense Amplifiers," IEEE Journal of Solid-State Circuits, Vol. 30, No. 4, pp. 480-486, April 1995.
[329] M. Bohus, "The Theory of Linear Controls," (Linearis Szabalyozasok Elmelete) Technical University Budapest, Tankonyvkiado, 1966.
[330] G. Fodor, "Analysis of Linear Systems," (Linearis Rendszerek Analizise) Muszaki Konyvkiado, 1967.
[331] H. Yamauchi et al., "A Circuit Design to Suppress Assymetrical Characteristics in High Density DRAM Sense Amplifiers," IEEE Journal of Solid-State Circuits, Vol. 25, No. 1, pp. 36-41, February 1990.
[332] T. P. Haraszti, "Associative Control for Fault-Tolerant CMOS/SOS RAMs," European Solid-State Circuit Conference, Digest of Technical Papers, pp. 194-198, September 1981.
[333] P. R. Gray and R. G. Meyer, "The Analysis and Design of Analog Integrated Circuits," Wiley, 1977.
[334] M. Annaratone, "Digital CMOS Circuit Design," Kluwer Academic, pp. 198-200 1986.
[41] R. L. Liboff and G. C. Dalman, "Transmission Lines, Waveguides, and Smith Charts," Macmillan, 1985.
[42] T. Sakurai, "Approximation of Wiring Delay in MOSFET LSI," IEEE Journal of Solid-State Circuits, Vol. SC-18, No. 4, pp. 418-426, August 1983.
[43] M. Shoji, "CMOS Digital Circuit Technology," Prentice Hall, pp. 357-359, 1988.
[44] C-Y. Wu, "The New General Realization Theory of FET-Like Integrated Voltage-Controlled Negative Differential Resistance Devices," IEEE Transactions on Circuits and Systems, Vol. CAS28, pp. 382-390, May 1981.
[45] M. Shoji and R. M. Rolfe, "Negative Capacitance by Terminator for Improving the Switching Speed of a Microcomputer Power Bus," IEEE Journal of Solid-State Circuits, Vol. 50-20, pp. 828-832, August 1985.
[46] K. Simonyi, "Theory of Electricity," (Elmeleti Villamossagtan) Tankonyvkiado, pp. 407-447, 1960.

[47] G. Bilardi, M. Pracchi and F. P. Preparata, "A Critique of Network Speed in VLSI Models of Computation," IEEE Journal of Solid-State Circuits, Vol. SC-17, No. 4, pp. 696-702, August 1982.

[48] P. R. Brent and H. T. Kung, "The Chip Complexity of Binary Arithmetic," Proceedings of the 12th Symposium on Theory of Computing, pp. 190-200, April 1980.

[49] C. D. Thompson, "A Complexity Theory for VLSI," Ph.D. Thesis, Department of Computer Science, Carnegie-Mellon University, August 1980.

[410] C. L. Seitz, "System Timing," Chapter 7 in C. Mead and L. Conway, "Introduction to VLSI Systems," Addison-Wesley 1979.

[411] R. L. Street, "The Analysis and Solution of Partial Differential Equations," Brooks and Cole, 1973.

[412] T. P. Haraszti, "Radiation Hardened CMOS Memory Circuits" Technical Information, Rockwell International, Y78-758/501, July 1978.

[413] D-S. Min et al., "Temperature-Compensation Circuit Techniques for High Density CMOS DRAMs," IEEE Journal of Solid-State Circuits, Vol. 27, No. 4, pp. 626-631, April 1992.

[414] D. L. Fraser, "High Speed MOSFET IC Design," International Electronic Devices Meeting, Seminar Guidebook, pp. 372-376, December 1986.

[415] K. Nakamura et al., "A 500-MHz 4-Mb CMOS Pipeline-Burst Cache SRAM with Point-to-Point Noise Reduction Coding I/O," IEEE Journal of Solid State Circuits, Vol. 32, No. 11, pp. 1758-1765, November 1997.

[416] K. Nagaraj and M. Satyan, "Novel CMOS Schmidt Trigger," Electronic Letters, Vol. 17, pp. 693-694, September 1981.

[417] E. L. Hudson and S. L. Smith, "An ECL Compatible 4K CMOS RAM," ISSCC82, Digest of Technical Papers, pp. 248-249, February 1982.

[418] R. E. Miller, "Switching Theory," Chapter 10, Review of D. Muller's Work, Wiley, 1965.

[419] B. Razovi, "Monolithic Phase-Locked Loops and Clock Recovery Circuits, IEEE Press, 1996.

[420] P. R. Gray and R. G. Meyer, "Analysis and Design of Analog Integrated Circuits," John Wiley, 1977.

[421] R. E. Best, "Phase-Locked Loops," McGraw Hill, 1993.

[422] F. M. Gardner, "Phaselock Techniques," John Wiley, 1979.

[423] H. B. Bakoglu and J. D. Meindl, "Optimal Interconnect Circuits for VLSI," International Solid-State Circuit Conference, Digest of Technical Papers, pp. 164-165, February 1984.

[424] J. R. Black, "Electromigration-A Brief Survey and Some Recent Results," IEEE Transactions on Electron Devices, Vol. Ed-16, No. 4, pp. 338-339, 1969.

[425] M. L. Cortes et al., "Modeling Power-Supply Disturbances in Digital Circuits," International Solid-State Circuit Conference, Digest of Technical Papers, pp. 164-165, February 1986.

[426] M. Shoji, "Reliable Chip Design Method in High Performance CMOS VLSI," Digest ICCD86, pp. 389-392, October 1986.

[51] A. K. Sharma, Semiconductor Memories, Technology, Testing and Reliability, IEEE Press, pp. 249-320, 1997.

[52] A. Reibman, R. M Smith, and K. S. Trivedi, "Markov and Markov Reward Model Transient Analysis: An Overview of Numerical Approaches," European Journal of Operation Ressearch, North-Holland, pp. 256-267, 1989.

[53] C. Hu, "IC Reliability Simulation," IEEE Journal of Solid-State Circuits, Vol. 27, No. 3, pp. 241-246, March 1992.

[54] F. A. Applegate, "A Commentary on Redundancy," General Electric, Technical Publiions, Spartan, 1962.

[55] K. S. Trivedi, "Probability and Statistics with Reliability, Queueing, and Computer Science Applications," Prentice Hall, 1982.

[56] A. Goyal et al., "The System Availability Estimator," Proceedings 16th International Symposium on Fault Tolerant Computing," CS Press, pp. 84-89, July 1986.

[57] P. A. Layman and S. G. Chamberlain, "A Compact Thermal Noise Model for the Investigation of Soft Error Rates in MOS VLSI Digital Circuits," IEEE Journal of Solid-State Circuits, Vol. 24, No. 1, pp. 78-89, February 1989.

[58] T. C. May and M. H. Woods, "A New Physical Mechanisms for Soft Errors in Dynamic Memories," Proceedings Reliability Physics Symposium, pp. 2-9, April 1978.

[59] D. Binder, C. E. Smith, and A. B. Holman, "Satellite Anomalies from Galactic Cosmic Rays," IEEE Transactions on Nuclear Science, NS-22, No. 6, pp. 2675-2680, December 1975.

[510] E. J. Kobetich and R. Katz, "Energy Deposition by Electron Beams and Delta Rays," Physics Review, No. 170, pp. 391-396, 1968.

[511] J. C. Pickel and J. T. Blandford, "Cosmic-Ray-Induced Errors in MOS Memory Cells," IEEE Annual Conference on Nuclear and Space Radiation Effects, Albuquerque, New Mexico, Rockwell Technical Information No. X78-317/501, July 1978.

[512] L. C. Northcliffe and R. F. Schilling, "Nuclear Data," A7, Academic Press, 1970.

[513] T. L. Turtlinger and M. V. Davey, "Understanding Single Event Phenomena in Complex Analog and Digital Integrated Circuits," IEEE Transactions on Nuclear Science, Vol. 37, No. 6, December 1990.

[514] T. Toyabe and T. Shinada, "A Soft-Error Rate Model for MOS Dynamic RAMs," IEEE Journal of Solid-State Circuits, Vol. SC-17, pp. 362-367, April 1982.

[515] E. Diehl et al., "Error Analysis and Prevention of Cosmic Ion-Induced Soft Errors in CMOS Static RAMs," IEEE Transactions on Nuclear Science, NS-29, pp. 1963-1971, 1982.

[516] R. J. MacPartland, "Circuit Simulations of Alpha-Particle-Induced Soft Errors in MOS Dynamic RAMs," IEEE Journal of Solid State Circuits, Vol. SC-16, No. 1, pp. 31-34, February 1981.

[517] C. Stapper, A. McLaren, and M. Dreckman, "Yield model for Productivity Optimization of VLSI Memory Chips with Redundancy and Partially Good Product," IBM Journal of Research and Development, Vol. 24, No. 3, pp. 398-409, May 1980.

[518] D. Moore and H. Walker, "Yield Simulation for Integrated Circuits," Kluwer Academic, 1987.

[519] J. Wallmark, "Design Considerations for Integrted Electronic Devices," Proceedings of the IRE, Vol. 48, No. 3, pp. 293-300, March 1960.

[520] R. Petritz, "Current Status of Large Scale Integration Technology," IEEE Journal of Solid State Circuits, Vol. 4, No. 2, pp. 130-147, December 1967.

[521] J. Price, "A New Look at Yield of Integrated Circuits," Proceedings of the IEEE, Vol. 58, No. 8, pp. 1290-1291, August 1970.

[522] B. Murphy, "Cost-Size Optima of Monolithic Integrated Circuits," Proceedings of the IEEE, Vol. 52, No. 12, pp. 1537-1545, December 1964.

[523] S. Hu, "Some Considerations in the Formulation of IC Yield Statistics, "Solid-State Electronics, Vol. 22, No. 2, pp. 205-211, February 1979.

[524] B. Murphy, Comments on "A New Look at Yield of Integrated Circuits," Proceedings of the IEEE, Vol. 59, No. 8, pp. 1128-1132, July 1971.

[525] V. Borisov, "A Probability Method for Estimating the Effectiveness of Redundancy in Semiconductor Memory Structures," Microelectronika, Vol. 8, No. 3, pp. 280-282, May-June 1979.

References 537

[526] T. Okabe, M. Nagata and S. Shimada, "Analysis on Yield of Integrated Circuits and New Expression for the Yield," Electrical Engineering in Japan, No. 92, pp. 135-141, December 1972.

[527] W. Maly, "Modeling of Point Defect Related Yield Losses for CAD of VLSI Circuits," IEEE International Conference on Computer-Aided Design, Digest of Technical Papers, pp. 161-163, November 1984.

[528] C. H. Stapper, "Defect Density Distribution for LSI Yield Calculations," IEEE Transactions on Electron Devices, Vol. ED-20, No. 7, pp. 655-657, July 1973.

[529] M B. Ketchen, "Point Defect Yield Model for Wafer Scale Integration," IEEE Circuits and Devices Magazine," Vol. 1, No. 4, pp. 24-34, July 1985.

[530] V. P. Nelson and B. D. Carroll, "Tutorial: Fault-Tolerant Computing," CS Press, Los Alamitos, Order No. 677, Chapters 1-2, 1986.

[531] M. E. Zaghloul and D. Gobovic, "Fault Modeling of Physical Failures in CMOS VLSI Circuits," IEEE Transactions on Circuits and Systems," Vol. 37, No. 12, pp. 1528-1543, December 1990.

[532] R. T. Smith, "Using a Laser Beam to Substitute Good Cells for Bad," Electronics, McGraw-Hill Publications, pp. 131-134, July 28, 1981.

[533] E. Hamdy et al., "Dielectric Based Antifuse for Logic and Memory ICs," International Electron Devices Meeting, Technical Digest, pp. 786-789, December 1988.

[534] J. Birkner et al., "A Very High-Speed Field Programmable Gate Array Using Metal-to-Metal Antifuse Programmable Elements," Custom Integrated Circuits Conference, Technical Digest, May 1991.

[535] V. G. McKenny, "A 5V 64K EPROM Utilizing Redundant Circuitry," IEEE International Solid-State Circuit Conference Digest of Technical Papers, pp. 146-147, February 1980.

[536] T. P. Haraszti, "A Novel Associative Approach for Fault-Tolerant MOS RAMs," IEEE Journal of Solid-State Circuits, Vol. SC-17, No. 3, June 1982.

[537] R. P. Cenker et al., "A Fault-Tolerant 64K Dynamic RAM," IEEE International Solid-State Circuit Conference, Digest of Technical Papers, pp. 150-151, February 1979.

[538] T. P. Haraszti et al., "Novel Fault-Tolerant Integrated Mass Storage System," European Solid-State Circuit Conference, Proceedings, pp. 141-144, September 1990.

[539] J. C. Kemp, "Redundant Digital Systems," Symposium on Redundancy Techniques for Computing Systems, Proceedings, pp. 285-293, 1962.

[540] F. J. MacWilliams and N.J.A. Sloane, "The Theory of Error Correcting Codes, Vol. I and II, North-Holland, 1977.

[541] C. E. Shannon, "A Mathematical Theory of Communication," Bell System Technique Journal, No. 27, pp. 379-423, 623-656, 1948.

[542] S. P. Lloyd," Binary Block Coding," Bell System Technique Journal, No. 36, pp. 517-535, 1957.

[543] A. M. Michelson and A. H. Levesque, "Error-Control Techniques for Digital Communication," Wiley-Interscience, pp. 234-269, 1985.

[544] R. C. Bose and D. K. Ray-Chaudhuri, "On a Class of Error Correcting Binary Group Codes," Information and Control, No. 3, pp. 68-79, 279-290, 1960.

[545] S. Lin and E. J. Weldon, Jr., "Long BCH Codes are Bad," Information Control, No. 11, pp. 445-451, 1967.

[546] E. R. Berlekamp, "Goppa Codes," IEEE Transactions on Information Theory, Vol. IT-18, pp. 415-426, 1972.

[547] W. W. Peterson and E. J. Weldon, Jr., "Error Correcting Codes," MIT Press, 1972.

[548] I. S. Reed and G. Solomon, "Polynomial Codes over Certain Finite Fields," Journal SIAM, No. 8, pp. 300-304, 1960.

[549] G. D. Forney, Jr., "Burst Correcting Codes for the Classic Burst Channel," IEEE Transactions on Communication Techniques, Vol. COM-19, pp. 772-781, 1971.
[550] T. Kasami and S. Lin, "On the Probability of Undetected Error for Maximum Distance Separable Codes," IEEE Transactions on Communication, Vol. COM-32, pp. 998-1006, 1984.
[551] G. Birkhoff and S. MacLane, "A Survey of Modern Algebra," Macmillan, 1965.
[552] T. P. Haraszti, "Intelligent Fault-Tolerant Memories for Mass Storage Devices," United States Air Force, Project F04701-85-C-0075, Report, pp. 69-78, January 1986.
[553] J. M. Berger, "A Note on Error Detection Codes for Assymetric Channels," Information and Control, No. 4, pp. 68-73, 1961.
[554] A. Hocquenghem, "Error Corrector Codes" (Codes Correctoeurs d'Erreurs), Chiffres, No. 2, pp. 147-156, 1959.
[555] R. W. Hamming, "Error Detecting and Error Correcting Codes, "Bell System Technique Journal, No. 29, pp. 147-160, 1950.
[556] R. T. Chien, "Memory Error Control: Beyond Parity," IEEE Spectrum, Vol. 10, No. 7, pp. 18-23, July 1973.
[557] A. C. Singleton, "Maximum Distance q-nary Codes," IEEE Transactions on Information Theory, Vol. IT-10, pp. 116-118, 1964.
[558] R. E. Blahut, "Theory and Practice of Error Control Codes, "Addison-Wesley, 1983.
[559] P. Elias, "Coding for Noisy Channels," IRE Convention Records, Part 4, pp. 37-46, 1955.
[560] T. P. Haraszti and R. P. Mento, "Novel Circuits for Radiation Hardened Memories," IEEE Nuclear Science Symposoium and Medical Imaging Conference, Proceedings, November 1991.
[561] J. A. Fifield and C. H. Stapper, "High-Speed On-Chip ECC for Synergistic Fault-Tolerant Memory Chips," IEEE Journal of Solid-State Circuits, Vo. 26, No. 10, pp. 1449-1452, October 1991.
[61] T. P. Ma and P. V. Dressendorfer," Ionizing Radiation Effects in MOS Devices and Circuits," John Wiley and Sons, 1989.
[62] W. Poch and A. G. Holmes-Siedle, "Permanent Radiation Effects in Complementary-Symmetry MOS Integrted Circuits," IEEE Transactions on Nuclear Science, Vol. NS-16, pp. 227-234, 1969.
[63] A. H. Johnston, "Super Recovery of Total Dose Damage in MOS Devices," IEEE Transactions on Nuclear Science, Vol. NS-31, No. 6, pp. 1427-1431, 1984.
[64] P. J. McWhorter and P. S. Winokur, "Simple Technique for Separating the Effects of Interface Traps and Trapped-Oxide Charge in Metal Oxide Semiconductor Transistors," Applied Physics Letters, Vol. 48, No. 2, pp. 133-135, 1986.
[65] F. W. Sexton and J. R. Schwank, "Correlation of Radiation Effects in Transistors and Integrated Circuits," IEEE Transactions on Nuclear Science, Vol. NS-32, No. 6, pp. 3975-3981, 1985.
[66] W. A. Dawes, G. F. Derbenwick, and B. L. Gregory, "Process Technology for Radiation-Hardened CMOS Integrated Circuits," IEEE Journal of Solid-State Circuits, Vol. SC-11, No. 4, pp. 459-465, August 1976.
[67] J. H. Yuan and E. Harrari, "High Performance Radiation Hard CMOS/SOS Technology," IEEE Transactions on Nuclear Science, Nol. NS-74, No. 6, pp. 2199, 1977.
[68] M. A. Xapsos et al., "Single-Event Upset, Enhanced Single-Event and Dose-Rate Effects with Pulsed Proton Beams," IEEE Transactions on Nuclear Science, Vol. NS-34, pp. 1419-1425, 1987.
[69] L. W. Massengill and S. E. Diehl, "Transient Radiation Upset Simulations of CMOS Memory Circuits," IEEE Transactions on Nuclear Science, Vol. NS-31, pp. 1337-1343, 1984.

[610] J. M. Aitken, "1μm MOSFET VLSI Technology: Part III - Radiation Effects," IEEE Journal of Solid-Stte Circuits, Vol. SC-14, No. 2, pp. 294-302, 1979.

[611] A. R. Knudson, A. B. Campbell, and E. C. Hammond, "Dose Dependence of Single Event Upset Rate in MOS dRAMs," IEEE Transactions on Nuclear Science, Vol. NS-30, pp. 4240-4245, 1983.

[612] B. L. Bhuva et al., "Quantification of the Memory Imprint Effect for a Charged Particle Environment," IEEE Transactions on Nuclear Science, Vol. NS-34, pp. 1414-1417, 1987.

[613] G. J. Bruckner, J. Wert, and P. Measel, "Transient Imprint Memory Effect in MOS Memories," IEEE Transactions on Nuclear Science, Vol. NS-33, pp. 1484-1486, 1986.

[614] G. E. Davis et al., "Transient Radiation Effects in SOI Memories," IEEE Transactions on Nuclear Science, Vol. NS-32, pp. 4432-4437, 1985.

[615] C. R. Troutman, "Latchup in CMOS Technology," Kluwer Academic, 1986.

[616] R. J. Hospelhorn and B. D. Shafer, "Radiation-Induced Latch-up Modeling of CMOS ICs," IEEE Tranactions on Nuclear Science, Vol. NS-34, pp. 1396-1401, 1987.

[617] A. H. Johnston and M. P. Boze, "Mechanisms for the Latchup Window Effect in Integrated Circuits," IEEE Transactions on Nuclear Science, Vol. NS-32, pp. 4018-4025, 1987.

[618] A. Ochoa et al., "Snap-back: A Stable Regenerative Breakdown Mode of MOS Devices," IEEE Transactions on Nuclear Science, Vol. NS-30, pp. 4127-4130, 1983.

[619] F. Najm, "Modeling MOS Snapback and Parasitic Bipolar Action for Circuit-Level ESD and High-Current Simulations," Circuits and Devices, Vol. 13, No. 2, pp. 7-10, March 1997.

[620] P. S. Winokur et al., "Implementing QML for Radiation Hardness Assurance," IEEE Transactions on Nuclear Science, Vol. 37, No. 6, pp. 1794-1805, December 1990.

[621] H. Borkan, "Radiation Hardening of CMOS Technologies - An Overview," IEEE Transactions on Nuclear Science, Vol. NS-24, No. 6, pp. 2043, 1977.

[622] T. P. Haraszti, "Radiation Hardened CMOS/SOS Memory Circuits," IEEE Transactions on Nuclear Science, Vol. NS-25, No. 6, pp. 1187-1201, June 1978.

[623] T. P. Haraszti et al., "Novel Circuits for Radiation Hardened Memories," IEEE Transactions on Nuclear Science, Vol. 39, No. 5, pp. 1341-1351, October 1992.

[624] C. C. Chen et al., "A Circuit Design for the Improvement of Radiation Hardness in CMOS Digital Circuits," IEEE Transactions on Nuclear Science, Vol. 39, No. 2, pp. 272-277, April 1992.

[625] T. P. Haraszti, R. P. Mento and N. Moyer, "Spaceborne Mass Storage Device with Fault-Tolerant Memories," IEEE/AIAA/NASA Digital Avionics Systems Conference, Proceedings, pp. 53-57, October 1990.

[626] J. P. Colinge, "Silicon-on-Insulator Technology, Kluwer Academic, 1991 and 1997.

[627] R. A. Kjar, S. N. Lee and R. K. Pancholy, "Self-Aligned Radiation Hard CMOS/SOS," IEEE Transactions on Nuclear Science, Vol. NS-23, No. 6, pp. 1610-1619, 1976.

[628] J-W. Park et al., "Performance Characteristics of SOI DRAM for Low Power Application," 1999 IEEE International Solid State Circuit Conference, Digest of Technical Papers, pp. 434-435, February 1999.

[629] V. Ferlet-Carrois et al., "Comparison of the Sensitivity to Heavy Ions of SRAMs in Different SIMOX Technologies," IEEE Electron Device Letters, Vol. 15, No. 3, pp. 82-84, March 1994.

[630] A. Gupta and P. K. Vasudev, "Recent Advances in Hetero-Epitaxial Silicon-on-Insulator Technology," Solid State Technology, pp. 104-109, February 1983.

[631] D. Yachou, J. Gautier, and C. Raynaud, "Self-Heating Effects on SOI Devices and Implications to Parameter Extraction," IEEE SOI Conference, Proceedings, pp. 148-149, 1993.

[632] V. Szekely and M. Rencz, "Uncovering Thermally Induced Behavior of Integrated Circuits with the SISSI Simulation Package," Collection of Papers presented at the International Workshop on Thermal Investigations on ICs and Microstructures, pp. 149-152, September 1997.

[633] S. Krisham and J. G. Fossum, "Grasping SOI Floating-Body Effects," Circuits and Devices, Vol. 14, No. 4, pp. 32-37, July 1998.

[634] G. G. Shahidi et al., "Partially Depleted SOI Technology for Digital Logic," 1999 IEEE International Solid State Circuit Conference, Digest of Technical Papers, pp. 426-427, February 1999.

[635] W-G. Kang et al., "Grounded Body SOI (GBSOI) nMOSFET by Wafer Bonding," IEEE Electron Device Letters, Vol. 16, No. 1, pp. 2-4, January 1997.

[636] D. H. Allen et al., "A 0.2μm 1.8V SOI 550MHz 64b Power PC Microprocessor with Copper Interconnects," 1999 IEEE International Solid State Circuit Conference, Slide Supplement to the Digest of Technical Papers, pp. 524-527, February 1999.

[637] S. N. Lee, R. A. Kjar and G. Kinoshita, "Island Edge Effects in CMOS/SOS Transistors," Rockwell International, Tewchnical Information X76-367/501, March 1976.

[638] D. W. Flatly and W. E. Ham, "Electrical Instabilities in SOS/MOS Transistors," Meeting Records, Electrochemical Society Meeting, pp. 487-491, 1974.

[639] J. L. Peel and R. K. Pancholy, "Investigation on Radiation Effects and Hardening Procedures for CMOS/SOS," Presented at the IEEE Annual Conference on Nuclear and Space Radiation Effects, July 1975.

[640] D. H. Phillips and G. Kinoshita, "Silicon-on-Sapphire Device Photoconducting Predictions," presented at the IEEE Annual Confernce on Nuclear and Space Radiation Effects, July 1974.

[641] S. Veeraraghavan and J. G. Fossum, "A Physical Short-Channel Model for Thin-Film SOI MOSFET Applicable to Device and Circuit CAD," IEEE Transactions on Electron Devices, Vol. 35, pp. 1866-1874, 1988.

[642] K. Shimomura et al., "A 1V 46ns 16Mb SOI-DFRAM with Body Control Technique," International Solid-State Circuit Conference, Digest of Technical Papers, pp. 68-69, February 1997.

[643] S. Cristolovenau and S.S.Li, "Electrical Characterization of Silicon-On-Insulator Materials and Devices," Kluwer Academic Publishers; pp. 209-273, 1999.

[644] K. Bernstein and N.J. Rohrer, "SOI Circuit Design Concepts," Kluwer Academic Publishers, pp. 119-192, 2000.

Index

Access Time, 6, 14-16, 18-31, 38-40
Active Load, 207
Address Activated (Address Change Detector)
Address Change Detector, 340, 341, 344, 345
Addressing Time, 15, 18, 16, 30-31, 38-40
Alpha Particle Impact (Charged Atomic Particle Impact)
Amplitude Attenuation, 297
Antifuse, 425-427, 430
Array Wiring, 278-311
Artificial Intelligence, 466, 467
Asynchronous Operation (Self-timed Operation)
Atomic Particle Impact:
    Causes, 388
    Characterization, 388, 389
    Effects, 388-390
    Error Rate, 390-398
    Induced Signals, 391
    Modelling, 390-398
    Space, 388-402
    Terrestrial, 388-402

Back-Channel Effect:
    Allay, 526, 527
    Drain Current, 525
    Mechansim, 524
    Photocurrent, 525, 526
Bandwidth, 7, 11-54, 61-79

Barkhausen Criteria, 114, 213
Bathtub Curve, 367
BCH Code (Error Control Code, Bose-Chaudhuri-Hocquenghem)
BEDO DRAM (Burst Extended Data Output Random Access Memory)
Bitline Decoupler, 221-224
Bitline:
    Clamping, 283-285
    Dummy, 286, 287
    Interdigitized, 379-382
    Model, 279-283, 296-311
    Signal, 280-287
    Termination, 282, 283, 296-307
    Twisted, 381, 382
Body-Tie, 517, 518
Burst Extended Data Output Dynamic Random Access Memory, 23-25

Cached Dynamic Random Access Memory, 72, 73
Cache-Memory:
    Copy-Back, 64
    Direct Mapped, 66, 67
    Fully Associative, 65
    Fundamentals, 61-65
    Hit-Miss, 63, 67
    Set Associative, 68, 69
    Write-Through, 64
CAM (Content Addressable Memory)

CAM Cell (Content Addressable Memory Cell)
CCD (Charge Coupled Device)
CDRAM (Cached Dynamic Random Access Memory)
Cell Capacitor:
  Effective Area, 105-109
  Ferroelectric, 100-102
  Granulated, 106-107
  Material, 98-102
  Paraelectric, 98-100
  Parasitic, 103, 104
  Poly-Poly, 104, 107, 109
  Poly-Semiconductor, 103, 105, 106, 109
  Stack, 106, 107, 109
  Thickness, 97-98
  Trench, 105-109
Characteristic Impedance, 282, 297-299
Charge Amplifier (Charge Transfer Sense Amplifier)
Charge Coupled Device, 154-156
Charge Coupling, 140-142, 176-178, 271-273
Charge Reference, 321-323
Charge Transfer Sense Amplifier, 271- 273
Charged Atomic Particle Impact (Atomic Particle Impact)
Clamp Circuit, 283-285
Clock Circuit, 341-354
Clock Impulse:
  Delay Control, 352-354
  Generator, 344-347
  Recovery, 347-352
  Timing, 341-344
  Transient Control, 354, 355
CMOS Memory (Complementary-Metal-Oxide-Semiconductor Memory)
Commercial-Off-The-Shelf, 483
Complementary-Metal-Oxide-Semiconductor Memory:
  Application Area, 7
  Architecture, 80-84
  Characterization, 1,2, 6-9
  Classification, 1, 2, 6-8
  Combination, 70-79
  Content Addressable, 54-61
  Hierarchical, 82-84
  Nonranked, 80-82
  Random Access, 10-42
  Sequential Access, 42-54
  Special, 61-79

Content Addressable Memory Cell:
  Associative Access, 146-148
  Circuit Implementations, 148
  Dynamic, 150, 151
  Static, 148-149, 151
Content Addressable Memory:
  All-Parallel, 56-58
  Basics, 54-46
  Cell, 146-151, 340
  Word-Parallel-Bit-Serial, 59-61
  Word-Serial-Bit-Parallel, 58, 59
Cosmic Particle Impact (Charged Atomic Particle Impact)
COTS (Commercial-Off-The-Shelf)
Counter:
  Binary, 346
  Johnson, 346, 347
  Nonlinear, 346, 347
Critical Charge, 390-392
Crosstalk:
  Array-Internal, 374-382
  Model, 375-379
  Reduction, 379-382
  Signal, 378, 379
Current Amplifier (Current Sense Amplifier)
Current Mirror (Current Source)
Current Reference:
  Current Mirror, 318, 319
  Regulated, 320, 321
  Widlar, 319
Current Sense Amplifier:
  Crosscoupled, 245-248
  Current-Mirror, 238-240
  Current-Voltage, 243, 244
  Damping, 256, 257
  Negative Feedback, 249, 260, 263-265
  Offset Reduced, 260-263
  Parallel Regulated, 252-255
  Positive Feedback, 240-243
  Sample-and-Feedback, 263-265
  Stability, 256, 257
Current Sensing, 232
Current Source, 226, 227, 238-240
Current Versus Voltage Sensing, 232-236
Cycle Time, 7, 14-16, 18-31, 38-40

Data Rate, 6, 7, 15, 18, 19-31, 38-40, 44-48
DDR (Double Data Rate)

Index 543

Decoder:
   Full-Complementary, 326
   NAND, 324, 325
   NOR, 324, 325
   Rectangular, 323-328
   Tree, 326
Decoupling Bitline Loads, 221-224
Defect:
   Clustering, 411, 412
   Density, 402-405
   Types, 402
Delay Mimicking:
   Bitline, 286, 287
   General, 348, 349
   Wordline, 295, 296
Differential Sense Amplifier:
   Charge Coupled, 271-273
   Current Mode, 232-256
   Voltage Mode, 192-231
Diode-Like Nonlinear Element, 527, 529
Dose-Rate, 475-477
Double Data Rate, 27, 79
DRAM (Dynamic Random Access Memory)
DRAM Cell (Memory Cell, DRAM)
DRAM-Cache Combination (Dynamic Random Access Memory and Cache Combination)
Dummy:
   Bitline, 286, 287
   Memory Cell, 321-323
   Wordline, 295, 296
Dynamic Four-Transistor Random Access Memory Cell, 158
Dynamic One-Transistor-One-Capacitor Random Access Memory Cell:
   Capacitor, 97-109
   Design Goals, 96
   Design Trade-Offs, 96, 97
   Designs, 97, 100, 103-109
   Insulator, 97-103
   Storage, 88, 90
   Read Signal, 93-96
   Refresh, 90-92
   Write Signal, 92, 93
Dynamic Random Access Memory:
   Basic Architecture, 12-14
   Cached, 70
   Characterization, 14
   Dynamic Storage, 11, 12
   Fundamentals, 14-20

Dynamic Random Access Memory *(Continued)*:
   Operation Modes, 12, 14-20
   Pipelining, 20-31
   Refresh, 12, 40, 41, 83-94
   Sense Amplifier, 163-265, 486-490
   Timing, 13-19
   Wordline, 287-311
Dynamic Random Access Memory and Cache Combination, 70-83
Dynamic Random Access Memory Cell (Memory Cell, DRAM)
Dynamic Three-Transistor Random Access Memory Cell:
   Derivative, 158
   Description, 110, 111, 158
   Read Signal, 112, 113
   Write Signal, 111, 112

ECC (Error Control Code)
Eccles-Jordan Circuit, 114
EDO DRAM (Extended Data Output Dynamic Random Access Memory)
EDRAM (Enhanced Dynamic Random Access Memory)
Eight-Transistor Shift-Register Cell, 141, 142
Eight-Transistor-Two-Resistor Content Addressable Memory Cell, 51
Electrical Programming, 425-428
Electromigration, 363
Enhanced Dynamic Random Access Memory, 70-72
Equitime Regions, 342, 343
Error:
   Categories, 413, 414, 419, 420
   Effects, 415
   Hard, 473
   Soft, 413
   Types, 419, 420
Error Control Code:
   Berger, 455, 456
   Bidirectional, 462, 463
   Binary, 440, 441
   Bose-Chaudhuri-Hocquenghem, 444, 445, 453, 457-463
   Convolutional, 491
   Cyclic, 447, 448
   Decoding, 447-467
   Efficiency, 446-453

Error Control Code *(Continued)*:
   Encoding, 447-467
   Family, 440, 441, 453
   Fundamentals, 438-441
   Gilbert-Varshamov, 445
   Goppa, 444
   Hamming, 457-461
   Linear Systematic, 453-463
   Multidirectional, 463
   Noncyclic, 447
   Parity Check, 453-455
   Performance, 442-446
   Read-Solomon, 461, 462, 445, 446
   Shortening, 457
Error Correction Code (Error Control Code)
Error Detection Code (Error Control Code)
Error Inducing Particle Flux, 393
Error-Control-Coding and Fault-Repair
   Combination, 464-467
Esaki-Diode Based Memory Cell (Memory
   Cell, Tunnel-Diode)
Extended Data Output Dynamic Random
   Access Memory, 20-23

Failure, 413
Failures in Time (FIT), 368, 389, 400, 401
Fast Page Mode, 21
Fault:
   Classification, 412, 413
   Effects, 414, 415, 416, 417
Fault Masking, 434-436
Fault Masking, 436-438
Fault Repair:
   Associative, 434-436
   Hierarchical, 435, 436
   Masking, 436-438
   Principle, 421-423
   Programming, 413-425
   Row/Column Replacement, 428-434
Fault Tolerance:
   Categories, 412-415
   Faults/Errors to Repair, 415-420
   Strategies, 420, 422
FDRAM Cell (Ferroelectric Dynamic
   Random Access Memory Cell)
Feedback:
   General, 237, 256
   Improvements, 252-256
   Junction, 250-252

Feedback *(Continued)*:
   Negative, 237, 260-263
   Positive, 237, 273-280
   Sampled, 263-265
   Separation, 224, 225
   Stability, 256, 257
   Types, 236, 237
Ferroelectric Dynamic Random Access
   Memory Cell, 100-102
FIFO Memory (First-In-First-Out Memory)
Fill Frequency, 7, 8
First-In-First-Out Memory, 51-54
FIT (Failures in Time)
Floating Body Effect (Floating Substrate
   Effect)
Floating Substrate Effect:
   History Dependency, 509-515
   Kink, 512, 513
   Memory Specific, 515
   Model, 510, 511
   Passgate Leakage, 513-515
   Premature Breakdown, 513
   Relieves, 516
Four-Transistor-Two-Capacitor Content
   Addressable Memory Cell, 150, 151
Four-Transistor-Two-Diode Shift-Register
   Cell, 142, 143
Fuse, 423-425, 427, 430

Granularity, 7, 9, 64, 68

Hierarchical Memory Organization, 80, 81, 83

Impulse Flattening, 304, 306
Input Buffer (Input Receiver)
Input Receiver:
   Differential, 337, 338
   Level-Sensitive Latch, 339
   Schmidt Trigger, 337-339

Kink Phenomena, 512, 513, 522, 525

Laser Annealing Effect, 118
Laser Programming, 423-425

Last-In-First-Out Memory, 52
Leakage Currents, 173-176, 473, 513-515, 521, 522, 525, 526
LET (Linear Energy Transfer)
LIFO Memory (Last-In-First-Out Memory)
Line Buffer, 353-355
Linear Energy Transfer, 396, 397, 506
Logic Gate Circuit:
    Address Change Detector, 340, 341, 344, 345
    Clock Generator, 344-347
    Counter, 346, 347
    Decoder, 325-328
    Error Contol, 454-467
    Fault-Repair, 428-438
    Input, 336-341, 344, 345
    Majority Decision, 436-438
    Output, 328-335
    Radiation Hardened, 495-499
    Timing, 341-355
LPF (Phase Locked Loop, Low Pass Filter)

Mean Time Between Errors, 368, 461
Mean Time Between Failures, 366-370
Memory Architecture (Memory Organization)
Memory Capacity, 5-7
Memory Cell:
    Basics, 85, 86
    CCD, 154-156
    Classification, 86-88
    Content Addressable, 146, 154
    Derivative, 158-161
    DRAM, 89-125, 158, 159, 413, 491-493
    Multiport, 156, 157
    Objectives, 88-89
    ROM, 132, 136
    Shift-Register, 136, 146
    SRAM, 125-131, 158, 159, 390-393, 493, 495
    State Retention, 398, 399, 493-495
    Tunnel-Diode, 152-154
Memory Organiztion, 184, 464-467, 499-501
Memory Subcircuit:
    Array, 278-311
    Bitline, 278-287, 296-311
    Clock, 341-355
    Decoder, 323-328
    Input Receiver, 336-341

Memory Subcircuit *(Continued)*:
    Logic, 323-355, 428-438, 493-495
    Memory Cell, 85-161
    Output Buffer, 328-335
    Power Line, 355-363
    Radiation Transferance, 398, 399, 481-501
    Reference, 311-323
    Sense Amplifier, 163-275
    Wordline, 287-311
Memory Timing:
    BEDO, 23, 24
    CAM, 57, 59, 61
    DDR, 21
    DRAM, 13-19
    EDO, 21, 22
    Fast Page Mode, 21
    Hierarchical, 83
    Nonranked, 82
    Page Mode, 17, 18
    Psuedo SRAM, 40, 41
    SDRAM, 25-28, 30
    Shift-Register SAM, 46-48
    Shuffle-Register SAM, 50, 51
    SRAM, 38, 39
    Static Column Mode, 19
MTBE (Mean Time Between Errors)
MTBF (Mean Time Between Failures)
Muller C Circuit, 348
Multiport Memory Cell (Memory Cell, Multiport)

Negative Capacitance, 294
Negative Resistance, 292-294, 153, 154
Neutron Fluence (Radioactive Radiation, Neutron)
Nibble Mode, 19
Nine-Transistor Shift-Register Cell, 160
Noise:
    Chip External, 388-468
    Chip Internal, 373-388
    Crosstalk, 374-379
    Particle Impact, 388-462
    Power Supply, 358-367, 382-385
    Source, 373, 374
    Thermal, 385-388
Noise Margin, 115, 129
Nondifferential Sense Amplifier:
    Basics, 256, 260
    Common-Drain, 273-275

Nondifferential Sense Amplifier *(Continued)*:
  Common-Gate, 269-273
  Common-Source, 266-269
Nonranked Memory Organization, 80, 82

Offset, 179-181, 257-265, 480-490
Offset Reduction:
  Sense Amplifiers, 257-265, 480-490
  Layout Design, 259
  Negative Feedback, 249, 250, 260-265, 480-488
  Sample-and-Feedback, 263-265, 489, 490
Omnidirectional Flux, 394
Operation Margin:
  Effecting Terms, 166-184
  Radiation Dependency, 170, 486, 487
  Temperature Dependency, 170
Optimized Voltage Swing, 229-231
Output Buffer:
  Coded, 333-335
  Digital Controlled, 333, 334
  Impedance Controlled, 333, 334
  Level Conversion, 328
  Low-Power, 332-334
  Reflection Reduced, 333, 334
  Simple Scaled, 330
  Tri-State, 331

Page Mode, 17-19
Parameter Tracking Circuits, 314, 315, 491-493
Parasitic Bipolar Device, 511, 513-515
Parasitic Diode (Diode-Like Nonlinear Element)
Particle Sensitive Cross Section, 396, 397, 505
PD (Phase Locked Loop, Phase Detector)
Performance Gap, 8, 9
Permanent Ionization (Total-Dose)
Phase Factor, 297
Phase Locked Loop:
  General, 349-352
  Low-Pass Filter, 350
  Phase Detector, 351, 352
  Transfer Function, 351, 352
  Voltage Controlled Oscillator, 352
Pipelining, 20-31, 39, 42
PLL (Phase Locked Loop)

Positive Feedback, 114-117, 126, 129
Power:
  Bounce, 355-363
  Bounce Reduction, 359-363
  Circuit, 357, 358, 360-361
  Current Density, 363
  Distribution, 355-359
  Line, 358-367, 382-385
  Model, 357, 358
  Noise, 358-367, 382-385
  Switching Current, 358-359
Power-Bounce Reduction:
  Architecture, 360
  Differential, 362
  Local Loop, 361
Power-Line Noise:
  Array-Internal, 382-385
  Model, 383-385
  Reduction, 385
  Signal, 384, 385
Precharge, 14, 17, 168, 182-188, 198, 413, 491-493
Precharge, 14, 17, 39, 168, 182-188, 198, 413, 491-493
Predecoder, 327, 328
Processing Hardening (Radiation Hardened Processing)
Propagation Coefficient, 297

Radiation Effected:
  Characteristics, 471, 475
  Data Upset, 473
  Latchup, 479, 480
  Leakage Current, 473
  Mobility, 473, 474
  Operation Margin, 486, 487
  Photocurrents, 475, 477
  Sense Amplifier, 486-490
  Snapback, 479-481
  Subthreshold Current, 472, 473
  Threshold Voltage, 471, 472, 485
Radiation Environments, 470, 471
Radiation Hardened Circuit (Radiation Hardening by Circuit Technique)
Radiation Hardened Memory Circuit:
  Architecture, 495-501
  Logic Gate, 495-499
  Memory Cell, 398-399, 493-495

Index

Radiation Hardened Memory Circuit *(Continued)*:
  Reference, 314-315, 481, 493
  Sense Amplifier, 486-490
Radiation Hardened Processing, 484-486
Radiation Hardening:
  General Methods, 484, 486
  Grades, 482, 483
  Requirements, 481, 484
Radiation Hardening by Circuit Technique:
  Combined, 464-467, 499-501
  Error Control Coding, 438-467
  Error Purging, 467
  Fault Masking, 434-436, 500
  Fault Repair, 421-438
  Global, 499-501
  Parameter Tracking, 314, 315, 491-493
  Self-Adjustment, 495-499
  Self-Compensation, 486-490
  State Retention, 398, 399, 493-495
  Voltage Limitation, 314, 315
Radioactive Environments (Radiation Environmentals)
Radioactive Radiation:
  Combined, 478-481
  Cosmic (Atomic Particle Impact)
  Electron, 478, 479
  Fabrication Induced, 477
  Neutron, 478
  Package (Atomic Particle Impact)
  Permanent Total-Dose, 471-475
  Proton, 478, 479
  Transient Dose-Rate, 475-477
RAM (Random Access Memory)
RAM Cell (Random Access Memory Cell)
Rambus Dynamic Random Access Memory, 73-76
Ramlink, 75
Random Access Memory:
  Basic, 10
  Categories, 11
  Dynamic, 11-36
  Fundamentals, 10, 11
  Pseudo Static, 40, 41
  Special, 61-83
  Static, 36-39
Random Access Memory Cell:
  Dynamic, 89-113, 154-161
  State Retention, 398, 399, 493-495
  Static, 113-128, 154-161

RDRAM (Rambus Dynamic Random Access Memory)
Read-Only Memory:
  Architecture, 41, 42
  Cell, 132, 136
Read-Only Memory Cell:
  Bilevel, 132-136
  Design, 134-136
  Multi-Level, 135
  NAND Array, 132, 133
  NOR Array, 132-134
  Programming, 134-136
  Storage, 132
Redundancy:
  Duplicated, 369, 370, 410, 411
  Effects, 363-374, 442-453
  Error Control Coding, 438-467
  Majority Decision, 436-438
  Optimization, 409-412, 446-453
  Triplicated, 370, 371, 410, 411
Reference Circuit:
  Charge, 321-323
  Current, 318-321
  Parameter Tracking, 314, 315, 491-493
  Regulated, 316-318, 320, 321
  Voltage, 311-318
Reflected Signal:
  Capacitive, 302-304
  Inductive, 305-307
  Open, 300, 301
  Resistive, 288, 305
  Short, 301
Reflection (Transmission Line, Signal Reflection)
Reflection Coefficient, 297
Regulated Reference Circuit, 320, 321
Reliability:
  Memory Circuit, 366-369
  Redundancy Effected, 369-374
Ring Oscillator, 345
ROM (Read-Only Memory)
ROM Cell (Read-Only Memory Cell)
RS Code (Error Control Code, Reed-Solomon)

SAM (Sequential Access Memory)
SAM Cell (Sequential Access Memory Cell)
Sample-and-Feedback Amplifier, 263-265, 489, 490

Scaled Inverter (Tapered Inverter)
Schmidt Trigger, 337-339
SDRAM (Synchronous Dynamic Random
    Access Memory)
Search Time, 6, 61
Self-Compensating Sense Amplifier, 486-490
Self-Compensation, 260-265, 480-490
Self-Repair, 464-467
Selftest, 464-467
Self-timed Operation, 13, 14, 341
Sense Amplifier:
    Charge-Translator, 271-273
    Circuit, 164
    Classification, 21, 190, 191
    Current, 232-257
    Differential, 192-265, 261-263, 480-490
    Enhanced, 220-231
    General, 184
    Nondifferential, 256-275
    Offset Reduction, 257-265
    Voltage, 192-234, 261-263, 480-490
Sense Circuit:
    Data Sensing, 164-166
    Operation Margin, 166-184
    Sense Signal, 164, 166
Sensitive Cross Section (Parasitic Sensitive
    Cross Section)
SEP (Single Event Phenomena)
Sequential Access Memory:
    First-In-First-Out, 51-54
    Generic Architecture, 43
    Last-In-Last-Out, 42
    Principle, 42-44
    Random Access Memory Based, 44, 45, 48
    Shift-Register Based, 45-48
    Shuffle: 48-51
Sequential Access Memory Cell:
    Dynamic, 138-143
    Static, 143-146
SER (Soft Error Rate)
SEU (Single Event Upset)
Seven-Transistor Shift-Register Cell, 143,
    144
Shannon's Theorem, 439, 440
Shift-Register Cell:
    Charge Distribution, 140-142
    Data Shifting, 136-138
    Derivative, 158, 160
    Diode-Transistor Combined, 142, 143
    Dynamic, 138-149, 160

Shift-Register Cell *(Continued)*:
    Feedback, 143-146
    Four-Phase, 142, 143
    Static, 143-146, 160
    Three-Phase, 143-145
    Transistor-Only, 138-142
    Two-Phase, 136-142
Shift-Register Memory, 45-48
Shuffle Memory. 48-51
Side-Channel Effect:
    Allay, 526, 527
    Breakdown, 523
    Kink, 522
    Leakage, 521, 522
    Mechanism, 520
Signal Accelerator Circuit:
    Negative Capacitance, 294
    Negative Resistance, 292-294
    Pull-Up-Pull-Down, 291, 292
    Sense Amplifier, 163-275
Signal Limiter, 229-231, 283-285
Silicon-On-Insulator:
    Design, 501-529
    Devices, 501-504
    Features, 505-509
    Special Effect, 509-529
Silicon-On-Sapphire, 501-529
Similarity Measure, 54, 55
Single Event Phenomena, 389
Single Event Upset, 389
Six-Transistor Shift-Register Cell, 138-140,
    160
Soft Error Rate Reduction:
    CMOS SOI (SOS), 501-529
    General, 398
    Sense Amplifier, 399, 400
    Special Memory Cell, 398-400
    Special Fabrication, 401
Soft Error Rate:
    Array, 394
    Decoder, 396
    Estimate, 390-398
    Memory Cell, 393
    Memory Chip, 396
    Reduction, 398-402
    Sense Amplifier, 395
SOI (Silicon-On-Insulator)
SOS (Silicon-on-Sapphire)
Spare:
    Block, 434-436

Spare *(Continued)*:
  Column, 430, 431, 432
  Decoder, 428-434
  Row, 429, 432
SR Cell (Shift-Register Cell)
SR Memory (Shift-Register Memory)
SRAM (Static Random Access Memory)
Stability Criteria (Feedback, Stability)
State Retention Memory Cell, 398, 399
State-Retention, 398, 399, 493-495
Static Column Mode, 17-19
Static Five-Transistor Random Access
  Memory Cell, 158
Static Four-Transistor-Two-Resistor Random
  Access Memory Cell:
  Design, 125-131
  Laser Annealed Poly, 127, 128
  Noise Margin, 128
  Positive Feedback, 126, 129
Static Random Access Memory:
  Basic Architecture, 36-38
  Pseudo, 40, 41
  Static Storage, 36, 38
  Timing, 38, 99
Static Random Access Memory Cell
  (Memory Cell, SRAM)
Static Six-Transistor Random Access
  Memory Cell:
  Bitline Termination, 123-125
  Derivative, 158, 159
  Design Objectives, 121
  Design Trade-Offs, 111, 122
  Designs, 122-125, 160
  Flipping Voltage, 114-117
  Positive Feedback, 114-117
  Read Signal, 119-121
  Stack-Transistor, 123
  Storage, 113-116
  Noise Margin, 115
  Static Retention, 398, 399, 493-495
  Write Signal, 116-119
Stopping Power, 393
Synchronous Dynamic Random Access
  Memory:
  Dual Bank, 25-28
  Double Date Rate, 27, 79
  Multi Bank, 29, 31
  Pipelined, 27-28
  Prefetched, 27-28
Synchronous Operation, 13, 14, 25-31, 341

Synclink, 76

Tapered Inverter, 331, 332, 353
Ten-Transistor Content Addressable Memory
  Cell, 148, 149
Ten-Transistor-Two-Resistor Shift-Register
  Cell, 146
Terms Effecting Operation Margins:
  Atomic Particle Impacts, 182, 388-402
  Bitline Droop, 182
  Charge Couplings, 176-178
  Incomplete Restore, 182
  Imbalance, 179-181, 257-265, 480-490
  Leakage Currents, 173-176, 473, 513-515,
    521, 522, 525, 526
  Noise, 181, 373-388
  Precharge Level Variation, 182-184, 198,
    413, 491-493
  Radioactive Radiation, 170, 182-184, 388-
    402, 470-501
  Supply Voltage Ranges, 171
  Threshold Voltage Shifts, 171-173, 471,
    472, 495
Thermal Noise:
  Amplitude, 385
  Model, 385-388
  Signal/Noise, 387
Threshold Voltage Ranges, 472, 485
Threshold Voltage Shift, 171-173, 471, 472,
  495
Time-to-Failure, 368
Timing (Memory Timing)
Total-Dose, 471-475
Transfer Function, 250-252
Transient Damping (Feedback, Damping)
Transient Ionization (Dose-Rate)
Transmission Line:
  Lossless, 299-304
  Lossy, 304-306
  Model, 296-311
  Model Validity Region, 308-311
  Signal Distortion, 197-307
  Termination, 296-307
  Transient, 301-305
TTF (Time-to-Failure)
Tunnel-Diode Based Memory Cell (Memory
  Cell, Tunnel-Diode)
Twelve-Transistor Shift-Register Cell, 145,
  146

550 CMOS Memory Circuits

VCM (Virtual Channel Memory)
VCO (Phase Locked Loop, Voltage Controlled Oscillator)
Video DRAM (Video Dynamic Random Access Memory)
Video Dynamic Random Access Memory, 33-36
Virtual Channel Memory, 76-79
Voltage Divider, 311-313
Voltage Limited Sense Circuit, 229-231, 314, 315, 486-490
Voltage Reference:
  Divider, 311, 312
  Parallel Regulated, 317, 318
  Series-Regulated, 316-318
  Temperature Stabilized, 315, 316
  Threshold-Drop, 314, 315, 491-493
  Threshold Voltage Tracking, 314, 315
Voltage Sense Amplifier:
  Active Load, 207-220
  Backgate Bias Reduction, 210
  Basic, 192-199
  Bisected, 193-199, 203
  Bitline Decoupled, 221-224
  Current Source, 226, 227
  Differential, 191-231
  Enhanced, 220, 231
  Feedback Separated, 224, 225
  Full-Complementary, 207-211
  Full-Complementary-Positive Feedback, 217, 220
  Negative Feedback, 260-265
  Nondifferential, 265-275
  Offset Reduced, 261-263, 480-490
  Optimum Voltage-Swing, 229-231
  Positive-Feedback, 211-217
  Sample-and-Feedback, 263-265, 489, 490
  Simple, 200-207
  Voltage Swing Limitation, 230, 231
Voting Circuit, 437, 438

Waterfall Curve, 444, 445
Wave Impedance (Characteristic Impedance)
Wave Velocity, 302, 304-307
Well Separation, 210
Wide DRAM (Wide Dynamic Random Access Memory)
Wide Dynamic Random Access Memory, 31-33

Winston Bridge, 356
Wordline:
  Divided, 291
  Dummy, 295, 296
  Model, 187-290, 296-311
  Signal, 290-294
  Signal Accelerator, 291-294
  Termination, 296-307

Yield:
  Effective, 409
  Estimate, 403-406, 410-412
  Fabrication, 402, 403
  Improvement, 403, 406-412
  Issues Effecting, 402-404, 416
  Maturity Influences, 407, 408
  Memory, 402-412
  Models, 404, 412
  Optimization, 408-412
  Redundancy Effected, 406-412

1T1C DRAM Cell (Dynamic One-Transistor-One-Capacitor Memory Cell)
3T DRAM Cell (Dynamic Three-Transistor Memory Cell)
4T DRAM Cell (Dynamic Four-Transistor Random Access Memory Cell)
4T2C CAM Cell (Four-Transistor-Two Capacitor Content Addressable Memory Cell)
4T2D SR Cell (Four-Transistor-Two-Diode Shift-Register Cell)
4T2R SRAM Cell (Static Four-Transistor-Two-Resistor Random Access Memory Cell)
5T SRAM Cell (Static Five-Transistor Random Access Memory Cell)
6P Memory, 487, 486
6T SR Cell (Six-Transistor Shift-Register Cell)
6T SRAM Cell (Static Six-Transistor Random Access Memory Cell)
7T SR Cell (Seven-Transistor Shift-Register Cell)
8T SR Cell (Eight-Transistor Shift-Register Cell)

Index

8T2R CAM Cell (Eight-Transistor-Two-Resistor Content Addressable Memory Cell)
9T SR Cell (Nine-Transistor Shift-Register Cell)
10T CAM Cell (Ten-Transistor Content Addressable Memory Cell)
10T2R SR Cell (Ten-Transistor-Two-Resistor Shift-Register Cell)
12T SR Cell (Twelve-Transistor Shift-Register Cell)